COLLEGE ALGEBRA
GUIDED NOTEBOOK

Christopher Schroeder

Author:
Chris Schroeder

Editors:
Daniel Breuer,
Kara Roché,
Ashish Samudre

Designers:
Trudy Gove,
Patrick Thompson,
Tee Jay Zajac

Cover Design:
Bryan Mitchell

VP Research & Development: Marcel Prevuznak

Director of Content: Kara Roché

A division of Quant Systems, Inc.

546 Long Point Road
Mount Pleasant, SC 29464

Copyright © 2019, 2016 by Hawkes Learning / Quant Systems, Inc. All rights reserved.

No part of this publication may be reproduced, stored in a retrieval system, or transmitted in any form or by any means, electronic, mechanical, photocopying, recording, or otherwise, without the prior written consent of the publisher.

Printed in the United States of America

10 9 8 7 6 5 4 3

ISBN: 978-1-941552-34-6

Contents

A Message to the Student

Have you ever been working on your homework and knew you had some notes somewhere for a problem you were stuck on, but didn't know where your notes were? Have you ever tried to enter an answer on a homework program by guessing and didn't know for sure why you got it wrong (or right)? Have you ever been working on math and wanted to throw your computer out the window? If you answered "yes" to any of these questions, then this notebook is for you. In my experiences as a College Algebra instructor, I have seen many students struggle with the material because of a lack of organization and motivation. This notebook was written to help students that struggle with these problems.

This notebook is designed to be used in conjunction with the Hawkes Learning Systems' *College Algebra* courseware. After a brief introduction to the concepts and objectives, each section of this workbook contains three main components: 1) **Learn**; 2) **Video Examples**; and 3) **Practice**. **Learn** will guide you through the topics that will be covered in the section and corresponds to the *Learn* mode on your courseware. **Video Examples** allows you the opportunity to solve problems that are similar to ones found on www.hawkestv.com. The examples correspond to the videos posted by Chris Schroeder from Morehead State University. In **Practice**, you will work through additional examples that are similar to problems you will find in the *Practice* mode of the courseware. At this point, you will be ready to work through the additional *Practice* problems, putting yourself in a great position to succed when attempting to *Certify* on that section in the courseware.

When used in this way, this notebook can be a valuable asset in helping you stay organized, motivated, and on track to succeed in your College Algebra course.

Chris Schroeder
Morehead State University

Important Information

Class Time: _____

Office Hours: _____

Teacher's phone: _____

Teacher's email: _____

Exam Dates: _____

Final Exam Date: _____

Hawkes Website: `learn.hawkeslearning.com`

Videos Website: `tv.hawkeslearning.com`
(Select "College Algebra: Chris Schroeder - Morehead State University")

Notes:

Schedule

Month:						
S	**M**	**T**	**W**	**T**	**F**	**S**

Notes:

Month:						
S	**M**	**T**	**W**	**T**	**F**	**S**

Notes:

Month:						
S	M	T	W	T	F	S

Notes:

Month:						
S	M	T	W	T	F	S

Notes:

Month:						
S	M	T	W	T	F	S

Notes:

My Habits for Success in College Algebra

Students who have consistent and effective study habits are far more likely to succeed in their academic work. Research shows that writing down what these habits will be prior to using them greatly increases the chances that you will utilize these habits effectively. With that in mind, take some time to write out what habits you will develop and use to help you succeed in this course. You can do this!

My Study Habits

1. I will begin working on each lesson at least _____ days before the Certification for that lesson is due. Specifically, when working through a lesson, I will use the following habits:

2. I will begin studying for exams at least _____ days before taking the exam. Specifically, when studying for an exam, I will use the following habits:

My Habits When Things Get Rough

1. I will do my best to attend every class. However, if I have to miss a class, to catch up on the material missed I will use the following habits:

2. If I am stuck on a lesson, I will use the following habits to get help and figure out the material:

Chapter 1

Number Systems and Fundamental Concepts of Algebra

Before we can embark on a journey into algebra, we need to be sure that we know the players and the rules of the game. The players are the numbers, and there are many different types of numbers. The rules tell us how to work with these numbers. Most of these you probably already know, but it's worth going over them again to make sure you're ready for algebra.

1.1   The Real Number System

The real numbers contain within them many different types of numbers. Our goal in this section is to be able to classify the different types of real numbers according to their properties.

Objectives:

- Absolute value and distance on the real-number line

- Common subsets of real numbers

- Order on the real-number line

- Set-builder notation and interval notation

- The real-number line

<u>Learn:</u>

1. Give an example of:

   (a) a number which is an integer but *not* a natural number: _____

   (b) a number which is a rational number but *not* an integer: _____

   (c) two irrational numbers: _____   _____

2. Fill in the blank with one of the symbols $<$ or $>$:

   (a) -4_____ -14

   (b) 7_____ -8

   (c) $\sqrt{2}$_____$\sqrt{5}$

3. List three numbers which are in the set $\{x|x$ is an integer and $-1 \le x < 3\}$.

4. Write the set $\{x|x \ge 3\}$ in interval notation.

5. Given two real numbers a and b, the _____ between then is defined to be $|a - b|$.

<u>Video Examples:</u>

Watch the videos for Section 1.1 on `www.hawkestv.com` and then solve the following problems.

- Write the following set as an interval using interval notation: $\{x|-5 \le x < 3\}$.

- Find the distance between the following numbers on the real-number line.

$$a = -3, b = -10$$

Practice:

1. Identify the following number. Choose all that apply.

$$-\frac{3}{4}$$

   Natural Number; Whole Number; Integer; Rational Number; Irrational Number; Real Number; Undefined

2. Evaluate the expression.

$$-|2 - 5|$$

3. Plot the set $\{-2.5, 3, 1.5, 0\}$ on the real-number line.

Remember to go through all of the problems in the Practice section before attempting to Certify!

1.2   The Arithmetic of Algebraic Expressions

In this section we look at the terminology and the rules involved in working with algebraic expressions. The material in this section is fundamental to algebra.

<u>Objectives:</u>

- Basic set operations and Venn diagrams

- Components and terminology of algebraic expressions

- The field properties and their use in algebra

- The order of mathematical operations

<u>Learn:</u>

1. **Algebraic expressions** are made up of *constants* and _____, combined by the operations of addition, subtraction, multiplication, division, exponentiation, and the taking of _____.

2. Write the terms separately for the algebraic expression $3xy^2 - 7x^2z^3 + 4(xy - z)$.

3. Write the additive and multiplicative versions of the Commutative Property.

   Additive:

   Multiplicative:

4. In the expression $3 - 2 \cdot 4 + 7$, which operation should be performed first: subtraction, multiplication or division?
   Simplify the expression.

5. Simplify each of the following set expressions, if possible.

    (a) $(-3, 2] \cup [1, 6] =$

    (b) $(-3, 2] \cap [1, 6] =$

Video Examples:

Watch the videos for Section 1.2 on www.hawkestv.com and then solve the following problems.

- Evaluate the following expression for the given values of the variables.

$$|x + 9y| - (5z + 6) \text{ for } x = -2, y = -1, \text{ and } z = 5$$

- Evaluate the following expression.

$$\frac{8 - 3 \cdot 4 - 5}{-2(-7 - 6 \div (1 + 2))}$$

Practice:

1. Evaluate the following expression for the given values of the variables.

$$3\sqrt{x - 3} + 7y^3 \text{ for } x = 28 \text{ and } y = -3$$

2. Simplify the following intersection of intervals.

$$(-\infty, -13] \cap [-13, 6)$$

3. Identify the property that justifies the following statement.

$$-4(6y + 1) = -24y - 4$$

Remember to go through all of the problems in the Practice section before attempting to Certify!

1.3 Properties of Exponents

Being able to evaluate expressions with exponents is an important skill to have to be successful in algebra. In the first part of this section, we'll review the properties we use to work with exponents. Then, we'll apply those properties to scientific notation and geometric problems.

1.3.a Properties of Exponents

The properties of exponents are a set of rules that allow us to evaluate and simplify exponential expressions. It's important that we have a solid working knowledge of these properties.

Objectives:

- Integer exponents

- Natural number exponents

- Properties of exponents and their use

Learn:

1. In the expression a^n, a is called the _____, and n is the _____.

2. Complete the following list of properties of exponents.

    (a) $a^n \cdot a^m =$

    (b) $\dfrac{a^n}{a^m} =$

    (c) $a^{-n} =$

    (d) $(a^n)^m =$

    (e) $(ab)^n =$

    (f) $\left(\dfrac{a}{b}\right)^n =$

3. If $a \neq 0$, then $a^0 =$ _____.

<u>Video Examples:</u>

Watch the videos for Section 1.3a on `www.hawkestv.com` and then solve the following problems.

- Simplify the expression, writing your answer with only positive exponents.

$$\frac{x^4 \cdot x^7}{x^3}$$

- Simplify the expression, writing your answer with only positive exponents.

$$\left[\frac{y^6 \left(xy^2\right)^{-3}}{3x^{-3}z} \right]^{-2}$$

Practice:

1. Simplify the expression, writing your answer with only positive exponents.

$$\left(2y^{-1}\right)^{-1}$$

2. Simplify the expression, writing your answer with only positive exponents.

$$\left(\frac{2a}{b^{-3}}\right)^2$$

3. Use the properties of exponents to simplify the expression, writing your answer with only positive exponents.

$$\left(\left(3x^{-2}y^4\right)^3\right)^{-1}$$

Remember to go through all of the problems in the Practice section before attempting to Certify!

1.3.b Scientific Notation and Geometric Problems Using Exponents

Scientific notation allows scientists to more easily work with very large or very small numbers using properties of exponents. Of course, geometric problems occur in many areas and exponents are often involved in these formulas.

<u>Objectives:</u>

- Interlude: working with geometric formulas

- Scientific notation

<u>Learn:</u>

1. For a number in scientific notation of the form $a \times 10^n$, between what two numbers must the absolute value of a be?

$$\underline{\hspace{1cm}} \leq a < \underline{\hspace{1cm}}$$

2. To convert a number in scientific notation to a number in decimal notation, we move the decimal point n units to the _____ if n is positive and to the _____ if n is negative.

3. Write the geometric formula for the following:

   (a) Volume of a sphere: $V =$

   (b) Surface area of a box: $S =$

   (c) Surface area of a right circular cylinder: $S =$

   (d) Volume of a right trapezoidal cylinder: $V =$

Video Examples:

Watch the videos for Section 1.3b on `www.hawkestv.com` and then solve the following problems.

- Evaluate the following expression, using the properties of exponents. Write your answer in standard notation.

$$\frac{\left(15.3 \times 10^{15}\right)\left(3 \times 10^{-11}\right)}{\left(5.1 \times 10^{9}\right)}$$

- Determine the volume of a right trapezoidal cylinder whose bases are $B = 18\,\text{m}$ and $b = 9\,\text{m}$, height is $h = 4\,\text{m}$, and length is $l = 33\,\text{m}$.

Practice:

1. Convert the following number from scientific notation to standard notation.

$$7.32 \times 10^5$$

2. Evaluate the following expression, using the properties of exponents. Write your answer in scientific notation.

$$\left(4 \times 10^{-6}\right)\left(3.7 \times 10^{-7}\right)\left(7 \times 10^{-11}\right)$$

3. Determine the surface area of a box whose length l is 11 feet, width w is 5 feet, and height h is 7 feet.

Remember to go through all of the problems in the Practice section before attempting to Certify!

1.4 Properties of Radicals

In this section, we investigate radicals and their properties. In the first part, we'll look at some of the methods used to simplify radical expressions. In the second part, we'll look at these same methods but when the radical is represented as a rational number exponent.

1.4.a Properties of Radicals

Simplifying radicals means more than evaluating square roots of numbers. We also want to be able to simplify algebraic expressions that contain radicals. We call these *radical expressions*. There are numerous properties that we'll look at which can be used to simplify these radical expressions.

Objectives:

- Roots and radical notation

- Simplifying radical expressions

Learn:

1. $\sqrt[n]{a} = b \iff a = $ _____

2. A radical expression is in **simplified form** when:

    (a) The _____ contains no factor with an exponent greater than or equal to the index of the radical.

    (b) The radicand contains no _____.

    (c) The denominator, if there is one, contains no _____.

    (d) The greatest common factor of the _____ and any exponents occurring in the radicand is 1.

3. $\sqrt[n]{a^n} = \begin{cases} |a| & \text{if } n \text{ is } \rule{2cm}{0.4pt} \\ a & \text{if } n \text{ is } \rule{2cm}{0.4pt}. \end{cases}$

4. Complete the following Properties of Radicals.

    (a) $\sqrt[n]{ab} = \qquad .$

    (b) $\sqrt[n]{\dfrac{a}{b}} = $ _____

    (c) $\sqrt[m]{\sqrt[n]{a}} = $

5. To simplify the expression $\dfrac{3 + \sqrt{2}}{5 - \sqrt{3}}$, we would multiply numerator and denominator by the **conjugate** of the denominator. What is the conjugate of the denominator?

Video Examples:

Watch the videos for Section 1.4a on www.hawkestv.com and then solve the following problems.

- Simplify the following radical expression.

$$\sqrt[4]{\frac{y^{12}z^8}{81}}$$

- Simplify the following radical by rationalizing the denominator.

$$\frac{\sqrt{z} + \sqrt{x}}{\sqrt{z} - \sqrt{x}}$$

Practice:

1. Determine if the following radical expression is a real number. If it is, evaluate the expression.

$$\sqrt[3]{-27}$$

2. Simplify the following radical expression.

$$\sqrt[3]{8x^4y^{17}}$$

3. Simplify the radical expression by rationalizing the denominator.

$$\frac{y}{\sqrt{5x}}$$

Remember to go through all of the problems in the Practice section before attempting to Certify!

1.4.b Rational Number Exponents

At times, it is useful to represent radical expressions as expressions with rational number exponents. Then, we can use the properties of exponents to simplify these same radical expressions. We'll get some practice doing just that in this section.

Objectives:

- Combining radical expressions

- Rational number exponents

- Simplifying radical expressions

Learn:

1. Radical expressions are called **like radicals**, if they have the same _____ and the same _____.

2. Meaning of $a^{1/n}$: If n is a natural number and if $\sqrt[n]{a}$ is a real number, then $a^{1/n} = $ _____.

3. Meaning of $a^{m/n}$: If m and n are natural numbers with $n \neq 0$, if m and n have no common factors greater than 1, and if $\sqrt[n]{a}$ is a real number, then $a^{m/n} = \sqrt[n]{\rule{1.5em}{0.4pt}} = (\quad)^m$.

4. $16^{3/4}$ can be written as $\sqrt[4]{16^3}$ or as $\left(\sqrt[4]{16}\right)^3$. Which form is easier to calculate? Simplify the expression.

<u>Video Examples:</u>

Watch the videos for Section 1.4b on `www.hawkestv.com` and then solve the following problems.

- Simplify the following expression.

$$\sqrt[3]{\sqrt[7]{x^{21}}}$$

- Simplify the following expression.

$$(3x^3 + 2)^{11/3}(3x^3 + 2)^{-10/3}$$

Practice:

1. Simplify the following expression. Assume all variables are positive.

$$\left(x^{1/2} \cdot y^{-1} \cdot z^{-1/3}\right)^{-2/3}$$

2. Combine the radical expressions, if possible.

$$\sqrt[5]{-729y^6} + 3y\sqrt[5]{3y}$$

3. Simplify the following expression, writing your answer using the same notation as the original expression.

$$\frac{x^{2/3}z^{-1/2}}{x^{-1/3}z}$$

Remember to go through all of the problems in the Practice section before attempting to Certify!

1.5 Polynomials and Factoring

Polynomials are an important type of algebraic expression and will appear often in this course. With that in mind, let's get some practice in identifying types of polynomials and factoring polynomials.

Objectives:

- Common factoring methods

- The algebra of polynomials

- The terminology of polynomial expressions

Learn:

1. Polynomials consisting of a single term are called _____, those consisting of two terms are called _____, and those consisting of three terms are called _____.

2. What is the degree of the following polynomial? What is the leading coefficient?

$$3x^2 - 7x^3 + 5x + 2$$

   Degree: Leading coefficient:

3. Which terms in the following two polynomials are **like** terms?

   $$-2a^3 - 4ab^2 + 7ab + 5b^2 \qquad\qquad 10ab - 6a^2b + 8b^2 - 2a^2$$

   Like terms:

4. What is the greatest common factor in the following polynomial?

   $$10a^2b - 6ab + 4a^2b^2$$

   GCF:

5. Complete the special binomial factoring formulas.

   (a) $A^2 - B^2 = ($ $)(A + B)$

   (b) $A^3 - B^3 = (A - B)($ $)$

   (c) $A^3 + B^3 = ($ $)(A^2 - AB + B^2)$

6. Complete the following steps which are used to factor a trinomial $a^2 + bx + c$ where $a \neq 1$.

   (a) Multiply a and _____.

   (b) Factor ac into two integers whose sum is _____. If no such factors exist, the trinomial is irreducible over the integers.

   (c) Rewrite b in the trinomial with the sum found in step 2, and _____. The resulting polynomial of four terms may now be factored by grouping.

Video Examples:

Watch the videos for Section 1.5 on `www.hawkestv.com` and then solve the following problems.

- Multiply the following polynomials.

$$\left(4x^2 + y\right)\left(x^2 - 5y\right)$$

- If possible, factor the following trinomial.

$$12y^2 - 17y + 6$$

Practice:

1. Add or subtract the following polynomials, as indicated.

$$\left(-7y^3 + 10 - 9y^4\right) - \left(2 - 4y^3\right)$$

2. If possible, factor the following polynomial by factoring out the greatest common factor.

$$15x^2y + 5x^4 - 15x^3y$$

3. If possible, factor the following polynomial by grouping.

$$ax - 4bx + 4ay - 16by$$

Remember to go through all of the problems in the Practice section before attempting to Certify!

1.6 The Complex Number System

In this section, we'll study *complex numbers*. Just like real numbers, we can add, subtract, multiply, and divide complex numbers. In addition, we will encounter equations in this class which do not have a real number solution, but do have a complex number solution.

Objectives:

- Roots and complex numbers

- The algebra of complex numbers

- The imaginary unit i and its properties

Learn:

1. The imaginary unit i is defined as $i =$ _____. In other words, i has the property that $i^2 =$ _____.

2. For any two real numbers a and b, the sum _____ is a **complex number**.

3. Identify the real part and the imaginary part of the complex number $-7 + 5i$.
   Real part: Imaginary part:

4. Determine the complex conjugate of the complex number $3 - 2i$.
   Conjugate:

5. What is the product of any complex number $a + bi$ and its conjugate $a - bi$.

$$(a + bi)(a - bi) =$$

<u>Video Examples:</u>

Watch the videos for Section 1.6 on `www.hawkestv.com` and then solve the following problems.

- Simplify the following square root expression.

$$\sqrt{-20}\sqrt{-2}$$

- Simplify the quotient.

$$\frac{3-i}{2+4i}$$

<u>Practice:</u>

1. Simplify the following expression.

$$(16 - 6i)(10 - 11i)$$

2. Simplify the following expression.

$$i^{43}$$

3. Simplify the following expression.

$$(6 - 8i)(6 + 8i)$$

Remember to go through all of the problems in the Practice section before attempting to Certify!

Chapter 2

Equations and Inequalities of One Variable

All things being equal, equations are one of the most important topics in mathematics. Developing a set of skills to be able to solve a wide variety of equations and inequalities will serve you well not only in this class, but in almost any career you may go into.

2.1 Linear Equations in One Variable

One of the most basic types of equations are **linear** equations. You probably solve these types of equations often without even writing anything down. For these problems, however, we'll probably need to write some things down. We'll break this section into two parts: solving linear equations and solving application problems involving linear equations.

2.1.a Linear Equations in One Variable

Let's start by getting some practice solving linear equations.

Objectives:

- Equivalent equations and the meaning of solutions

- Solving linear equations

- Solving absolute value equations

<u>Learn:</u>

1. Is the following equation an **identity**, a **contradiction**, or a **conditional** equation? Can you give an example of each?

$$x + 3 = 5$$

2. A **linear equation in one variable** is an equation that can be transformed into the form _____, where a and b are real numbers and $a \neq 0$.

3. To solve an equation with fractions, we can get rid of the fractions by multiplying each side of the equation by the least common _____ or LCD. To get rid of the fractions in the equation below, by what number should each side of the equation be multiplied?

$$\frac{x}{3} + \frac{x-2}{6} = \frac{x+5}{2} \qquad \text{LCD} = \text{_____}$$

4. What two values for x would make the following equation true?

$$|x| = 6$$

$$x = \text{_____} \quad \text{and } x = \text{_____}$$

5. Without solving, how many solutions would the following equation have? Why?

$$|x - 2| + 10 = 0$$

<u>Video Examples:</u>

Watch the videos for Section 2.1a on `www.hawkestv.com` and then solve the following problems.

- $0.5t + 1.4 = 3t$

- $|3x - 7| = 13$

<u>Practice:</u>

1. Solve the following linear equation.

$$2t + 1 = 3(t + 2) - 5$$

2. Solve the following linear equation. Remember it may be easier to multiply each side of the equation by the LCD first.

$$\frac{2y+3}{4} + \frac{3}{2} = \frac{3y-1}{2}$$

3. Solve the following absolute value equation. Don't forget to check your answers.

$$|x+5| = |x-4|$$

Now complete the rest of the problems in the Practice section on your Hawkes courseware and then you'll be ready to Certify!

2.1.b   Applications of Linear Equations in One Variable

What good is learning all of this mathematics if we can't apply it? In this section, we'll look at some application problems that can be solved with linear equations.

Objectives:

- Interlude: discount, tax and tip problems

- Interlude: distance and interest problems

- Interlude: number problems

- Solving equations for one variable

- Solving linear equations for one variable

Learn:

1. Two equations that have the same solution set are called _____ **equations**.

2. **Solving for a variable** means to transform the equation into an equivalent one in which the specified variable is _____ on one side of the equation.

3. Write the basic distance formula.

4. In the simple interest formula $I = Prt$, what does the variable P represent?

<u>Video Examples:</u>

Watch the videos for Section 2.1b on `www.hawkestv.com` and then solve the following problems.

- Find three consecutive odd integers whose sum is 189.

- Suppose two ships leave a port at the same time. The first travels east at 24 miles per hour and the second travels west at 30 miles per hour. When will the ships be 81 miles apart?

Practice:

1. Solve the following formula for the indicated variable.

$$F = \frac{9}{5}C + 32, \text{ for } C$$

2. Suppose your bill for a meal comes to \$18.76. If you want to leave an 18% tip, how much should you pay?

3. If $5000 is invested in a savings account which pays 2.4% annually, how much money will be in the account after 1 year?

Now complete the rest of the problems in the Practice section on your Hawkes courseware and then you'll be ready to Certify!

2.2 Linear Inequalities in One Variable

Oftentimes, we want to find all of the numbers that make a particular inequality true. Since there will usually be infinitely many solutions to such a problem, we'll need to represent our answers either graphically or using interval notation.

Objectives:

- Solving absolute value inequalities

- Solving compound linear inequalities

- Solving linear inequalities

- Solving linear inequality word problems

Learn:

1. Fill in the blanks with $<$ or $>$.

   (a) If $C > 0$, $A < B \iff A \cdot C$ ___ $B \cdot C$.

   (b) If $C < 0$, $A < B \iff A \cdot C$ ___ $B \cdot C$.

2. If the solution to a linear inequality is $x < 5$, graph this solution and write it using interval notation.

3. A **compound inequality** is a statement containing two inequality symbols, and can be interpreted as two distinct inequalities joined by the word "___".

4. To solve the absolute value inequality $|2x + 6| > 3$, what two independent inequalities would you solve?

   or

Video Examples:

Watch the videos for Section 2.2 on www.hawkestv.com and then solve the following problems. Write your solutions using interval notation.

- $2|a + 4| \leq 16$

- $-3(x + 2) \leq 12$ or $15 + x < 17$

Practice:

1. Solve the following compound inequality.

$$\frac{6}{8} < \frac{y+2}{4} < \frac{18}{8}$$

2. Solve the following absolute value inequality.

$$4|2 - r| \le 8$$

3. In a class in which the final course grade depends entirely on the average of four equally weighted 100-point tests, Jason has scored 87, 91, and 83 on the first three. What range of scores on the fourth test will give Jason a B for the semester (an average between 80 and 89, inclusive)? Assume that all test scores have a non-negative integer value.

Now complete the rest of the problems in the Practice section on your Hawkes courseware and then you'll be ready to Certify!

2.3 Quadratic Equations in One Variable

One of the most common types of equation is the quadratic equation. There are numerous methods that can be used to solve quadratic equations. Determining which method is best for a particular equation is part of what you'll learn in this section.

Objectives:

- Interlude: gravity problems

- Solving "perfect square" quadratic equations

- Solving quadratic equation word problems

- Solving quadratic equations by completing the square

- Solving quadratic equations by factoring

- The quadratic formula

Learn:

1. A **quadratic equation in one variable**, say the variable x, is an equation that can be transformed into the form _____, where a, b, and c are real numbers and $a \neq 0$.

2. If A and B are algebraic expressions and $AB = 0$, then _____ $= 0$ or _____ $= 0$.

3. If you wanted to solve the equation $x^2 + 6x - 4 = 0$ by **completing the square**, what would the equation look like after the first step?

4. According to the **quadratic formula**, the solutions of the equation $ax^2 + bx + c = 0$ are

$$x = \qquad\qquad .$$

5. If a ball is thrown upward with a velocity of $20 \, \text{ft/s}$ from a height of $42 \, \text{ft}$, what values would be substituted for v_0 and h_0 in the formula $h = -\frac{1}{2}gt^2 + v_0 t + h_0$?

$$v_0 = \underline{\qquad\qquad}, h_0 = \underline{\qquad\qquad}$$

<u>Video Examples:</u>

Watch the videos for Section 2.3 on `www.hawkestv.com` and then solve the following problems.

- Solve the equation $x^2 + 6x + 9 = 64$ by the square root method.

- Solve the equation $x^2 + 8x + 7 = -8$ by completing the square.

Practice:

1. Solve the following quadratic equation by factoring.

$$y^2 + 3y - 28 = 0$$

2. Solve the following quadratic equation using the quadratic formula.

$$2x^2 - 4x - 3 = 0$$

3. How long would it take for a stone dropped from the top of a 196-foot building to hit the ground?

Now complete the rest of the problems in the Practice section on your Hawkes courseware and then you'll be ready to Certify!

2.4 Higher Degree Polynomial Equations

Of course there are numerous equations with a degree of more than 2 that can be solved. In this section, we will look at some of the methods that are used to solve these higher degree polynomial equations.

Objectives:

- Solving general polynomial equations by factoring

- Solving polynomial-like equations by factoring

- Solving quadratic-like equations

Learn:

1. An equation is *quadratic-like*, or quadratic in form, if it can be written in the form _____ where a, b, and c are constants, $a \neq 0$, and A is an algebraic expression.

2. In the quadratic-like equation below, determine the value of A so that the equation can be written as $aA^2 + bA + c = 0$.

$$(x^3 + x)^2 - (x^3 + x) - 12 = 0 \qquad A = \underline{\qquad\qquad}$$

3. To solve $8x^3 - 27 = 0$ by factoring, we would write the left side of the equation as a difference of two cubes. What are the two things being cubed?

$$(\qquad)^3 - (\qquad)^3 = 0$$

4. In the equation $x^{5/2} + 2x^{3/2} - 8x^{1/2} = 0$, what power of x can be factored out on the left side of the equation?

<u>Video Examples:</u>

Watch the videos for Section 2.4 on `www.hawkestv.com` and then solve the following problems.

- Solve the following equation by grouping.

$$2x^3 + x^2 + 2x + 1 = 0$$

- Solve the following equation.

$$3x^{13/5} + 2x^{8/5} - 5x^{3/5} = 0$$

Practice:

1. Solve the following polynomial equation.

$$2y^3 + 2y^2 = 4y$$

2. Solve the following quadratic-like equation.

$$\left(z^2 - 7\right)^2 + 7\left(z^2 - 7\right) - 18 = 0$$

3. Solve the following polynomial equation.

$$8x^3 - 1 = 0$$

Now complete the rest of the problems in the Practice section on your Hawkes courseware and then you'll be ready to Certify!

2.5 Rational Expressions and Equations

Fractions that involve algebraic expressions are called rational expressions. In this section, we will look at ways to combine and simplify these expressions as well as solving equations involving these expressions.

<u>Objectives:</u>

- Combining rational expressions

- Finding the restricted values of a rational expression

- Interlude: work-rate problems

- Simplifying complex rational expressions

- Simplifying rational expressions

- Solving rational equations

<u>Learn:</u>

1. A **rational expression** is an expression that can be written as a _____ of two polynomials $\dfrac{P}{Q}$ (with $Q \neq 0$).

2. In the following rational expression, how can we rewrite the denominator to cancel a common factor?

$$\frac{x^2 - 3x - 10}{5 - x} = \frac{(x+2)(x-5)}{-(\quad\quad)}$$

3. In order to add or subtract two rational expressions, we must first find the _____ _____ _____ or LCD.

4. In the following complex rational expression, what is the LCD of all the fractions?

$$\frac{\dfrac{1}{x} - \dfrac{1}{y}}{\dfrac{1}{x} + \dfrac{1}{y^2}} \qquad \text{LCD} = \underline{\quad\quad}$$

5. In the following rational equation, what is the LCD that should be multiplied by both sides of the equation?

$$\frac{x}{x-1} + \frac{2}{x-3} = \frac{2}{x^2 - 4x + 3} \qquad \text{LCD} = \underline{\quad\quad\quad}$$

Video Examples:

Watch the videos for Section 2.5 on `www.hawkestv.com` and then solve the following problems.

- Simplify the following rational expression.

$$\frac{x^2 + 2}{x - 3} - \frac{x + 4}{x + 5}$$

- Solve the following rational equation.

$$\frac{2}{2x + 1} - \frac{x}{x - 4} = \frac{-3x^2 + x - 4}{2x^2 - 7x - 4}$$

Practice:

1. Add or subtract the following rational expressions.

$$\frac{x+3}{x-3} - \frac{x-1}{x+6} - \frac{2}{x^2+3x-18}$$

2. Simplify the following complex rational expression.

$$\frac{8y+32}{2-\dfrac{32}{y^2}}$$

3. If Kendra were to clean her bathroom alone, it would take 2 hours. Her sister Angela could do the job in 4 hours. How long would it take them working together?

Remember to go through all of the problems in the Practice section before attempting to Certify!

2.6 Radical Equations

Some equations are a little more out there than others. These are called radical equations and there is a very specific set of steps we can use to solve them, which we'll look at in this section.

<u>Objectives:</u>

- Solving equations with positive rational exponents

- Solving formulas for a specific variable

- Solving radical equations

<u>Learn:</u>

1. Fill in the blanks for the Method for Solving Radical equations.

   Step 1: Begin by _____ the radical expression on one side of the equation. If there is more than one radical expression, choose one to isolate on one side.

   Step 2: Raise both side of the equation by the power necessary to "_____" the isolated radical. That is, if the radical is an nth root, raise both sides to the _____ power.

   Step 3: If any radical expressions remain, simplify the equation if possible and then repeat steps 1 and 2 until the result is a _____ equation. When a polynomial equation has been obtained, _____ the equation using polynomial methods.

   Step 4: _____ your solutions in the original equation! Any extraneous solutions must be discarded.

2. Isolate the radical in the following equation.

$$\sqrt{x-3}+4=x \qquad\qquad \sqrt{x-3}=$$

3. To what power should we raise both sides of the following equation?

$$\left(2x^2-4x+12\right)^{1/5}=3 \qquad\qquad \underline{}^{th} \text{ power}$$

<u>Video Examples:</u>

Watch the videos for Section 2.6 on `www.hawkestv.com` and then solve the following problems.

- Solve the following radical equation.

$$\sqrt{50 + 7s} - s = 8$$

- Solve the following equation.

$$x^{3/5} - 8 = 0$$

Practice:

1. Solve the following radical equation.

$$\sqrt{7x + 11} + 6 = x + 5$$

2. Solve the following radical equation.

$$\sqrt[3]{6y^2 + 27y} = \sqrt[3]{2y^2 + 40}$$

3. Solve the following equation.

$$z^{2/3} - \frac{49}{64} = 0$$

Now complete the rest of the problems in the Practice section on your Hawkes courseware and then you'll be ready to Certify!

Chapter 3

Linear Equations and Inequalities of Two Variables

In the previous chapter, we mainly considered equations and inequalities that contained only one variable. We were able to list our solutions as a single number or, in the case of inequalities, as a section of the real number line. In this chapter we turn our attention to equations and inequalities of two variables. In the case of equations, our solutions will consist of ordered pairs. In the case of inequalities, our solutions will be sections of a two-dimensional coordinate system.

3.1   The Cartesian Coordinate System

You've probably worked with the Cartesian coordinate system before, but in this section we will look at the basic properties of this system and familiarize ourselves with a way to find the distance between two points as well as the midpoint of two points.

Objectives:

- The components of the Cartesian coordinate system

- The distance and midpoint formulas

- The graph of an equation

<u>Learn:</u>

1. The **Cartesian coordinate system** consists of two _____ real number lines intersecting at the point _____ of each line.

2. In which quadrant would each of the following points be plotted?

   (a) $(-3, 2)$ Quadrant _____

   (b) $(3, -2)$ Quadrant _____

   (c) $(3, 2)$ Quadrant _____

   (d) $(-3, -2)$ Quadrant _____

3. Fill in the blanks in the following table so that the ordered pairs are solutions of the equation $x - y = 4$.

x	-1	0
y		1

4. The distance between two points, (x_1, y_1) and (x_2, y_2) is given by

$$d =$$

5. The midpoint between two points, (x_1, y_1) and (x_2, y_2) is given by

$$(\quad , \quad)$$

<u>Video Examples:</u>

Watch the videos for Section 3.1 on `www.hawkestv.com` and then solve the following problems.

- Fill in the table so that the ordered pairs are solutions of the equation $y = x^2 - x - 2$.

x	0	2
y		1

- Find the distance between $(4, 1)$ an $(-2, -1)$.

Practice:

1. Consider the following pair of points: $(8, -6)$ and $(6, -3)$.

    (a) Determine the distance between the two points.

    (b) Determine the midpoint of the line segment joining the pair of points.

2. Consider the following points: $(9, 4), (-4, 2)$ and $(-4, 4)$. Determine whether or not the given points form a right triangle. If the triangle is **not** a right triangle, determine if it is isosceles or scalene.

3. Find the perimeter of the triangle whose vertices are the following specified points in the plane.

$$(-9, 2), (-7, -1) \text{ and } (3, 1)$$

Now complete the rest of the problems in the Practice section on your Hawkes courseware and then you'll be ready to Certify!

3.2 Linear Equations in Two Variables

The solution set of a linear equation in two variables is a line in the Cartesian plane. Our job in this section is to recognize these types of equations and begin to understand what the graph of the solution will look like.

Objectives:

- Finding the intercepts of the coordinate axes

- Recognizing linear equations in two variables

Learn:

1. A **linear equation in two variables**, say the variables x and y, is an equation that can be written in the form _____, where a, b, and c are constants and a and b are not both zero. This form of such an equation is called the **standard form**.

2. To find the x-intercept of a linear equation, we let $y =$ _____ and solve for x. To find the y-intercept of a linear equation, we let $x =$ _____ and solve for y.

3. An equation of the form $ax = c$ represents a _____ line with an x-intercept of _____.

4. An equation of the form $by = c$ represents a _____ line with a y-intercept of _____.

Video Examples:

Watch the videos for Section 3.2 on `www.hawkestv.com` and then solve the following problems.

- Find the x- and y-intercepts of the equation $3y - 5x = -4(1 + x)$.

- Determine if the equation is linear.

$$\frac{6}{x} - \frac{5}{y} = 2xy$$

Practice:

1. Determine the x- and y-intercepts of the following linear equation, if possible.

$$11x + 8 = -11y + 8$$

2. Determine if the following equation is linear.

$$\frac{12x - y}{5} + 8y = 2$$

3. Determine if the following equation is linear. If the equation is linear, then write it in standard form: $ax + by = c$.

$$(10 + y)^2 - y^2 = 11x + 2$$

Now complete the rest of the problems in the Practice section on your Hawkes courseware and then you'll be ready to Certify!

3.3 Forms of Linear Equations

Linear equations can appear in a variety of forms. Depending on what you want to do with the equation, some forms are more useful than others. In this section, we'll look at a few of the different forms of linear equations and how to use certain ones to graph the corresponding line.

Objectives:

- Be able to graph a line using the slope-intercept form

- Be able to graph a line using the standard form

- Be able to graph a linear equation

- Be able to write the equation of a line in slope-intercept form

- Be able to write the equation of a line in standard form

- Identify the graph when given a point and the slope

- Understand the meaning of and be able to calculate the slope of a line

Learn:

1. Given a line L through the points (x_1, y_1) and (x_2, y_2), the **slope** of the line L is the ratio —————————.

2. The only lines with slope equal to 0 are ———————— lines.

3. The only lines for which the slope is undefined are ———————— lines.

4. The **slope-intercept form** of a line is an equation of the form ———————— where ——— is the slope of the line, and the y-intercept of the line is $(0, \_\_\_)$.

5. Given an ordered pair (x_1, y_1) and a real number m, an equation for the line passing through the point (x_1, y_1) with slope m is ————————.

<u>Video Examples:</u>

Watch the videos for Section 3.3 on www.hawkestv.com and then solve the following problems. Write your solutions using interval notation.

- Find an equation in slope-intercept form of the line through $(0, 3)$ with slope -4.

- Find an equation of the line, in slope-intercept form, that goes through the points $(-2, 3)$ and $(4, -1)$.

Practice:

1. Write the standard form of the equation for the line that passes through the points $(-1, 1)$ and $(-1, -1)$.

2. Find the slope of the line determined by the equation $4x - y = 22$.

3. Graph the linear equation:

$$y = \frac{-2}{3}x - 2$$

Now complete the rest of the problems in the Practice section on your Hawkes courseware and then you'll be ready to Certify!

3.4 Parallel and Perpendicular Lines

When we encounter a pair of linear equations, one of the things we will be interested in is whether or not these lines intersect. If they don't intersect, the two lines are parallel. If they do intersect, we then may want to determine if they do so at a right angle. If this is the case, we say that the two lines are perpendicular.

Objectives:

- Determine if lines are parallel given points

- Determine if two lines are parallel

- Determine if two lines are parallel, perpendicular, or neither

- Determine if two lines are perpendicular

- Find the equation of a line parallel to a given equation

- Find the equation of a line perpendicular to a given equation

Learn:

1. Two lines are parallel if an only if they have the same _____.

2. If m_1 and m_2 represent the slopes of two perpendicular lines, neither or which is vertical, then

$$m_1 = \qquad\qquad \text{and } m_2 = \qquad .$$

3. Given the line $3x - 2y = 9$, what is the slope of a line that is perpendicular to the given line?

$$m = \underline{\quad\quad}$$

4. Which of the following lines is perpendicular to the line $y = -5x + 3$?

$$y = \frac{1}{5}x - 2 \qquad y = -5x - 2 \qquad y = -\frac{1}{5}x - 2$$

Video Examples:

Watch the videos for Section 3.4 on www.hawkestv.com and then solve the following problems.

- Are the following lines parallel, perpendicular, or neither?

$$5 - (4y + 3x) = 5(x - y) \quad \text{and} \quad y + 4 = 3 + 8x$$

- Find the equation of the line through $(8, 5)$, perpendicular to $\dfrac{3x - y}{4} = \dfrac{4x - 5}{2}$.

Practice:

1. Find the equation, in slope-intercept form, for the line which is parallel to the following line and passes through the point $(-4, 12)$.

$$1 - \frac{3y + 6x}{3} = 3$$

2. Find the equation, in slope-intercept form, for the line which is perpendicular to the following line and passes through the point $(3, -3)$.

$$7x + 5y = 4y - 4$$

3. Determine if the two lines are parallel, perpendicular, or neither.

$$\frac{7x - 3y}{4} = x + 2 \quad \text{and} \quad -3y - 2x = x + 1$$

Now complete the rest of the problems in the Practice section on your Hawkes courseware and then you'll be ready to Certify!

3.5 Linear Inequalities in Two Variables

When we solved linear inequalities in one variable, one of the ways we represented our solution was by shading in a portion of the real-number line. To solve linear inequalities in two variables, we will need to represent our solutions graphically by shading in a portion of the Cartesian plane. In this section, we will look at the methods used to determine what part of the plane to shade in.

Objectives:

- Solving a system of linear inequalities

- Solving linear inequalities in two variables

- Solving linear inequalities joined by "and" or "or"

- Solving linear inequalities with absolute values

- Solving linear inequalities with absolute values by graphing

Learn:

1. A linear inequality in the two variables x and y is an inequality that can be written in the form

   _____, _____, _____, or _____,

   where a, b, and c are constants and a and b are not both 0.

2. In solving linear inequalities in two variables, we make the boundary line _____ if the inequality symbol is $\leq$ or $\geq$ and _____ if the symbol is $<$ or $>$.

3. For the inequality $y > \dfrac{3}{2}$, would you shade the portion of the graph that is *above* or *below* the line $y = \dfrac{3}{2}$?

4. The union of two set A and B, denoted $A \cup B$, is the set containing all elements that are in A _____ B, while the intersection of two sets A and B, denoted $A \cap B$, is the set containing all elements that are in both A _____ B.

5. What two inequalities must be solved to solve the absolute value inequality $|x + 3| > 1$?

   _____ or _____

Video Examples:

Watch the videos for Section 3.5 on www.hawkestv.com and then solve the following problems.

- Solve the inequality $3x + 3 < -3y + 1$.

- Solve the absolute value inequality $|4x - 2y| > 8$.

<u>Practice:</u>

1. Solve the system of two linear inequalities graphically.

$$2x + 2y < -4 \ \text{ and } \ x \geq 5$$

2. Solve the system of two linear inequalities graphically.

$$3y < -4x + 18 \ \text{ or } \ 4y \geq -3x + 18$$

3. Graph the solution set of the following linear inequality.

$$|4x + 8| \leq 18 \ \text{ or } \ |4y + 8| \leq 18$$

Now complete the rest of the problems in the Practice section on your Hawkes courseware and then you'll be ready to Certify!

3.6 Introduction to Circles

Another type of two variable equation is the equation of a circle. In order to graph a circle, we need to know two things: the center of the circle and the radius of the circle. We will look at forms of equations of circles that allow us to quickly identify these two important items.

<u>Objectives:</u>

- Graphing circles

- Standard form of a circle

<u>Learn:</u>

1. The **standard form** of the equation for a circle of radius r with center (h, k) is _____.

2. Determine the center and radius of the circle with equation $(x - 3)^2 + (y + 4)^2 = 16$.

   Center:(,), Radius:_____

3. What is the radius of a circle which is tangent to the line $x = 3$ and whose center is $(6, 2)$?

4. Fill in the blanks to complete the square for the following equation of a circle.

$$x^2 + y^2 + 4x - 6y = 3$$
$$\left(x^2 + 4x + \underline{}\right) + \left(y^2 - 6y + \underline{}\right) = 3 + \underline{} + \underline{}$$
$$(x + \underline{})^2 + (y - \underline{})^2 = \underline{}$$

Video Examples:

Watch the videos for Section 3.6 on www.hawkestv.com and then solve the following problems.

- Find the equation of the circle with diameter endpoints $(3, 5)$ and $(-1, -3)$.

- Find the center and radius of the circle with equation $x^2 + y^2 - 2x + 10y = -1$.

Practice:

1. Consider the equation below. Find the center and radius of this circle.

$$\left(x - \frac{3}{4}\right)^2 + \left(y + \frac{2}{7}\right)^2 = \frac{16}{25}$$

2. Find the standard form of the equation for the circle with radius $\sqrt{5}$ and center $(-3, 6)$.

3. Find the standard form of the equation for the circle with center $(5, -2)$ and tangent to the x-axis.

Remember to go through all of the problems in the Practice section before attempting to Certify!

Chapter 4

Relations, Functions, and Their Graphs

We've seen that the solutions to two variable equations can be represented as a graph in the Cartesian plane. In this chapter, we will explore that idea further. In particular, we will introduce the topic of *functions* - one of the most important concepts in all of mathematics. We will learn many different types of functions and familiarize ourselves with their respective graphs.

4.1   Relations and Functions

Before we can understand functions, we must first look at *relations*. As we'll see, functions are a type of relation. Recognizing which relations are also functions is one of the key skills we'll learn in this section.

Objectives:

- Determine the difference quotient

- Functional notation and function evaluation

- Functions and the vertical line test

- Implied domain of a function

- Relations, domains, and ranges

<u>Learn:</u>

1. A **relation** is a set of _____ pairs. The **domain** of a relation is the set of all the _____ coordinates, and the **range** of a relation is the set of all second _____.

2. For the relation $R = \{(0, 3), (2, -1), (4, 2), (-2, \sqrt{2})\}$, list the domain and the range. Remember to use curly brackets ({ }) to indicate these are sets.

> Domain: Range:

3. A **function** is a relation in which every element of the domain is paired with _____ one element of the range.

4. For the linear function $y = -3x + 1$, x is called the _____ variable and y is called the _____ variable.

5. Explain why $x = 2$ can not be plugged into the function $f(x) = \dfrac{x^2 - 3x}{x - 2}$.

<u>Video Examples:</u>

Watch the videos for Section 4.1 on `www.hawkestv.com` and then solve the following problems.

- Solve for y as a function of x.

$$\frac{2y + x}{3} = 4$$

- For the function $f(x) = \sqrt{3 + x} - 2$, find $f(1)$ and the domain of the function.

Practice:

1. Given the relation $3y - 5x = -3$, describe the domain and range of the relation and determine if the relation is a function.

2. Given the function $f(x) = \sqrt{x+9} - 3$, find $f(x-1)$.

3. Given the following function, determine the difference quotient, $\dfrac{f(x+h) - f(x)}{h}$.

$$f(x) = x^2 - 3x + 2$$

Now complete the rest of the problems in the Practice section on your Hawkes courseware and then you'll be ready to Certify!

4.2 Linear and Quadratic Functions

Two of the most common types of function are linear functions and quadratic functions. In part (a) of this section, we'll consider some of the properties of each of these types of functions. In part (b), we'll look at how quadratic functions can be used to solve problems involving maximization or minimization.

4.2.a Linear and Quadratic Functions

We've already looked at many of the properties of linear functions in Chapter 3, although we were not considering them to be functions at that time. If we can solve a linear equation in two variables for the variable y, then we can think of our equation as a linear function. Quadratic functions have graphs which are parabolas. One of the skills we'll learn is how to find the vertex of a parabola for a given quadratic function.

Objectives:

- Linear functions and their graphs

- Quadratic functions and their graphs

Learn:

1. A **linear function** f of one variable, say the variable x, is any function that can be written in the form _____, where a and b are real numbers.

2. A **quadratic, or second-degree, function** f of one variable, say the variable x, is any function that can be written in the form _____, where a, b, and c are real numbers and $a \neq 0$.

3. The graph of $g(x) = (x-3)^2$ will be the graph of $f(x) = x^2$ shifted to the _____ 3 units.

4. The graph of $h(x) = x^2 - 3$ will be the graph of $f(x) = x^2$ shifted _____ 3 units.

5. The graph of the quadratic function $f(x) = -(x+2)^2 + 4$ has its vertex at the point (,   ).

Video Examples:

Watch the videos for Section 4.2a on `www.hawkestv.com` and then solve the following problems.

- Find the slope and the y-intercept of the linear function $r(x) = 2(4 - 3x) + 5x$.

- Find the vertex of the quadratic function $q(x) = x^2 + 3x + 1$.

Practice:

1. Find the slope and the y-intercept of the following linear function.

$$a(x) = (4x - 18) - (-14 + 6x)$$

2. Find the vertex of the following quadratic function.

$$f(x) = (x - 9)(x + 3) + 36$$

3. Consider the quadratic function $f(x) = (x - 2)^2 - 5$.

   (a) Find the vertex of this function.

   (b) Find the x-intercepts of this function, if any.

   (c) Determine two points on the graph of the parabola other than the vertex and the x-intercepts.

   Now complete the rest of the problems in the Practice section on your Hawkes courseware and then you'll be ready to Certify!

4.2.b   Max/Min Applications of Quadratic Functions

There are numerous situations that can arise in which we are interested in maximizing or minimizing a quantity. For example, you may want to maximize the area inside of a set amount of fencing or minimize the amount of money spent in a business. In this section, we'll learn how quadratic functions can help us solve many of these types of problems.

Objectives:

- Interlude: maximization/minimization problems

Learn:

1. Given a quadratic function in the form $f(x) = ax^2 + bx + c$, the x-coordinate of the vertex will be given by _____ and the y-coordinate of the vertex will be given by $f(\_\_\_\_\_)$.

2. For the function $f(x) = 3x^2 - 2x + 4$, find a, b, and c so that the equation is in the form $f(x) = ax^2 + bx + c$.

$$a = \_\_\_\_\_, \quad b = \_\_\_\_\_, \quad \text{and } c = \_\_\_\_\_$$

3. If $A = xy$ and $y = \dfrac{400 - 3x}{2}$, write the function $A(x)$ by substituting for y in $A = xy$.

$$A(x) =$$

Video Examples:

Watch the videos for Section 4.2b on www.hawkestv.com and then solve the following problems. Write your solutions using interval notation.

- Of all rectangles with a perimeter of 36, find the one with the largest area.

- Of all pairs of numbers (x, y) that have a sum of 22, find the pair with maximum product.

Practice:

1. Determine whether the following function has a maximum, a minimum, or neither. If it has either a maximum or minimum, find what that value is and where it occurs.

$$f(x) = -4x^2 + 24x - 38$$

2. The total cost of producing a type of car is given by $C(x) = 19000 - 30x + 0.04x^2$, where x is the number of cars produced. How many cars should be produced to incur minimum cost?

3. Among all pairs of numbers (x, y) such that $2x + y = 14$, find the pair for which the sum of squares, $x^2 + y^2$, is minimum.

Now complete the rest of the problems in the Practice section on your Hawkes courseware and then you'll be ready to Certify!

4.3 Other Common Functions

Of course, there are *many* more types of functions besides linear and quadratic functions. In this section, we'll look at some other types of functions which are common in mathematics. In part (a), we'll investigate the properties and graphs of many of these functions. In part (b), we'll look at problems dealing with direct and inverse variation.

4.3.a Other Common Functions

Objectives:

- Evaluating piecewise-defined functions

- Identify graphs of piecewise-defined functions

- Understand shapes of common functions

Learn:

1. For functions of the form $f(x) = ax^n$, what is the domain and range when

   (a) n is even?　　　　　　Domain:　　　　　　　　　Range:

   (b) n is odd?　　　　　　Domain:　　　　　　　　　Range:

2. For functions of the form $f(x) = \dfrac{a}{x^n}$, what is the domain and range when

   (a) n is even?　　　　　　Domain:　　　　　　　　　Range:

   (b) n is odd?　　　　　　Domain:　　　　　　　　　Range:

3. For functions of the form $f(x) = ax^{1/n}$, what is the domain and range when

   (a) n is even?　　　　　　Domain:　　　　　　　　　Range:

   (b) n is odd?　　　　　　Domain:　　　　　　　　　Range:

4. For the greatest integer function, $f(x) = [[x]]$, find the following:

   (a) $f(3.2) = [[3.2]] = $ _____

   (b) $f(-3.2) = [[-3.2]] = $ _____

Video Examples:

Watch the videos for Section 4.3a on www.hawkestv.com and then solve the following problems.

- Choose the function that corresponds with each of the following graphs.

$$f(x) = x^4, \quad f(x) = \frac{1}{x^4}, \quad f(x) = \sqrt[4]{x}, \quad f(x) = 4|x|$$

1.

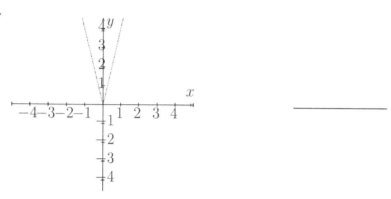

_____

2.

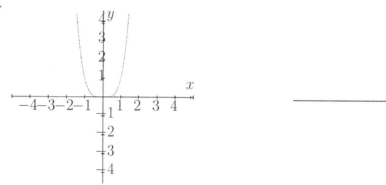

_____

3.

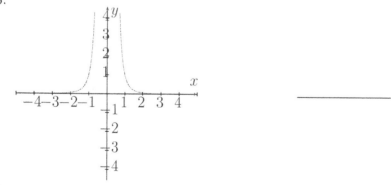

_____

4.

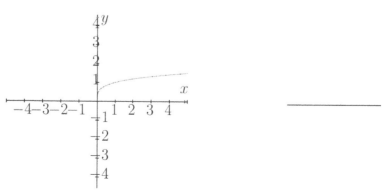

_____

Practice:

1. Consider the function $f(x) = \dfrac{9}{8x^3}$. Sketch a graph of this function. Find and plot at least two ordered pairs.

2. Consider the function $t(x) = 6|x|$. Sketch a graph of this function. Find and plot at least two ordered pairs.

3. Consider the piecewise-defined function $\begin{cases} 3x + 1 & \text{if } x \leq -4 \\ \frac{2}{3}x - 7 & \text{if } x > -4 \end{cases}$.

    (a) Evaluate the function at $x = -4$.

    (b) Evaluate the function at $x = -8$.

    (c) Evaluate the function at $x = 0$.

Now complete the rest of the problems in the Practice section on your Hawkes courseware and then you'll be ready to Certify!

4.3.b   Direct and Inverse Variation

There are a lot of instances of one quantity changing as the result of another quantity changing. This is called *variation*. If increasing one quantity causes another quantity to increase, we call it *direct variation*. If increasing one quantity causes another quantity to decrease, we call it *inverse variation*. We'll look at both types of variations in this section.

Objectives:

- Interlude: variation problems

Learn:

1. We say that y **varies directly as the nth power of** x if there is a non-zero constant k such that $y =$ _____.

2. We say that y **varies inversely as the nth power of** x if there is a non-zero constant k such that $y =$ _____.

3. If y varies inversely as the square root of x, then $y = \dfrac{k}{\rule{2em}{0.4pt}}$.

4. If y varies directly as the square of x, then $y = k \cdot \rule{3em}{0.4pt}$.

Video Examples:

Watch the videos for Section 4.3b on `www.hawkestv.com` and then solve the following problems.

- Suppose y varies directly as the square root of x and $y = 10$ when $x = 4$. What is y when $x = 12$?

- Suppose y varies inversely as the square of x and $y = 7$ when $x = 2$. What is x when $y = 9$?

Practice:

1. Suppose y varies directly as the square of x, and that $y = 7$ when $x = 3$. What is y when $x = 5$?

2. The distance that an object falls from rest, when air resistance is negligible, varies directly as the square of the time that it falls (before it hits the ground). A stone dropped from rest travels 249 feet in the first 5 seconds. How far will it have fallen at the end of 7 seconds?

3. The volume of a gas in a container varies inversely as the pressure on the gas. If a gas has a volume of 221 cubic inches under a pressure of 5 pounds per square inch, what will be its volume if the pressure is increased to 6 pounds per square inch?

Now complete the rest of the problems in the Practice section on your Hawkes courseware and then you'll be ready to Certify!

4.4 Transformations of Functions

Now that we have a catalog of many different types of functions, we can begin to investigate how making changes to the formula of one of those functions will change the corresponding graph of the function. This section will consider a number of these transformations.

Objectives:

- Determine if a function is even, odd, or neither

- Determine if a given equation is even, odd, or neither

- Intervals of monotonicity

- Shifting, stretching, and reflecting graphs

- Symmetry of functions and equations

Learn:

1. Suppose the graph of $y = f(x)$ is known.

    (a) The graph of $g(x) = f(x - 2)$ will shift the graph of f _____ units to the

       _____.

    (b) The graph of $g(x) = f(x) - 4$ will shift the graph of f _____ units

       _____.

    (c) The graph of $g(x) = -f(x)$ will reflect the graph of f over the _____-axis.

    (d) The graph of $g(x) = 3f(x)$ will _____ the graph of f by a factor

       of _____.

2. Is $f(x) = x^4 - 3x^2 + 2$ an *even* function, an *odd* function, or neither?

3. If replacing x with $-x$ and y with $-y$ in an equation results in an equivalent equation, is the equation symmetric with respect to the x-axis, y-axis, or origin?

4. Consider the graph of $f(x) = (x-2)^2 + 1$ shown below.

   (a) On which interval(s) is f increasing?

   (b) On which interval(s) is f decreasing?

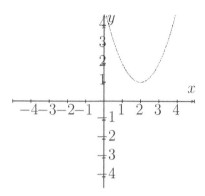

Video Examples:

Watch the videos for Section 4.4 on `www.hawkestv.com` and then solve the following problems.

- Consider the function $f(x) = -\dfrac{1}{(x-3)^2} + 4$.

  1. What is the more basic function that is being transformed?

  2. Does the graph of f shift the basic function horizontally? If so, which direction and how many units?

  3. Does the graph of f reflect the basic function over an axis? If so, which one?

  4. Does the graph of f shift the basic function vertically? If so, which direction and how many units?

- Consider the function $g(x) = (x+3)^2 - 2$.

    1. Where is the vertex of the graph of g located?

    2. Which way is the parabola opening?

    3. On which interval(s) is the function decreasing?

Practice:

1. Determine if the equation $y = \left| \dfrac{x}{3} \right|$ has x-axis symmetry, y-axis symmetry, origin symmetry, or none of these.

2. Consider the function $h(x) = \sqrt{x+3} - 5$.

    (a) Identify the more basic function that has been shifted, reflected, stretched, or compressed.

    (b) Identify the direction and number of units of the horizontal shift.

    (c) Identify the direction and number of units of the vertical shift.

3. Find and identify all of the intervals where the following function is increasing, decreasing, or constant.

$$m(x) = |x - 3|$$

    Remember to go through all of the problems in the Practice section before attempting to Certify!

4.5 Combining Functions

We know that we can combine the real numbers with the operations of addition, subtraction, multiplication, and division. Wouldn't it be great if we could combine functions using the same operations? As a matter of fact, we can. We'll look at those four operations and throw an extra one in (called *composition*) free of charge.

Objectives:

- Combining functions arithmetically

- Composing functions

- Decomposing functions

Learn:

1. Suppose $f(2) = 3$ and $g(2) = -4$. Find the values of the following functions.

$$
\begin{aligned}
(f + g)(2) &= f(2) + g(2) = \\
(f - g)(2) &= f(2) - g(2) = \\
(fg)(2) &= f(2)g(2) = \\
\left(\frac{f}{g}\right)(2) &= \frac{f(2)}{g(2)} =
\end{aligned}
$$

2. Let f and g be two functions. The **composition** of f and g, denoted $(f \circ g)(x)$, is the function defined by $(f \circ g)(x) = $ _____.

3. Suppose $g(-2) = 4$ and $f(4) = 10$. Find $(f \circ g)(-2)$.

$$(f \circ g)(-2) = f(g(-2)) = f(\quad) =$$

4. Suppose you want to decompose the function $f(x) = \sqrt[3]{2x^3 - 7}$ into the composition $g(h(x))$. What would a possible function be for $h(x)$?

Watch the videos for Section 4.5 on www.hawkestv.com and then solve the following problems.

- Let $f(x) = x^3 + 4$ and $g(x) = \sqrt{x}$.

  1. Find $(f + g)(x)$ and its domain.

  2. Find $\left(\dfrac{f}{g}\right)(x)$ and its domain.

- Let $f(x) = \sqrt{2x - 5} + 3$. Find $g(x)$ and $h(x)$ so that $f(x) = (g \circ h)(x) = g(h(x))$.

Practice:

1. For $f(x) = 3 + \sqrt{x}$ and $g(x) = \sqrt{x+3}$, determine $(f \circ g)(1)$.

2. For $f(x) = \sqrt{x-1}$ and $g(x) = \dfrac{x+3}{3}$, determine the formula and domain for $(f \circ g)(x)$.

3. Consider the functions $f(x) = x - 4$ and $g(x) = x^{2/3}$. Find the formula for $\left(\dfrac{f}{g}\right)(x)$.

Remember to go through all of the problems in the Practice section before attempting to Certify!

4.6 Inverses of Functions

We can think of functions as a rule for sending one number to another number. For example, the function $f(x) = x^3$ sends the number $x = 2$ to the number 8 (since $2^3 = 8$). In this section, we'll develop methods for finding a function that sends the number back. This is called the *inverse* function.

Objectives:

- Finding inverse function formulas

- Inverse functions and the horizontal line test

- Inverse of relations

Learn:

1. Let R be a relation. The **inverse of R**, denoted R^{-1}, is the set
   $R^{-1} = \{(\quad , \quad) | (a, b) \in R\}$.

2. A function f is **one-to-one** if for every pair of distinct elements x_1 and x_2 in the domain of f, $f(x_1) \quad f(x_2)$.

3. Let f be a one-to-one function, and assume that f is defined by a formula. To find a formula for f^{-1}, perform the following steps:

   (a) Replace $f(x)$ in the definition of f with the variable _____.

   (b) _____ x and y in the equation.

   (c) Solve the new equation for _____.

   (d) Replace y in the remaining equation with _____.

4. Given a function f and its inverse f^{-1}, what would $f(f^{-1}(x))$ simplify to?

<u>Video Examples:</u>

Watch the videos for Section 4.6 on `www.hawkestv.com` and then solve the following problems.

- Let $R = \{(2,5), (-3,1), (0,1), (-7,4)\}$.

  1. Find R^{-1}.

  2. Find the domain and range of R^{-1}.
     Domain: Range:

  3. Is R^{-1} a function?

- Let $h(x) = \sqrt[3]{3x + 6}$. Find a formula for $h^{-1}(x)$.

Practice:

1. Given the relation $y = 3x - 3$:

    (a) Find four points contained in the inverse of the relation.

    (b) Define the domain and range of the inverse.

2. Consider the function $h(x) = 2\sqrt[5]{x} + 3$. Find a formula for the inverse of the function, if possible.

3. Consider the function $k(x) = \dfrac{-2}{x+2}$. Find a formula for the inverse of the function, if possible.

Remember to go through all of the problems in the Practice section before attempting to Certify!

Chapter 5

Polynomial Functions

Polynomial functions are a very common type of function that can occur in a wide variety of mathematical applications. In this chapter we take an in-depth look at these functions, including graphing, finding zeros, and the Fundamental Theorem of Algebra which, as its name implies, is pretty important.

5.1   Introduction to Polynomial Equations and Graphs

We'll begin our study of polynomial functions by looking at some of the properties of their graphs, such as behavior at infinity and factoring. We'll also take a look at polynomial inequalities, which just sound like fun.

Objectives:

- Determine the behavior of a function as x goes to positive or negative infinity

- Graphing polynomials by factoring

- Solving polynomial inequalities

- Zeros of polynomials and solutions of polynomial equations

Learn:

1. The number k is said to be a **zero** of the polynomial function
   $f(x) = a_n x^n + a_{n-1} x^{n-1} + \cdots + a_1 x + a_0$ if _____ $= 0$.

2. Fill in the blanks for **The Leading Coefficient Test**.

	Degree, n, is even	Degree, n, is odd
Leading coefficient, a_n, is positive	$f(x) \to$ _____ as $x \to -\infty$ $f(x) \to$ _____ as $x \to \infty$	$f(x) \to$ _____ as $x \to -\infty$ $f(x) \to$ _____ as $x \to \infty$
Leading coefficient, a_n is negative	$f(x) \to$ _____ as $x \to -\infty$ $f(x) \to$ _____ as $x \to \infty$	$f(x) \to$ _____ as $x \to -\infty$ $f(x) \to$ _____ as $x \to \infty$

3. Consider the polynomial function $f(x) = (x - 4)(2x + 3)(x + 5)$.

    (a) Determine the x-coordinates of the x-intercepts of f.

$$\{\underline{\hspace{1cm}}, \underline{\hspace{1cm}}, \underline{\hspace{1cm}}\}$$

    (b) Determine the y-intercept of f.

$$(0, \underline{\hspace{1.5cm}})$$

    (c) If the three linear factors were multiplied out, what would the leading term of the polynomial be?

4. A portion of the graph of $f(x) = x(x - 3)(x + 2)$ is shown below. On what interval(s) is $x(x - 3)(x + 2) \geq 0$?

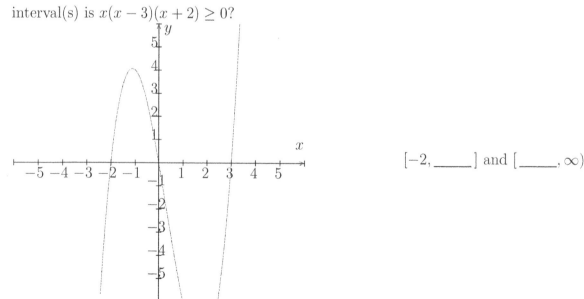

$[-2, \underline{\hspace{1cm}}]$ and $[\underline{\hspace{1cm}}, \infty)$

<u>Video Examples:</u>

Watch the videos for Section 5.1 on `www.hawkestv.com` and then solve the following problems.

- Sketch the graph of $r(x) = (x - 2)(x + 4)(1 - x)$ by finding the x-intercepts, the y-intercepts, and determining the behavior at infinity.

- Solve the polynomial inequality $x^2(x + 3)(x - 5) \leq 0$.

<u>Practice:</u>

1. Does the given value of x solve the polynomial equation?

$$-53x^2 + 21x = 7x^3 + 24; \quad x = -8$$

2. Solve the polynomial equation $x^3 - 15x^2 + 56x = 0$ by factoring or using the quadratic formula, making sure to identify all the solutions.

3. Consider the polynomial $f(x) = -4x^2(x+5)(x-8)$.

    (a) Determine the degree and leading coefficient of $f(x)$.

    (b) Describe the behavior of the graph of $f(x)$ as $x \to \pm\infty$.

Now complete the rest of the problems in the Practice section on your Hawkes courseware and then you'll be ready to Certify!

5.2 Polynomial Division and the Division Algorithm

It can be very difficult to determine the zeros of some higher-degree polynomial equations. As we have seen, however, once a polynomial is in factored form it becomes much easier to recognize the zeros of that polynomial. In this section, we will investigate how using *polynomial division* can help us find factors of a given polynomial.

Objectives:

- Constructing polynomials with given zeros

- Polynomial long division and synthetic division

- The Division Algorithm and the Remainder Theorem

Learn:

1. Let $p(x)$ and $d(x)$ be polynomials such that $d(x) \neq 0$ and with the degree of d less than or equal to the degree of p. Then there are unique polynomials $q(x)$ and $r(x)$, called the **quotient** and the **remainder**, respectively, such that

$$p(x) = \underline{\hspace{1cm}} d(x) + \underline{\hspace{1cm}}.$$

2. The number k is a zero of a polynomial $p(x)$ if and only if the linear polynomial _____ is a factor of p.

3. Suppose you want to divide $-x^3 + 3x^2 - 7x + 2$ by $x + 4$. Set this up as a *synthetic division* problem.

4. Complete the synthetic division in the previous problem. Based on your answer, if $p(x) = -x^3 + 3x^2 - 7x + 2$, then $p(-4) = \underline{\hspace{1cm}}$.

5. Suppose a polynomial has zeros of $-1, -6$, and 5. What three factors must be included in the factorization of this polynomial?

<u>Video Examples:</u>

Watch the videos for Section 5.2 on `www.hawkestv.com` and then solve the following problems.

- Divide. Write your answer in the form $\dfrac{p(x)}{d(x)} = q(x) + \dfrac{r(x)}{d(x)}$.

$$\frac{5x^2 + 9x + 6}{x + 2}$$

- Use synthetic division to determine if k is a zero of $p(x)$. If it is not, what is $p(k)$?

$$p(x) = 2x^3 - 8x^2 - 23x + 63; \quad k = 2$$

Practice:

1. Use polynomial long division to write the fraction in the form $q(x) + \dfrac{r(x)}{d(x)}$, where $d(x)$ is the denominator of the original fraction, $q(x)$ is the quotient, and $r(x)$ is the remainder.

$$\frac{12x^4 - 18x^3 + 26x^2 + 3x + 3}{3x^2 + 2}$$

2. Use synthetic division to write the fraction in the form $q(x) + \dfrac{r(x)}{d(x)}$, where $d(x)$ is the denominator of the original fraction, $q(x)$ is the quotient, and $r(x)$ is the remainder.

$$\frac{5x^3 - 23ix^2 + 5x + (4 - 4i)}{x - 4i}$$

3. Construct a polynomial which is third-degree, with zeros of $-3, -1$, and 2, and passes through the point $(3, 9)$.

Now complete the rest of the problems in the Practice section on your Hawkes courseware and then you'll be ready to Certify!

5.3 Locating Real Zeros of Polynomial

If we know one of the zeros of a given polynomial, we can use division to factor that polynomial. However, sometimes it may be difficult to even locate one zero of a given polynomial. In this section, we will look at some methods designed to help us determine what numbers might possibly be zeros of a polynomial.

Objectives:

- Descartes' Rule of Signs

- The Intermediate Value Theorem

- The Rational Zero Theorem

- Use Descartes' Rule of Signs to find the real zeros of a polynomial

Learn:

1. If $f(x) = a_n x^n + a_{n-1} x^{n-1} + \cdots + a_1 x + a_0$ is a polynomial with integer coefficients, then any *rational zero* of f must be of the form $\dfrac{p}{q}$, where p is a factor of the constant term _____ and q is a factor of the leading coefficient _____.

2. Consider the polynomial $f(x) = 2x^3 - 9x^2 + 11x - 6$.

    (a) List the factors of a_0:

    (b) List the factors of a_3:

    (c) List the possible rational zeros:

3. How many *variations in sign* occur in each of the following polynomials.

    (a) $g(x) = 7x^3 + 4x^2 - 5x + 13$ Variations in sign:

    (b) $h(x) = -11x^4 + 5x^3 - 7x^2 + 3x + 1$ Variations in sign:

4. Use synthetic division to determine if -3 is a lower bound for the function $f(x) = 2x^3 - x^2 - 7x + 6$.

5. Suppose $f(x)$ is a polynomial with real coefficients and that $f(2) = -5$ and $f(3) = 2$. According to the **Intermediate Value Theorem**, there is at least one point c with _____ $< c <$ _____ and $f(c) = 0$.

Video Examples:

Watch the videos for Section 5.3 on www.hawkestv.com and then solve the following problems.

- Given the polynomial $D(x) = -2x^3 + 9x^2 + 11x - 30$:

    1. List all of the potential rational zeros.

    2. Use polynomial division and the quadratic formula, if necessary, to identify the actual zeros.

- Solve: $x^3 - 3 + 3x^2 - x = 0$.

Practice:

1. Given the polynomial $f(x) = 2x^3 - x^2 + 8x - 4$:

   (a) List all of the possible rational zeros.

   (b) Use polynomial division and the quadratic formula, if necessary, to identify the actual zeros.

2. Given the function below, does the Intermediate Value Theorem guarantee that a real zero exists between the indicated values?

$$g(x) = F(x) = x^4 - 9x^3 + 64x^2 + 156x - 144; \quad \frac{11}{2} \text{ and } \frac{13}{2}$$

3. Given the polynomial $g(x) = x^3 - x^2 - 14x + 24$:

    (a) Use Descartes' Rule of Signs to determine the possible number of positive and negative zeros.

    (b) Use synthetic division to identify the best integer upper and lower bounds of the real zeros.

    (c) Using your answers to the preceding steps, polynomial division, and the quadratic formula, if necessary, list all of the zeros.

      Now complete the rest of the problems in the Practice section on your Hawkes courseware and then you'll be ready to Certify!

5.4 The Fundamental Theorem of Algebra

Well, this certainly sounds important! We conclude this chapter by combining all of the methods we learned to be able find the zeros of a multitude of polynomials. We'll show that this is never a vain attempt. In fact, the Fundamental Theorem of Algebra guarantees the existence of at least one zero for *any* polynomial of degree 1 or more.

Objectives:

- Conjugate pairs of zeros

- Constructing polynomials with given properties

- Find the real zeros of a given polynomial

- Multiple zeros and their geometric meaning

Learn:

1. If f is a polynomial of degree n, with $n \geq 1$, then f has at least one _____.

2. The graph of an nth-degree polynomial function has *at most* _____ x-intercepts at *at most* _____ turning points.

3. Consider the polynomial function $f(x) = x(x-2)^3(x+1)$. List the zeros of f and the multiplicity of each zero.
   Zero: _____ Multiplicity: _____   Zero: _____ Multiplicity: _____
   Zero: _____ Multiplicity: _____

4. Suppose that $3 - 5i$ is a zero of a given polynomial function. According to the **Conjugate Roots Theorem**, what complex number is also a zero?

Watch the videos for Section 5.4 on www.hawkestv.com and then solve the following problems.

- Find the zeros and sketch the graph of the polynomial function
  $f(x) = x^3(x+3)(x-1)^2$.

- Construct a polynomial which is third degree, 3 is a zero of multiplicity 2, -1 is the only other zero, and the leading coefficient is 2.

Practice:

1. Consider the polynomial function $f(x) = 2x^3 + 13x^2 + 24x + 9$.

   (a) Use all available methods to factor the polynomial function completely.

   (b) Determine the degree and the y-intercept as an ordered pair.

   (c) Determine the x-intercepts at which f crosses the axis.

   (d) Determine the number of zero(s) of f at which it "flattens out".

2. Use all available methods (in particular, the Conjugate Roots Theorem, if applicable) to factor the following polynomial function completely, making use of the given zero.

$$f(x) = x^4 - 10x^3 + 28x^2 - 40x + 96; \quad -2i \text{ is a zero.}$$

3. Construct a polynomial function with the following properties: fifth degree, 5 is a zero of multiplicity 3, -4 is the only other zero, leading coefficient is 3.

Now complete the rest of the problems in the Practice section on your Hawkes courseware and then you'll be ready to Certify!

Chapter 6

Rational Functions and Conic Sections

In this chapter we consider another type of function: rational functions. In the first section, we'll look at the properties of rational functions and the methods that are available to draw their graphs. For the remainder of the chapter, we turn our attention to investigating and classifying the conic sections: the ellipse, the parabola, and the hyperbola.

6.1 Rational Functions

We'll look at this section on rational functions in two parts: the first will deal with graphing the rational functions through a series of steps including finding intercepts and *asymptotes*. Then, we'll look at how being able to graph rational functions will allow us to more easily solve rational inequalities.

6.1.a Rational Functions

In many ways, graphing rational functions is similar to graphing polynomial functions: we want to find the x-intercepts, the y-intercepts, and determine the behavior of the graph at infinity. However, for rational functions, the behavior at infinity is often determined by an asymptote. Finding asymptotes will be one of the main points of focus in this section.

Objectives:

- Determine the asymptotes of a given rational function

- Graphing rational functions

- Horizontal and oblique asymptotes

- Using asymptotes to graph rational functions

- Vertical asymptotes

Learn:

1. Consider the rational function $f(x) = \dfrac{x}{x+2}$. What is the domain of $f(x)$?

2. The vertical line _____ is a **vertical asymptote** of a function f if $f(x)$ increases in magnitude without bound as x approaches c.

3. The horizontal line _____ is a **horizontal asymptote** of a function f if $f(x)$ approaches the value c as $x \to -\infty$ or as $x \to \infty$.

4. Determine the vertical asymptote and horizontal asymptote for the rational function whose graph is shown below.

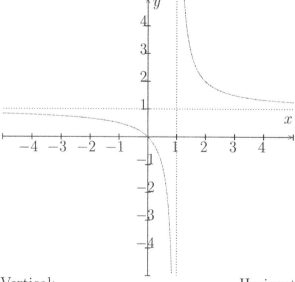

Vertical: Horizontal:

5. Determine the vertical asymptote(s) and horizontal asymptote, if they exist, for the functions below.

   (a) $f(x) = \dfrac{x^2 + 3x + 1}{x^2 - 4}$

   Vertical: Horizontal:

   (b) $g(x) = \dfrac{3x + 5}{2x - 6}$

   Vertical: Horizontal:

<u>Video Examples:</u>

Watch the videos for Section 6.1a on `www.hawkestv.com` and then solve the following problems.

- Consider the rational function $f(x) = \dfrac{-x + 3}{x - 4}$.

  1. Find any vertical asymptotes.

  2. Find any horizontal asymptotes.

  3. Find three points on the graph of f.

- Consider the rational function $g(x) = \dfrac{x^3 + 4}{x^2 - 9}$.

  1. Find any vertical asymptotes.

  2. Find the *oblique* asymptote by using long division.

Practice:

1. Given the rational function: $f(x) = \dfrac{-16x^2 - 8x + 8}{4x + 7}$. Find equations for the oblique asymptotes, if any, for the rational function.

2. Given the rational function: $g(x) = \dfrac{-x^2 - 6x - 2}{x^2 - 13x + 30}$.

    (a) Find equations for the vertical asymptotes, if any.

    (b) Find equations for the horizontal or oblique asymptotes, if any.

    (c) Find four ordered pairs for the graph of the rational function.

3. Consider the following rational function: $f(x) = \dfrac{-8}{x^2 - 25}$.

    (a) Find the vertical asymptotes, if any, of this function.

    (b) Find the horizontal asymptotes, if any, of this function.

    (c) Sketch the graph of the function by finding x-intercepts, the y-intercept, plotting additional points, and making use of symmetry.

Now complete the rest of the problems in the Practice section on your Hawkes courseware and then you'll be ready to Certify!

6.1.b   Rational Inequalities

Now that we can graph rational functions, we possess all of the skills we need to solve rational inequalities. As was the case for linear inequalities, we well need to indicate our solutions using interval notation. Let's get started.

Objectives:

- Solving rational inequalities

Learn:

1. Consider the graph of the rational function $f(x) = \dfrac{x^2 + 2}{x^2 - x - 6}$ shown below, and use it to determine the solution of the rational inequality $\dfrac{x^2 + 2}{x^2 - x - 6} > 0$.

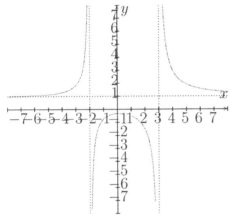

2. Consider the rational inequality $0 \leq \dfrac{2x + 4}{x - 3}$.

   (a) Set the numerator of the right side equal to zero and solve for x.

   (b) Set the denominator of the right side equal to zero and solve for x.

(c) Determine the intervals which need to be tested with a test point based on your answers above.

$$(-\infty, \underline{\hspace{1cm}}], [\underline{\hspace{1cm}}, \underline{\hspace{1cm}}), \text{ and } (\underline{\hspace{1cm}}, \infty)$$

(d) Testing a point from each of these intervals, we see that the solution set of the inequality is:

$$(-\infty, \underline{\hspace{1cm}}] \text{ and } (\underline{\hspace{1cm}}, \infty)$$

3. For the rational inequality $\dfrac{1}{x+4} < \dfrac{1}{x-2}$, rewrite the inequality in standard form and simplify.

$$\frac{1}{x+4} < \frac{1}{x-2}$$

$$\frac{1}{x+4} - \hspace{1cm} < 0$$

$$\frac{(x - \underline{\hspace{1cm}}) - (x + \underline{\hspace{1cm}})}{(x+4)(x-2)} < 0$$

$$\frac{\overline{\hspace{2cm}}}{(x+4)(x-2)} < 0$$

Video Examples:

Watch the videos for Section 6.1b on www.hawkestv.com and then solve the following problems.

- Solve the rational inequality. Write your answer in interval notation.

$$\frac{-3}{x-2} \leq \frac{5x}{x-2}$$

- Solve the rational inequality. Write your answer in interval notation.

$$\frac{3}{x^2 - 16} \geq \frac{x}{x^2 - 16}$$

Practice:

1. Solve the rational inequality. Write your answer in interval notation.

$$\frac{x - 6}{4} \geq 0$$

2. Solve the rational inequality. Write your answer in interval notation.

$$\frac{x+3}{x+1} < \frac{x}{x+4}$$

3. Solve the rational inequality. Write your answer in interval notation.

$$4x > \frac{16}{x+3}$$

Now complete the rest of the problems in the Practice section on your Hawkes courseware and then you'll be ready to Certify!

6.2 The Ellipse

We turn our attention to conic sections for the rest of this chapter and we begin with *ellipses*. One of the most common examples of ellipses is the orbit of planets. We'll investigate this phenomenon after we get a working understanding of the standard form of an ellipse.

Objectives:

- Interlude: planetary orbits

- Using the standard form of an ellipse

Learn:

1. A conic section described by an equation of the form
   $Ax^2 + Cy^2 + Dx + Ey + F = 0$, where at least one of the two coefficients A and C is not equal to 0 is:

   (a) an **ellipse** if the product AC is _____;

   (b) a **parabola** if the product AC is _____;

   (c) a **hyperbola** if the product AC is _____.

2. Consider the ellipse given by the equation $\dfrac{x^2}{25} + \dfrac{y^2}{9} = 1$.

   (a) The x-intercepts of the ellipse are at the points (_____, 0) and (_____, 0).

   (b) The y-intercepts of the ellipse are at the points (0, _____) and (0, _____).

   (c) Using the formula $c^2 = a^2 - b^2$, we can determine that the foci of the ellipse are at the point (_____, 0) and (_____, 0).

3. Consider the ellipse given by the equation $\dfrac{(x-2)^2}{4} + \dfrac{(y+3)^2}{16} = 1$.

   (a) The center of the ellipse is at the point (_____, _____).

   (b) The major axis is vertical and has length _____.

   (c) The minor axis is horizontal and has length _____.

4. Given an ellipse with major and minor axes of lengths $2a$ and $2b$, respectively, the **eccentricity** of the ellipse, often given the symbol e, is defined by:

$$e = \frac{c}{a} = \frac{\sqrt{\underline{\quad\quad}}}{\underline{\quad\quad}}.$$

<u>Video Examples:</u>

Watch the videos for Section 6.2 on `www.hawkestv.com` and then solve the following problems.

- Consider the following equation of an ellipse. Graph the equation by finding the center, the major axis endpoints, the minor axis endpoints, and the foci.

$$x^2 + \frac{(y-3)^2}{49} = 1$$

- Consider the following equation of an ellipse. Rewrite the equation in standard form, then graph the equation by finding the center, the major axis endpoints, the minor axis endpoints, and the foci.

$$16x^2 + 64y^2 + 160x + 512y + 400 = 0$$

<u>Practice:</u>

1. Find the equation, in standard form, for the ellipse which has vertices at $(-8, -5)$ and $(-2, -5)$ and eccentricity $e = \dfrac{2}{3}$.

2. Find the equation, in standard form, for the ellipse which is centered at $(-4, -5)$, has major axis of length 18 oriented vertically, and minor axis of length 6.

3. The planet Uranus moves in an elliptical orbit with the sun at one focus. Given that Uranus' closest approach to the sun is approximately 2741.3 million kilometers and that the eccentricity of Uranus' orbit is approximately 0.046, estimate this planet's maximum distance from the sun.

Now complete the rest of the problems in the Practice section on your Hawkes courseware and then you'll be ready to Certify!

6.3   The Parabola

The next conic section we will look at is the *parabola*. While we have studied parabolas some when graphing quadratic functions, in this section we take a more in-depth look at the properties and equations of parabolas, including ones that open left or right.

Objectives:

- Using the standard form of a parabola

Learn:

1. Assume p is a fixed non-zero real number. The **standard form** of the equation for the parabola with vertex at (h, k) is:

    (a) $(x - h)^2 = 4p(y - k)$ if the parabola is _____ oriented (the focus is at (_____, _____) and the equation of the directrix is $y =$ _____);

    (b) $(y - k)^2 = 4p(x - h)$ if the parabola is _____ oriented (the focus it at (_____, _____) and the equation of the directrix is $y =$ _____).

2. Given the equation of the parabola $(x-2)^2 = 12(y+3)$, determine the following.

    (a) The vertex of the parabola: (_____, -3)

    (b) The orientation of the parabola (vertical or horizontal):

    (c) The value of p (*note:* $4p = 12$):

    (d) The focus of the parabola: $(2,$ _____$)$

    (e) The directrix of the parabola: $y =$ _____

3. Complete the steps necessary to put the equation in the standard form of a parabola.

$$
\begin{aligned}
y^2 + 4x + 6y + 25 &= 0 \\
y^2 + 6y &= -4y - \underline{} \\
y^2 + 6y + \underline{} &= -4y - 25 + \underline{} \\
(y + \underline{})^2 &= -4(y + \underline{}) \\
(y + 3)^2 &= 4(\underline{})(y + 4)
\end{aligned}
$$

<u>Video Examples:</u>

Watch the videos for Section 6.3 on `www.hawkestv.com` and then solve the following problems.

- Find the equation, in standard form, for the parabola with the given properties.

  Vertex at $(12, -13)$, focus at $(12, 2)$

- Given the parabola $x^2 - 6x + 4y = -9$, find the vertex, focus, and directrix.

Practice:

1. Find the equation, in standard form, for the parabola which is symmetric with respect to the line $y = -3$, directrix is the line $x = -1$, and $p = -1$.

2. Given the parabola $y^2 - 8y = 8x + 8$:

   (a) Find the focus of the parabola.

   (b) Find the directrix of the parabola.

   (c) Find the vertex and two points which lie on the line through the focus that is parallel to the directrix.

3. A spotlight is made by placing a strong light bulb inside a reflective paraboloid formed by rotating the parabola $x^2 = 5y$ around its axis of symmetry (assume that x and y are in units of inches). In order to have the brightest, most concentrated light beam, how far from the vertex should the bulb be placed?

Now complete the rest of the problems in the Practice section on your Hawkes courseware and then you'll be ready to Certify!

6.4   The Hyperbola

The final conic section to consider in this chapter is the *hyperbola*. Hyperbolas share some similarities with other conic sections, in particular with ellipses, but there are some key differences and we will explore those in this section.

Objectives:

- Using the standard form of a hyperbola

Learn:

1. For the points on a hyperbola, the magnitude of the _____ of the distances to two foci is a fixed constant.

2. The standard form of the equation for the hyperbola with center at (h, k) is:

   (a) $\dfrac{(x - h)^2}{a^2} - \dfrac{(y - k)^2}{b^2} = 1$ if the foci are aligned _____;

   (b) $\dfrac{(y - k)^2}{a^2} - \dfrac{(x - h)^2}{b^2} = 1$ if the foci are aligned _____.

3. Consider the hyperbola given by the equation $\dfrac{(x - 3)^2}{4} - \dfrac{(y + 1)^2}{16} = 1$.

   (a) The vertices of the hyperbola are at the point (_____, _____) and (_____, _____).

   (b) The foci of the parabola are aligned _____.

   (c) The foci of the hyperbola are at the points (_____, −1) and (_____, −1).

   (d) The slopes of the asymptotes of the hyperbola are ±_____.

<u>Video Examples:</u>

Watch the videos for Section 6.4 on `www.hawkestv.com` and then solve the following problems.

- For the hyperbola given by the equation $\dfrac{(x+3)^2}{49} - \dfrac{(y+4)^2}{16} = 1$, sketch a graph of the hyperbola indicating the vertices, foci, and asymptotes.

- Find the equation in standard form of the hyperbola with foci at $(-1, 3)$ and $(-1, 13)$ and asymptotes of $y - 8 = \pm\dfrac{3}{4}(x + 1)$.

Practice:

1. Consider the hyperbola given by the equation $\dfrac{(y-4)^2}{16} - (x+8)^2 = 1$.

   (a) Determine the coordinates of the vertices of the hyperbola.

   (b) Determine the coordinates of the foci of the hyperbola.

   (c) Determine the equations of the asymptotes of the hyperbola.

2. Find the equation in standard form of the hyperbola with the foci at $(4, -2)$ and $(12, -2)$ and the vertices at $(5, -2)$ and $(11, -2)$.

3. Find the equation in standard form of the hyperbola shown below.

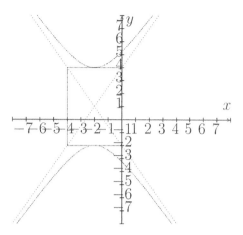

Now complete the rest of the problems in the Practice section on your Hawkes courseware and then you'll be ready to Certify!

Chapter 7

Exponential and Logarithmic Functions

One of the most commonly used types of functions in mathematics is exponential functions. There are numerous applications of these functions in various areas from investing to population to radioactive decay. We study logarithmic functions in this chapter as well, since logarithmic functions are the inverses of exponential functions. Like their exponential partners, logarithmic functions serve as an excellent function for modeling a variety of real-world phenomena.

7.1   Exponential Functions and Their Graphs

In this section we take an introductory look at exponential functions. We'll investigate the properties of these functions and the basic shape of their graph. In addition, we'll try our hand at solving some basic exponential equations.

Objectives:

- Graphing exponential functions

- Solving elementary exponential equations

Learn:

1. Let a be a fixed positive real number not equal to 1. The **exponential function with base a** is the function $f(x) = $ _____.

2. Recall that if $a \neq 0$, then a^0 is defined to be _____.

3. Given a positive real number a not equal to 1, the function $f(x) = a^x$ is:

   (a) A decreasing function if $0 < a < 1$, with $f(x) \to$ _____ as $x \to -\infty$ and
      $f(x) \to$ _____ as $x \to \infty$.

   (b) An increasing function if $a > 1$, with $f(x) \to$ _____ as $x \to -\infty$ and
      $f(x) \to$ _____ as $x \to \infty$.

4. In which direction and how many units would the graph of $f(x) = 3^x$ be shifted
   to form the graph of $g(x) = 3^{x-2} + 4$.

   Horizontal: _____ _____ units   Vertical: _____ _____ units

5. Fill in the blanks to solve the exponential equation.

$$
\begin{aligned}
27^x - 3 &= 0 \\
27^x &= \underline{\quad} \\
\underline{\quad 3x \quad} &= \underline{\quad}^1 \\
\underline{\quad} &= 1 \\
x &= \underline{\quad}
\end{aligned}
$$

<u>Video Examples:</u>

Watch the videos for Section 7.1 on www.hawkestv.com and then solve the following
problems.

- Sketch the graph of $g(x) = 2 - \left(\dfrac{3}{2}\right)^{x+3}$.

- Solve the following exponential equation.

$$6^{4x-5} = \frac{1}{36}$$

Practice:

1. Sketch the graph of $q(x) = \left(\frac{7}{2}\right)^{x-3}$.

2. Solve the following exponential equation.

$$\left(\frac{1}{2}\right)^{3x+2} = \left(\frac{1}{4}\right)^{2}$$

3. Solve the following exponential equation.

$$3^{-x} = 81$$

Now complete the rest of the problems in the Practice section on your Hawkes courseware and then you'll be ready to Certify!

7.2 Applications of Exponential Functions

As we mentioned, exponential functions lend themselves well to a wide variety of real-world problems. In this section, we'll take a look at just a few of these applications.

Objectives:

- Compound interest and the number e

- Models of cost and revenue

- Models of decay

- Models of depreciation

- Models of population growth

- Radioactive decay

Learn:

1. Suppose an isotope has a half-life of 3 years. If there are initially 26 g of this isotope, how many grams will there be in 6 years?

2. Suppose $1000 dollars is invested in a savings account at an annual interest rate of 3.5%, compounded weekly. Fill in the values for the variables in the formula $A(t) = P\left(1 + \dfrac{r}{n}\right)^{nt}$ that would be used to find the amount in the account after 3 years.

   $P = $ _____, $r = $ _____, $n = $ _____, $t = $ _____

3. If P dollars are invested for t years in a continuously compounded account with an annual interest rate of r, the accumulation is: $A(t) = $ _____.

4. Which exponential function formula would be used to determine the population of bacteria in 24 hours if the initial population and the doubling time of the bacteria are known?

Video Examples:

Watch the videos for Section 7.2 on www.hawkestv.com and then solve the following problems.

- The population of a certain inner-city area is estimated to be declining according to the model $P(t) = 466,000e^{-0.018t}$, where t is the number of years from the present. What does this model predict the population will be in 12 years? Round to the nearest person.

- Tessa has recently inherited $5500, which she wants to deposit into an IRA account. She has determined that her two best bets are an account that compounds daily at an annual rate of 3.5% and an account that compounds annually at an annual rate of 2.8%. Which account would pay Tessa more interest? How much would Tessa's balance be from the higher earning account after 4.5 years? Round to two decimal places.

Practice:

1. Mary invests $7800 in a new savings account which earns 5.1% annual interest, compounded monthly. What will be the value of the investment after 5 years?

2. The function $C(t) = C(1 + r)^t$ models the rise in the cost of a product that has a cost of C today, subject to an average yearly inflation rate of r for t years. If the average annal rate of inflation over the next 8 years is assumed to be 1.5%, what will the inflation-adjusted cost of a $32,000 motorcycle be in 8 years? Round to two decimal places.

3. The revenue function is given by $R(x) = x \cdot p(x)$ dollars where x is the number of units sold and $p(x)$ is the unit price. If $p(x) = 41(4)^{-x/4}$, find the revenue if 16 units are sold.

Now complete the rest of the problems in the Practice section on your Hawkes courseware and then you'll be ready to Certify!

7.3 Logarithmic Functions and Their Graphs

When we find the inverse of an exponential function, we get a logarithmic function. These two classes of functions are closely related and, since they are inverses, their basic graphs are reflections over the line $y = x$. We'll explore the graphs of logarithmic functions in more detail in this section. In addition, we'll use the properties of logarithms to evaluate logarithmic expressions and solve logarithmic equations.

Objectives:

- Definition of logarithmic functions

- Evaluating elementary logarithmic expressions

- Graphing logarithmic functions

- Solving logarithmic equations

Learn:

1. Let a be a fixed positive real number not equal to 1. The **logarithmic function with base a** is defined to be the _____ of the exponential function with base a, and is denoted $\log_a(x)$.

2. We know that the domain of $f(x) = 2^x$ is $(-\infty, \infty)$ and the range of $f(x) = 2^x$ is $(0, \infty)$. Using this information, what is the domain and range of $f^{-1}(x) = \log_2(x)$.
   Domain: Range:

3. The graph of $g(x) = \log_5(x + 3) - 2$ is the graph of $f(x) = \log_5(x)$ shifted 3 units to the _____ and _____ units down.

4. Use the fact that $\log_a(a^x) = x$ to evaluate the following logarithmic expressions.

   (a) $\log_3(27) = \log_3(3^3) = $ _____

   (b) $\log_5(\sqrt{5}) = \log_5(5^{1/2}) = $ _____

5. The function $\log_{10}(x)$ is called the _____ **logarithm**, and is usually written as _____.

6. The function $\log_e(x)$ is called the _____ **logarithm**, and is usually written as _____.

<u>Video Examples:</u>

Watch the videos for Section 7.3 on www.hawkestv.com and then solve the following problems.

- Evaluate $\log\left(\sqrt[3]{10^5}\right)$.

- Solve the logarithmic equation for x.

$$\log_3(5x + 2) = 3$$

<u>Practice:</u>

1. Evaluate the following logarithmic expression without the use of a calculator.

$$\log_{1/8}(64)$$

2. Use the elementary properties of logarithms to solve the following equation.

$$\log_{81}(x) = \frac{3}{4}$$

3. Use the elementary properties of logarithms to solve the following equation.

$$6^{\log_4(x)} = 36$$

Now complete the rest of the problems in the Practice section on your Hawkes courseware and then you'll be ready to Certify!

7.4 Properties and Applications of Logarithms

Just as was the case with exponential functions, logarithmic functions have numerous applications to real-world problems. In this section, we'll look at some of those applications in addition to studying some further properties of logarithms.

Objectives:

- Applications of logarithmic functions

- Evaluate logarithmic expressions

- Properties of logarithms

Learn:

1. Fill in the blanks to complete the properties of logarithms.

   (a) $\log_a(xy) = $ _____ $+$ _____

   (b) $\log_a\left(\dfrac{x}{y}\right) = $ _____ $-$ _____

   (c) $\log_a(x^r) = $ _____ $\log_a(x)$

2. Use the properties of logarithms to expand the logarithmic expression $\log\left(\dfrac{x^2 y}{z}\right)$.

$$
\begin{aligned}
\log\left(\frac{x^2 y}{z}\right) &= \log(\underline{\quad}) - \log(\underline{\quad}) \\
&= \log(\underline{\quad}) + \log(\underline{\quad}) - \log(\underline{\quad}) \\
&= \underline{\quad}\log(x) + \log(\underline{\quad}) - \log(\underline{\quad})
\end{aligned}
$$

3. Rewrite the logarithmic expression $\log_7(19)$ in terms of the natural logarithm (ln) using the change of base formula.

4. Write the formula for earthquake intensity according to the **Richter scale** using I_0 for the intensity of a just-discernible earthquake and I for the intensity of the earthquake being analyzed.

   $R = $

<u>Video Examples:</u>

Watch the videos for Section 7.4 on `www.hawkestv.com` and then solve the following problems.

- Expand the logarithmic expression $\ln\left(\dfrac{x^3\sqrt{y^6}}{e^2}\right)$.

- Simplify the logarithmic expression $\log\left(\log\left(100000^{2x}\right)\right)$.

Practice:

1. Use the properties of logarithms to condense the following expression as much as possible, writing the answer as a single term with a coefficient of 1. All exponents should be positive.

$$\ln(x) + \ln(14y)$$

2. Evaluate the following logarithmic expression. Round off your answer to two decimal places.

$$\log_{1/2}(0.478)$$

3. If a sample of a certain solution is determined to have a $[H_3O^+]$ concentration of 5.99×10^{-5} moles/liter, what is its pH? Round off your answer to one decimal place.

Now complete the rest of the problems in the Practice section on your Hawkes courseware and then you'll be ready to Certify!

7.5 Exponential and Logarithmic Equations

In the final section of this chapter, we focus on the interchangeability of logarithmic and exponential expressions. Somewhat ironically, to solve an exponential equation, we convert it to a logarithmic equation, and vice-versa.

<u>Objectives:</u>

- Further applications of exponential and logarithmic equations

- Solving exponential equations

- Solving logarithmic equations

<u>Learn:</u>

1. Rewrite the exponential equation as a logarithmic equation: $2^{3x} = 15$.

2. Use the properties of logarithms to rewrite the left side of the logarithmic equation as a single logarithm.

$$\log_3(x+1) + \log_3(x-2) = \log_3(x-4)$$

3. Use properties of logarithms to fill in the blanks.

   (a) $\ln(e^2) = $ _____

   (b) $\log_2(x) - \log_2(x+1) = $ _____

   (c) $\log_7(1) = $ _____

   (d) $\log_5(5^3) = $ _____

Video Examples:

Watch the videos for Section 7.5 on www.hawkestv.com and then solve the following problems.

- Solve the exponential equation $5^{3x-1} = 780$.

- Solve the logarithmic equation $\log_3(x + 1) + \log_3(x) = \log_3 2$.

Practice:

1. Solve the following exponential equation.

$$3e^{3x+14} = 111$$

2. Solve the following exponential equation.

$$e^{3x-6} = 9^{2x/11}$$

3. Solve the following logarithmic equation.

$$\log(x - 4) + \log(x + 2) = \log(x - 5)$$

Now complete the rest of the problems in the Practice section on your Hawkes courseware and then you'll be ready to Certify!

Chapter 8

Systems of Equations

Thus far, we have solved many equations of different types. However, there are quite a few situations where one may need to solve more than one equation to solve a problem. We call this a system of equations and we will learn a variety of methods to help us solve these systems in this chapter.

8.1 Solving Systems by Substitution and Elimination

One of the most basic types of systems of equations is a linear system. There are a few ways to solve a linear system of equations, but the two most common methods are substitution and elimination. We'll explore both of these methods in this section.

Objectives:

- Applications of systems of equations

- Definition and classifications of linear systems of equations

- Solving systems of equations using elimination

- Solving systems of equations using substitution

<u>Learn:</u>

1. Fill in the blanks to complete the statements which describe the three varieties
   of linear equations.

   (a) A linear system of equations with no solution is said to be _____.

   (b) A linear system of equations that has one solution is called _____.

   (c) A linear system of equations with more than one solution must have an
       infinite number of solutions, and is said to be _____.

2. Consider the system $\begin{cases} 2x + 4y = 8 \\ 3x - y = -9 \end{cases}$. If solving this system by substitution, it would be easiest to solve equation (2) for y.

   (a) Solve equation (2) for y: $y =$ _____

   (b) Now substitute your answer from the previous step for y in Equation (1).
       $2x + 4(\rule{2cm}{0.4pt}) = -2$

   (c) Now solve the remaining equation from the previous step for x.

       $x =$ _____

   (d) Finally, use that value for x to find the corresponding y value by substi-
       tuting into the equation you found in Step 1.
       $y = 3(\rule{1cm}{0.4pt}) + 9 =$ _____

3. Consider the system $\begin{cases} 3x - 2y &= -3 \\ 5x + 3y &= 8 \end{cases}$. Suppose you wanted to solve this system by elimination. In order to eliminate y, you could multiply Equation (1) by -3 and Equation (2) by 2. Perform these multiplications to get a new system.

$$\begin{cases} 3x - 2y &= -3 \\ 5x + 3y &= 8 \end{cases} \longrightarrow \begin{cases} \underline{} + 6y &= \underline{} \\ \underline{} + 6y &= \underline{} \end{cases}$$

Verify that $(1, 2)$ is the solution to this system by solving it.

4. The sum of three integers is 41. The sum of the first and third integers exceeds the second by 11. The first integer is 10 less than the second. Set up a system of three linear equations that would be used to solve the system. Use the variables x, y, and z for the first, second, and third integers, respectively.

$$\begin{cases} x + y + z &= \underline{} \\ x + z &= \underline{} + 11 \\ x &= \underline{} \end{cases}$$

Verify that $x = 5, y = 15$, and $z = 21$ is the solution to this system by solving it.

<u>Video Examples:</u>

Watch the videos for Section 8.1 on `www.hawkestv.com` and then solve the following problems.

- Solve the following system. Use either the substitution or elimination method.

$$\begin{cases} -2x & = & 14 \\ -4x + 5y & = & 13 \end{cases}$$

- Solve the following system. Use either the substitution or elimination method.

$$\begin{cases} 4x - 2y & = & -10 \\ -3x + y & = & 6 \end{cases}$$

<u>Practice:</u>

1. Solve the following system of equations. If the system is dependent, express the solution set in terms of one of the variables.

$$\begin{cases} 2x - 4y &= 18 \\ -10x + 20y &= -91 \end{cases}$$

2. Use a system of equations to solve the following problem. How many ounces of a 17% alcohol solution and a 34% alcohol solution must be combined to obtain 51 ounces of a 32% solution?

3. Use any convenient method to solve the following system of equations. If the system is dependent, express the solution set in terms of one of the variables.

$$\begin{cases} x + 3y & = & -11 \\ y - 6z & = & 3 \\ -2x - 5y + 2z & = & 21 \end{cases}$$

Now complete the rest of the problems in the Practice section on your Hawkes courseware and then you'll be ready to Certify!

8.2 Matrix Notation and Gaussian Elimination

In the last section, we used the elimination method to solve systems of linear equations. In this section, we will codify that method using matrix notation and Gaussian elimination. This is the preferred method of most mathematicians for solving systems of equations, especially ones that are larger in size.

Objectives:

- Elementary row operations

- Gauss-Jordan elimination and reduced row echelon form

- Gaussian elimination and row echelon form

- Linear systems, matrices, and augmented matrices

Learn:

1. Given the matrix $A = \begin{bmatrix} 3 & -2 & 11 \\ 0 & 6 & -5 \\ 4 & 15 & -21 \\ 31 & -16 & 1 \end{bmatrix}$, determine

   (a) the order of A;

   (b) the value of a_{32};

   (c) the value of a_{12}.

2. Convert the system of linear equations to an augmented matrix.

$$\begin{cases} 4x - 6y &= 18 \\ -2x + 5y &= 7 \end{cases} \longrightarrow \begin{bmatrix} -2 & \underline{\quad} & \underline{\quad} \\ \underline{\quad} & \underline{\quad} & 7 \end{bmatrix}$$

3. A matrix is in **row echelon form** if:

   (a) The first non-zero entry in each row is _____.

   (b) Every entry below each **leading 1** is _____, and each leading 1 appears farther to the _____ than the leading 1's in previous rows.

   (c) All rows consisting entirely of _____ (if there are any) appear at the bottom.

4. Perform the indicated row operation to the augmented matrix.

$$
\left[\begin{array}{ccc|c}
1 & -2 & 5 & 3 \\
2 & 4 & -4 & 0 \\
3 & -1 & 6 & -5
\end{array}\right]
\xrightarrow{-3R_1+R_3}
\left[\begin{array}{ccc|c}
1 & 2 & 5 & 3 \\
2 & 4 & -4 & 0 \\
 & & &
\end{array}\right]
$$

5. A matrix is said to be in **reduced row echelon form** if:

   (a) It is in _____ _____ _____.

   (b) Each entry _____ a leading 1 is also 0.

<u>Video Examples:</u>

Watch the videos for Section 8.2 on `www.hawkestv.com` and then solve the following problems.

- Solve the following system of equations using an augmented matrix and reducing it to row echelon form.

$$
\begin{cases}
x - 5y = -7 \\
4x + 3y = 18
\end{cases}
$$

- Solve the following system of equations using an augmented matrix and reducing to reduced row echelon form.

$$\begin{cases} x + 4y & = & 14 \\ y - 3z & = & 1 \\ -2x - 3y + 5z & = & -3 \end{cases}$$

<u>Practice:</u>

1. Use Gaussian elimination and back-substitution to solve the following system of equations.

$$\begin{cases} 3x - 4y &= 16 \\ -6x + 8y &= -34 \end{cases}$$

2. Use Gauss-Jordan elimination to solve the following system of equations.

$$\begin{cases} 2x + 7y &= -47 \\ 5x + 8y &= -70 \end{cases}$$

3. Use Gauss-Jordan elimination to solve the following system of equations.

$$\begin{cases} -3x + 2y + 2z &= -14 \\ x + 5y - 4z &= 27 \\ 4x + 3y - 6z &= 41 \end{cases}$$

Now complete the rest of the problems in the Practice section on your Hawkes courseware and then you'll be ready to Certify!

8.3   Determinants and Cramer's Rule

In this section, we investigate a method for solving linear systems of equations called Cramer's rule. In order to use this method, we'll first need to learn how to find the determinant of a matrix. Determinants play a very important role in the area of linear algebra, and we'll see how they can play a role in solving systems of equations.

Objectives:

- Determinants and their evaluation

- Determining cofactors

- Determining minors

- Using Cramer's rule to solve linear systems

Learn:

1. Find the determinant of the 2×2 matrix $A = \begin{bmatrix} 3 & 7 \\ -2 & 4 \end{bmatrix}$.

$$|A| = (3)(\underline{\hspace{0.6cm}}) - (-2)(\underline{\hspace{0.6cm}}) = \underline{\hspace{0.6cm}} - (-14) = \underline{\hspace{0.6cm}}$$

2. Let A be an $n \times n$ matrix and let i and j be two fixed numbers each between 1 and n.

   (a) The **minor** of the element a_{ij} is the determinant of the $(n-1) \times (n-1)$ matrix formed from A by deleting its _____ row and _____ column.

   (b) The **cofactor** of the element a_{ij} is _____ times the minor of a_{ij}.

3. In evaluating the determinant of the following matrix, along which row or column should you expand to make the calculations the easiest? Why?

$$A = \begin{bmatrix} -2 & 5 & 0 \\ 3 & 4 & 0 \\ 10 & -7 & 2 \end{bmatrix}$$

4. Consider the system of equations $\begin{cases} 2x + 3y &= 8 \\ -x + 4y &= 7 \end{cases}$. Find the following.

(a) $D = \begin{vmatrix} 2 & 3 \\ -1 & 4 \end{vmatrix} =$

(b) $D_x = \begin{vmatrix} & 3 \\ & 4 \end{vmatrix} =$

(c) $D_y = \begin{vmatrix} 2 & \\ -1 & \end{vmatrix} =$

(d) $x = \dfrac{D_x}{D} = \dfrac{}{} =$

(e) $y = \dfrac{D_y}{D} = \dfrac{}{} =$

Video Examples:

Watch the videos for Section 8.3 on www.hawkestv.com and then solve the following problems.

- Evaluate the determinant of the matrix.

$$A = \begin{bmatrix} -7 & 19 \\ 13 & 21 \end{bmatrix}$$

- Use Cramer's rule to solve the system $\begin{cases} x + 22y &= -140 \\ y + 15z &= -66 \\ 7x + 16z &= -120 \end{cases}$.

Practice:

1. Determine the cofactor a_{13} in the matrix.

$$A = \begin{bmatrix} 5 & -8 & 5 \\ 3 & 5 & 6 \\ 1 & 1 & -6 \end{bmatrix}$$

2. Evaluate the determinant of the matrix $A = \begin{bmatrix} 6 & 0 & 5 \\ -9 & 5 & 0 \\ 7 & -5 & 5 \end{bmatrix}$.

3. Use Cramer's rule to solve the system $\begin{cases} 7x - 3y & = & -19 \\ 8x + 2y & = & -38 \end{cases}$.

Now complete the rest of the problems in the Practice section on your Hawkes courseware and then you'll be ready to Certify!

8.4   The Algebra of Matrices

Now that we have worked with matrices to solve linear systems of equations, it's time that we take a more in-depth look at these objects. As is the case with numbers, we can perform many operations with matrices including addition, subtraction, and multiplication. These operations form an algebra of matrices. We will investigate these operations, as well as some of their restrictions, in this section.

Objectives:

- Matrix addition

- Matrix multiplication

- Operations on matrices

- Scalar multiplication

Learn:

1. Two matrices A and B can be added to form the new matrix $A + B$ only if A and B are of the same _____.

2. Note that the following matrices A and B are both of order 2×3, so they can be added together. Find $A + B$.

$$A = \begin{bmatrix} -5 & 7 & -10 \\ 3 & 8 & 4 \end{bmatrix}, \quad B = \begin{bmatrix} 15 & -9 & 1 \\ 0 & 14 & -6 \end{bmatrix}, \quad A + B = \begin{bmatrix} 10 & & \\ & 22 & \end{bmatrix}$$

3. Find the matrix $-2A$ for $A = \begin{bmatrix} 4 & -8 \\ 0 & 12 \\ -6 & 9 \end{bmatrix}$.

$$-2A = \begin{bmatrix} -8 & \\ & -24 \\ & -18 \end{bmatrix}$$

4. Two matrices A and B can be multiplied together, resulting in a new matrix denoted AB, only if the length of each _____ of A (the matrix on the left) is the same as the length of each _____ of B (the matrix on the right).

5. Consider the two matrices below. Note that A is of order 2×3 and B is of order 3×2 so the matrix AB is defined and will be of order 2×2. Find the entry in the first row, second column of AB.

$$A = \begin{bmatrix} 4 & -5 & 2 \\ -3 & 8 & 7 \end{bmatrix}, \qquad B = \begin{bmatrix} -6 & 9 \\ 0 & -4 \\ 1 & 10 \end{bmatrix}$$

$$AB_{12} = 4(\underline{\hspace{0.8cm}}) + (\underline{\hspace{0.8cm}})(-4) + (\underline{\hspace{0.8cm}})(10) =$$

Video Examples:

Watch the videos for Section 8.4 on www.hawkestv.com and then solve the following problems.

• Given the following matrices, if possible, determine $4A - 2B$.

$$A = \begin{bmatrix} 4 & -1 \\ 3 & 7 \end{bmatrix}, \quad B = \begin{bmatrix} -9 & -8 \\ -2 & -4 \end{bmatrix}$$

- Given the following matrices, if possible, determine $BA + B$.

$$A = \begin{bmatrix} -1 & 8 \\ -7 & 8 \end{bmatrix}, \quad B = \begin{bmatrix} 2 & -10 \end{bmatrix}$$

Practice:

1. Given the following matrices, if possible, determine $A + B + C$.

$$A = \begin{bmatrix} 7 & 1 \\ 6 & 8 \\ 1 & 0 \end{bmatrix}, \quad B = \begin{bmatrix} -4 & 6 \\ 8 & 2 \\ -3 & 5 \end{bmatrix}, \quad C = \begin{bmatrix} 4 & 8 \\ -6 & 3 \\ -4 & -5 \end{bmatrix}$$

2. Determine the values of the variables that will make the following equation true, if possible.

$$4 \begin{bmatrix} 2x & y & 3x \end{bmatrix} + 3 \begin{bmatrix} 2x & 3y & -z \end{bmatrix} = \begin{bmatrix} 98 & -104 & -45 \end{bmatrix}$$

3. Given the following matrices, if possible, determine $(BA)C$.

$$A = \begin{bmatrix} 8 & -10 \\ -4 & -8 \end{bmatrix}, \ B = \begin{bmatrix} -2 & -9 \end{bmatrix}, \ \text{and } C = \begin{bmatrix} -6 & -8 \\ 1 & -7 \end{bmatrix}$$

Now complete the rest of the problems in the Practice section on your Hawkes courseware and then you'll be ready to Certify!

8.5   Inverses of Matrices

We know that every real number has a multiplicative inverse. For example, the multiplicative inverse of 2 is $\frac{1}{2}$ since $2\left(\frac{1}{2}\right) = 1$. In the same way, we can define multiplicative inverses for certain types of matrices. Moreover, we can use these inverse matrices to solve certain systems of equations.

Objectives:

- Find the inverse of a matrix

- The matrix form of a linear system

- Using matrix inverses to solve linear systems

Learn:

1. Write the following linear system as a matrix equation: $\begin{cases} 2x - 3y + z &= -5 \\ -4x + 7y + 6z &= 11 \\ 5x - y + 3z &= -4 \end{cases}$.

$$\begin{bmatrix} 2 & & 1 \\ & 7 & 6 \\ 5 & & \end{bmatrix}\begin{bmatrix} x \\ \\ \end{bmatrix} = \begin{bmatrix} \\ 11 \\ \end{bmatrix}$$

2. The **n × n identity matrix**, denoted by I_n, is the $n \times n$ matrix consisting of _____ on the main diagonal and _____ everywhere else.

3. Verify that the two matrices are inverses by showing that $AA^{-1} = I$.

$$A = \begin{bmatrix} 2 & 5 \\ 3 & 8 \end{bmatrix}, \quad A^{-1} = \begin{bmatrix} 8 & -5 \\ -3 & 2 \end{bmatrix}$$

4. Let $A = \begin{bmatrix} a & b \\ c & d \end{bmatrix}$. Then $A^{-1} = \dfrac{1}{|A|}\begin{bmatrix} d & \\ -c & \end{bmatrix}$.

5. Solve the system by the inverse matrix method: $\begin{cases} 2x + 5y & = & 1 \\ 3x + 8y & = & 2 \end{cases}$.

(a) $\begin{bmatrix} 2 & \\ & 8 \end{bmatrix} \begin{bmatrix} x \\ y \end{bmatrix} = \begin{bmatrix} \\ 2 \end{bmatrix}$

(b) $\begin{bmatrix} \\ y \end{bmatrix} = A^{-1} \begin{bmatrix} 1 \\ 2 \end{bmatrix}$

(c) $\begin{bmatrix} x \\ y \end{bmatrix} = \begin{bmatrix} 8 & \\ & 2 \end{bmatrix} \begin{bmatrix} \\ 2 \end{bmatrix} = \begin{bmatrix} -1 \\ \end{bmatrix}$

Video Examples:

Watch the videos for Section 8.5 on www.hawkestv.com and then solve the following problems.

- Solve the following system by the inverse matrix method, if possible.

$$\begin{cases} -2x + 5y & = & 5 \\ 6x - 15y & = & -15 \end{cases}$$

- Solve the following system by the inverse method, if possible.

$$\begin{cases} -3y - 2z &= 2 \\ -x - y + 4z &= -24 \\ 4x - 2y - 17z &= 88 \end{cases}$$

Practice:

1. Find the inverse of the following matrix, if possible.

$$A = \begin{bmatrix} 6 & 7 \\ -1 & -3 \end{bmatrix}$$

2. Find the inverse of the following matrix, if possible.

$$A = \begin{bmatrix} -1 & -3 & -2 \\ -4 & -11 & -10 \\ 3 & 8 & 7 \end{bmatrix}$$

3. Solve the following system by the inverse matrix method, if possible.

$$\begin{cases} -2x + 3y + z & = & -15 \\ x + 7y - z & = & -8 \\ 3x + 4y - 2z & = & 7 \end{cases}$$

Now complete the rest of the problems in the Practice section on your Hawkes courseware and then you'll be ready to Certify!

8.6 Linear Programming

There are many situations that arise where constraints are given for certain variables and a solution must be found among all of the values that meet those constraints. Often we'll want to find the solution that not only satisfies the constraints, but also optimizes a value. These types of problems can be modeled by linear programming. We'll look at the methods used to set up and solve such problems in this section.

Objectives:

- Linear programming in two variables

Learn:

1. Fill in the blanks for the steps in solving a linear programming problem.

   (a) Identify the variables to be considered in the problem, determine all the constraints on the variables imposed by the problem, and sketch the _____ _____ described by the constraints.

   (b) Determine the function that is to be either maximized or minimized. Such a function is called the _____ _____.

   (c) Evaluate the function at each of the vertices of the feasible region and compare the values found. If the feasible region is bounded, the optimum value of the function will occur at a _____. If the feasible region is unbounded, the optimum value of the function may not exist, but if it does it will occur at a _____.

2. A _____ feasible region is one in which all the points lie within some fixed distance of the origin. An _____ region has points that are arbitrarily far away from the origin.

3. Find the maximum and minimum values of $f(x, y) = 3x - 2y$ subject to the feasible region shown. Remember that the maximum and minimum values must occur at a vertex.

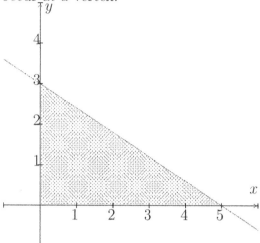

4. Sketch the feasible region for the following constraints.

$$\begin{cases} x \geq 0 \\ y \geq 0 \\ x + y \leq 3 \end{cases}$$

5. Suppose a shirt costs $10 and a pair of shorts costs $15. What would the constraint inequality be if a person could spend no more than $45 on shirts and shorts?

Video Examples:

Watch the videos for Section 8.6 on www.hawkestv.com and then solve the following problems.

- Find the minimum and maximum values of the objective function, and the points at which these values occur subject to the given constraints.

$$\text{Objective Function:} f(x, y) = 8x + 4y; \quad \text{Constraints:} \begin{cases} x \geq 0 \\ y \geq 0 \\ x + y \leq 9 \end{cases}$$

- To prevent pests, an orchard can have no more than 9 times as many apple trees as peach trees. Also, the number of apple trees plus 3 times the number of peach trees must not exceed 372. The revenue from a single apple tree is \$140 and the revenue from a single peach tree is \$365. Determine the number of each type of tree that will maximize revenue. What is the maximum revenue?

Practice:

1. Find the minimum and maximum values of the objective function, and the points at which these values occur subject to the given constraints.

$$\text{Objective Function:} f(x, y) = 11x + 13y; \quad \text{Constraints:} \begin{cases} x \geq 0; \ y \geq 0 \\ 11x + 3y \leq 66 \\ 11x + 3y \leq 132 \end{cases}$$

2. On your birthday your father-in-law gave you $19,000. You would like to invest at least $4000 of the money in municipal bonds yielding 5% and no more than $2000 in Treasury bills yielding 6%. How much should be placed in each investment in order to maximize the interest earned in one year? Assume simple interest applies. Let x represent the amount of money in municipal bonds and y represent the amount of money in Treasury bills.

Now complete the rest of the problems in the Practice section on your Hawkes courseware and then you'll be ready to Certify!

8.7 Nonlinear Systems of Equations

To this point, we have only considered systems of equations in which each equation was linear. In this section, we will look at systems of equations in which the equations are *nonlinear*. As you might expect, in this case we may have multiple points at which, say, a circle intersects a parabola. Luckily, the methods of elimination and substitution that we used to solve linear systems can also be used to solve nonlinear systems.

Objectives:

- Solving nonlinear systems algebraically

- Solving nonlinear systems graphically

Learn:

1. Consider the nonlinear system $\begin{cases} x + y & = & 2 \\ x^2 + y^2 & = & 4 \end{cases}$. The graphs of each equation are shown below. What are the points of intersection?

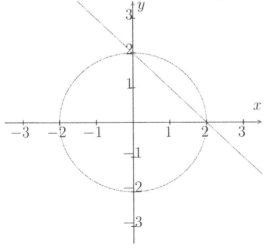

2. Consider the nonlinear system $\begin{cases} x^2 + y^2 &= 1 \\ y &= x^2 - 1 \end{cases}$. The y-coordinates of
the solutions to this system are $y = 0$ and $y = -1$. Find the corresponding
x-coordinates. (*Hint:* There are three solutions to the system.)

3. Consider the nonlinear system $\begin{cases} y - x^2 &= 4 \\ x^2 + y^2 &= 16 \end{cases}$. Solve this system algebraically.

(a) Solve the first equation for x^2: $x^2 = $ _____

(b) Solve the second equation for x^2: $x^2 = $ _____

(c) Equate the expressions and solve for y:

$$y - 4 = 16 - y^2$$
$$y^2 + y - \underline{\hspace{1cm}} = 0$$
$$(y + 5)(y - \underline{\hspace{0.6cm}}) = 0$$
$$y = -5, \underline{\hspace{0.8cm}}$$

(d) Find the corresponding values for x. (*Hint:* There are three solutions to
the system, two of which have imaginary values for x.)

Video Examples:

Watch the videos for Section 8.7 on `www.hawkestv.com` and then solve the following problems.

- Solve the nonlinear system $\begin{cases} x^2 + y^2 &= 12 \\ 2x^2 + 3y^2 &= 40 \end{cases}$.

- Carlos takes 10 minutes longer than Patricia does to make the 110-mile drive between two cities. Patricia drives 6 miles an hour faster. How fast do Carlos and Patricia drive?

Practice:

1. Solve the nonlinear system $\begin{cases} x^2 + y^2 &= 61 \\ 2x^2 - 4y^2 &= -28 \end{cases}$.

2. Solve the nonlinear system $\begin{cases} x^2 + y^2 &= 8 \\ y - 2 &= -x^2 \end{cases}$.

3. The product of two positive integers is 45, and their sum is 14. What are the integers?

Now complete the rest of the problems in the Practice section on your Hawkes courseware and then you'll be ready to Certify!

Chapter 9

An Introduction to Sequences, Series, Combinatorics, and Probability

As the title suggests, in this chapter we will cover a wide variety of topics. Sequences and series are used widely in mathematics and a solid working knowledge of these topics will be very beneficial to future mathematical studies. Probability and combinatorics form a foundation for the further study of statistics. We will be introduced to all of these topics in this chapter.

9.1 Sequences and Series

Sequences occur naturally in a wide variety of areas. Recognizing sequences and finding convenient ways to express them are skills that we will learn in this section. In addition, we'll investigate what happens when we add the entries in a sequence together. These are called series.

Objectives:

- Partial sums and series

- Recursively and explicitly defined sequences

- The formula for the general term of a sequence

<u>Learn:</u>

1. Determine the first three terms of the sequence $a_n = 2n + 3$.

    (a) $a_1 = 2(1) + 3 =$

    (b) $a_2 = 2(\underline{\qquad}) + 3 =$

    (c) $a_3 = 2(\underline{\qquad}) + 3 =$

2. A _____ formula is one that refers to one or more of the terms preceding a_n in the definition for a_n.

3. Suppose $a_n = 4a_{n-1} - 3$ and $a_1 = 1$. Find a_2.

$$a_2 = 4a_1 - 3 = 4(\underline{\qquad}) - 3 =$$

4. The sum $a_1 + a_2 + \cdots + a_n$ is expressed in summation notation as:

5. We say that an infinite series $\displaystyle\sum_{i=1}^{\infty} a_i$ _____ if the sequence of partial sums $S_n = \displaystyle\sum_{i=1}^{n} a_i$ approaches some fixed real number S. If the sequence of partial sums does not approach some fixed number, we say the series _____.

Video Examples:

Watch the videos for Section 9.1 on `www.hawkestv.com` and then solve the following problems.

- Determine the first five terms of the sequence whose n^{th} term is defined as follows.

$$a_1 = 2 \text{ and } a_n = \sqrt{(-2a_{n-1})^2 - 2} \text{ for } n \geq 2$$

- Translate the summation notation that follows into an expanded sum. Then use the formulas and properties from this section to evaluate the sums.

$$\sum_{i=3}^{7}(i - 5)$$

Practice:

1. Determine the first five terms of the sequence whose n^{th} term is defined as follows.

$$a_n = \frac{-2n^2}{-3n}$$

2. Find a possible formula for the general n^{th} term of the sequence that begins as follows.

$$-3, 9, -27, 81, -243, \ldots$$

3. Find a formula for the nth partial sum S_n of the following series. If the series is finite, determine the sum. If the series is infinite, determine if it converges or diverges, and if it converges, determine the sum.

$$\sum_{i=1}^{20} \left((-2)^i - (-2)^{i-1} \right)$$

Now complete the rest of the problems in the Practice section on your Hawkes courseware and then you'll be ready to Certify!

9.2 Arithmetic Sequences and Series

One of the more common types of sequences is called an *arithmetic* sequence. An arithmetic sequence is simply one in which each pair of successive terms has the same difference. We'll investigate these types of sequences as well as their associated series in this section.

<u>Objectives:</u>

- Evaluating partial sums of arithmetic sequences

- The formula for the general term of an arithmetic sequence

<u>Learn:</u>

1. A sequence $\{a_n\}$ is an **arithmetic sequence** if there is some constant d so that

   _____ − _____ $= d$ for each $n = 1, 2, 3, \ldots$.

2. Complete the formula for the partial sum of an arithmetic sequence $\{a_n\}$.

$$S_n = na_1 + d\sum_{i=1}^{n-1} i = na_1 + d\left(\text{———}\right)$$

3. Find the formula for the general n^{th} term of the arithmetic sequence $-3, 2, 7, 12, \ldots$.

$$a_n = a_1 + d(n-1) = \underline{} + \underline{}(n-1) = \underline{}$$

4. Use the formula $S_n = \left(\dfrac{n}{2}\right)(a_1 + a_n)$ to determine S_5 for the arithmetic sequence $3, 7, 11, 15, 19, \ldots$.

Video Examples:

Watch the videos for Section 9.2 on `www.hawkestv.com` and then solve the following problems.

- Find the explicit formula for the general n^{th} term of the arithmetic sequence described below.

$$a_{16} = \frac{79}{2} \text{ and } d = \frac{5}{2}$$

- Find the value of the following partial sum of an arithmetic sequence.

$$4 + 12 + \cdots + 164$$

Practice:

1. Find the explicit formula for the general n^{th} term of the arithmetic sequence described below.

$$a_{28} = -48 \text{ and } a_{36} = -64$$

2. Given that a is an arithmetic sequence, $a_1 = -9$ and $a_5 = 11$, what is a_{75}?

3. A person accepts a position with a company at a salary of $28,000 for the first year. The person is guaranteed a raise of $1700 per year for the first 4 years. Determine the person's total compensation through 5 full years of employment.

Now complete the rest of the problems in the Practice section on your Hawkes courseware and then you'll be ready to Certify!

9.3 Geometric Sequences and Series

Another common type of sequence is called a *geometric* sequence. In this section, we'll investigate the properties of these sequences and look at the conditions that will cause the associated infinite geometric series to converge to a finite number.

Objectives:

- Convert repeating decimals to rational fractions

- Evaluating infinite geometric series

- Evaluating partial sums and infinite geometric series

- The formula for the general term of a geometric sequence

Learn:

1. A sequence $\{a_n\}$ is a **geometric sequence** if there is some constant $r \neq 0$ so that $\dfrac{\rule{1cm}{0.4pt}}{} = r$ for each $n = 1, 2, 3, \ldots$.

2. Determine the formula for the general n^{th} term of the geometric sequence $2, -6, 18, \ldots$.

    (a) $r = \dfrac{-6}{2} = \underline{\hspace{1cm}}$

    (b) $a_1 = \underline{\hspace{1cm}}$

    (c) $a_n = a_1 r^{n-1} = \underline{\hspace{1cm}}(\underline{\hspace{1cm}})^{n-1}$

3. Find the sum of the first 6 terms of the geometric sequence $1, -2, 4, -8, \ldots$.

    $$S_6 = \frac{a_1(1 - r^6)}{1 - r} = \frac{1(1 - (-2)^6)}{1 - (-2)} =$$

4. If $|r| < 1$, the infinite geometric series $\displaystyle\sum_{n=1}^{\infty} a_1 r^{n-1} = a_1 + a_1 r + a_1 r^2 + a_1 r^3 + \cdots$ converges, and the sum of the series is given by $S = \underline{\hspace{1cm}}$.

5. Find the sum of the infinite geometric series $\displaystyle\sum_{n=1}^{\infty} 2\left(\frac{1}{3}\right)^{n-1}$.

    $$\sum_{n=1}^{\infty} 2\left(\frac{1}{3}\right)^{n-1} = \frac{3}{1 - \dfrac{1}{3}} = \frac{3}{\rule{1cm}{0.4pt}} =$$

Video Examples:

Watch the videos for Section 9.3 on `www.hawkestv.com` and then solve the following problems.

- Given that a is a geometric sequence and $a_1 = 1$, and $a_3 = \dfrac{1}{4}$, what is a_{12}?

- Determine if the following infinite geometric series converges. If the sum converges, find the sum.

$$\sum_{i=0}^{\infty} 2 \left(\frac{-3}{4} \right)^i$$

Practice:

1. Find the explicit formula for the general n^{th} term of the geometric sequence described below.

$$a_6 = \frac{46875}{2} \text{ and } r = 5$$

2. The following sum is a partial sum of a geometric sequence. Use this fact to evaluate the sum.

$$-1 + (-3) + \cdots + (-6561)$$

3. A rubber ball is dropped from a height of 81 feet, and on each bounce it rebounds up 48% of its previous height. How far has it traveled vertically at the moment when it hits the ground for the 22nd time?

Now complete the rest of the problems in the Practice section on your Hawkes courseware and then you'll be ready to Certify!

9.4 Mathematical Induction

Oftentimes, mathematicians are given the task of proving that a certain statement is true. When this statement is to apply to all of the natural numbers, one of the most common methods of proof is *proof by induction*. In this section, we'll look at the general methodology of mathematical induction and use it to prove some basic propositions about the natural numbers.

<u>Objectives:</u>

- Use mathematical induction to prove inequalities

- Use mathematical induction to prove summation formulas

- Use mathematical induction to prove various propositions

<u>Learn:</u>

1. Assume that $P(n)$ is a statement about each natural number n. Suppose that the following two conditions hold:

   (a) $P(\underline{\quad})$ is true.

   (b) For each natural number k, if $P(k)$ is true then $P(\underline{\quad\quad})$ is true.

   Then the statement $P(n)$ is true for every natural number n.

2. Consider the following statement $P(n)$: $1 + 2 + 4 + 8 + \cdots + 2^{n-1} = 2^n - 1$.

   (a) Show that $P(1)$ is true.

   (b) Write the statement $P(k)$.

   (c) Write the statement $P(k+1)$.

<u>Video Examples:</u>

Watch the videos for Section 9.4 on `www.hawkestv.com` and then solve the following problems.

- Verify the following statement for $n = 1$ by writing $P(1)$ and determining whether it is true or false.

$$P(n) : n(n + 1)(n + 2) \text{ is divisible by } 6$$

- Show that the statement $P(n)$ is true for all natural numbers n.

$$P(n) : 1(1!) + 2(2!) + \cdots + n(n!) = (n + 1)! - 1$$

Practice:

1. Verify the following statement for $n = 1$ by writing $P(1)$ and determining whether it is true or false.

$$P(n) : 11^n - 6 \text{ is divisible by } 5$$

2. Show that the statement $P(n)$ is true for all natural numbers n.

$$P(n) : \frac{1}{1 \cdot 2} + \frac{1}{2 \cdot 3} + \frac{1}{3 \cdot 4} + \cdots + \frac{1}{n(n+1)} = \frac{n}{n+1}$$

3. Show that the following statement $P(n)$ is true for all natural numbers n.

$$P(n) : \frac{1}{2} + \frac{1}{4} + \frac{1}{8} + \cdots + \frac{1}{2^n} = \frac{2^n - 1}{2^n}$$

Now complete the rest of the problems in the Practice section on your Hawkes courseware and then you'll be ready to Certify!

9.5 An Introduction to Combinatorics

In this section, we'll look at the basics of combinatorics. The general purpose of combinatorics is to be able to count complex things quickly and easily. The value of this skill should not be underestimated. We'll look at this topic in two parts: a) Counting, Permutations, and Combinations; and b) The Binomial and Multinomial Theorems.

9.5.a An Introduction to Combinatorics - Counting, Permutations, and Combinations

Here, we look at the basic procedures for counting. Included are two of the most basic of such methods: permutations and combinations. We'll study how to know whether something should be counted with a permutation or combination, and how to apply each.

Objectives:

- Combinations

- Permutations

- The multiplication principle of counting

Learn:

1. Suppose $E_1, E_2, \ldots, E_n$ is a sequence of events, each of which has a certain number of possible outcomes. Suppose event E_1 has m_1 possible outcomes, and that after event E_1 has occurred, event E_2 has m_2 possible outcomes. Similarly, after event E_2, event E_3 has m_3 possible outcomes, and so on. Then the total number of ways that all n events can occur is _____.

2. The number of permutations of n objects taken k at a time is $_nP_k = $ ——.

3. If order does not matter, should a combination or a permutation be used to count?

4. The number of combinations of n objects taken k at a time is $_nC_k = $ ——.

5. How many different arrangements are there of the letters in the word SOCCER?
$$\frac{6!}{\qquad} = \frac{720}{2} =$$

<u>Video Examples:</u>

Watch the videos for Section 9.5.a on `www.hawkestv.com` and then solve the following problems.

- A license plate must contain three numerical digits followed by three letters. if the first digit cannot be 0 or 1, how many different license plates can be created?

- Suppose you have 10 Physics texts, 8 Computer Science texts, and 12 Math texts. How many different ways can you select 2 of each to take with you on vacation?

Practice:

1. How many different 8-digit phone numbers do not contain the digit 8? Assume that any digit in the phone number can be any of the remaining numbers?

2. How many ways are there of choosing three cards from a standard 52-card deck and arranging them in a row? How many different three-card hands can be dealt from a standard 52-card deck?

3. A trade union asks its members to select 3 people, form a slate of 10, to serve as representatives at a national meeting. How many different sets of 3 can be chosen?

Now complete the rest of the problems in the Practice section on your Hawkes courseware and then you'll be ready to Certify!

9.5.b An Introduction to Combinatorics - The Binomial and Multinomial Theorems

As you probably know, raising algebraic expressions to powers can be a tedious task. For example, if we wanted to evaluate $(x+5)^4$, we would have to multiply four factors of $(x+5)$ together. This would be quite a bit of work. Wouldn't it be great if there was a formula for telling us exactly what the expanded form of this is? Luckily, there is and we'll study it here.

Objectives:

- The Binomial and Multinomial Theorems

Learn:

1. Given non-negative integers n and k, with $k \leq n$, we define $\dbinom{n}{k} = \dfrac{n!}{\rule{1cm}{0.4pt}}$.

2. Expand the expression $(x+5)^4$ using the binomial coefficient formula.

$$(x+5)^4 = \binom{4}{0}x^0 y^4 + \binom{4}{}\rule{1cm}{0.4pt}x^1 y^3 + \binom{4}{2}x^2 y^2 + \binom{4}{}x^3 y^1 + \rule{1cm}{0.4pt}$$

3. Given the positive integer n, and any two expressions A and B,
$$(A+B)^n = \sum_{k=0}^{n} \binom{n}{k} \rule{1cm}{0.4pt}.$$

4. Given the positive integer n and expression $A_1, A_2, \ldots, A_r$,

$$(A_1 + A_2 + \cdots + A_r)^n = \sum_{k_1+k_2+\cdots+k_r=n} \binom{n}{k_1, k_2, \ldots, k_r}\rule{2cm}{0.4pt}$$

Video Examples:

Watch the videos for Section 9.5.b on `www.hawkestv.com` and then solve the following problems.

- Expand the expression $(5x + y)^5$.

- Expand the expression $(3x + 6y - 3z)^3$.

1. What is the coefficient of the term containing x^7y in the expansion $(x + y)^8$?

2. Expand the expression $(x - 2y)^6$.

3. Expand the expression $(2x + 2y - 3z)^3$.

Now complete the rest of the problems in the Practice section on your Hawkes courseware and then you'll be ready to Certify!

9.6 An Introduction to Probability

The mathematical field of probability is one of the most applicable to real-world situations. It is used in everything from gambling to insurance. In this section, we'll introduce some of the basic language and concepts of this important topic.

Objectives:

- The language of probability

- Unions, intersections, and independent events

- Using combinatorics to compute probabilities

Learn:

1. An **experiment**, in probability theory, is any activity that results in well-defined _____.

2. If E is an event of an experiment with equally likely outcomes, and if S is the sample space of the experiment, then the probability of E is given by $P(E) = $ ———.

3. A fair die is rolled, and the number on the top is noted. Determine the probability that the number is even.

   (a) $E = \{2, 4, \_\_\_\_\}$ and $n(E) = \_\_\_\_$

   (b) $S = \{1, 2, 3, 4, \_\_\_\_, \_\_\_\_\}$ and $n(S) = \_\_\_\_$

   (c) $P(E) = \dfrac{3}{\_\_\_\_} = \dfrac{1}{2}$

4. Let $E = \{1, 3, 5\}$ and $F = \{1, 2\}$. Find

   (a) $E \cup F$

   (b) $E \cap F$

5. If $E_1, E_2, \ldots, E_n$ are pairwise disjoint, then

$$P(E_1 \cup E_2 \cup \cdots \cup E_n) = \underline{\hspace{0.7cm}} + \underline{\hspace{0.7cm}} + \cdots + \underline{\hspace{0.7cm}}$$

<u>Video Examples:</u>

Watch the videos for Section 9.6 on `www.hawkestv.com` and then solve the following problems.

- A card is drawn from a standard 52-card deck. Find the probability of drawing a face card (Jack, Queen, or King) in the suit of Spades.

- A coin is flipped 5 times. Find the probability of getting tails exactly 4 times.

Practice:

1. Determine the size of the sample space of the following experiment.
   A slot machine lever is pulled; there are 5 slots, each of which can hold 5 different values.

2. What is the probability that a 7-digit driver's license number chosen at random will not have 8 as a digit?

3. A pair of dice is rolled. Find the probability that the sum of the two dice is 5 or 8.

Now complete the rest of the problems in the Practice section on your Hawkes courseware and then you'll be ready to Certify!

C0-ALU-897

Adobe® Flash® CS6
The Professional Portfolio

Managing Editor: Ellenn Behoriam
Cover & Interior Design: Erika Kendra
Copy Editor: Angelina Kendra
Printer: Prestige Printers

Copyright © 2012 Against The Clock, Inc.
All rights reserved. Printed in the United States of
America. This publication is protected by copyright,
and permission should be obtained in writing from the
publisher prior to any prohibited reproduction, storage
in a retrieval system, or transmission in any form or by
any means, electronic, mechanical, photocopying,
recording, or likewise.

A portion of the images supplied in this book are copyright © PhotoDisc, Inc., 201 Fourth Ave., Seattle, WA 98121,
or copyright © PhotoSpin, 4030 Palos Verdes Dr. N., Suite 200, Rollings Hills Estates, CA. These images are the sole
property of PhotoDisc or PhotoSpin and are used by Against The Clock with the permission of the owners. The video
files for Project 7 are copyright © Digital Juice, Inc., 600 Technology Park, Lake Mary, FL 32746. These images and
videos may not be distributed, copied, transferred, or reproduced by any means whatsoever, other than for the comple-
tion of the exercises and projects contained in this Against The Clock training material.

Against The Clock and the Against The Clock logo are trademarks of Against The Clock, Inc., registered in the
United States and elsewhere. References to and instructional materials provided for any particular application program,
operating system, hardware platform, or other commercially available product or products do not represent
an endorsement of such product or products by Against The Clock, Inc.

Photoshop, Acrobat, Illustrator, InDesign, PageMaker, Flash, Dreamweaver, Premiere, and PostScript are trademarks
of Adobe Systems Incorporated. Macintosh is a trademark of Apple Computer, Inc. FrontPage, Publisher, PowerPoint,
Word, Excel, Office, Microsoft, MS-DOS, and Windows are either registered trademarks or trademarks of Microsoft
Corporation.

Other product and company names mentioned herein may be the trademarks of their respective owners.

10   9   8   7   6   5   4   3   2   1

Print ISBN: 978-1-936201-15-0

Ebook ISBN: 978-1-936201-16-7

AGAINST THE CLOCK
mastering graphic technology

4710 28th Street North, Saint Petersburg, FL 33714
800-256-4ATC • www.againsttheclock.com

Acknowledgements

ABOUT AGAINST THE CLOCK

Against The Clock, long recognized as one of the nation's leaders in courseware development, has been publishing high-quality educational materials for the graphic and computer arts industries since 1990. The company has developed a solid and widely-respected approach to teaching people how to effectively utilize graphics applications, while maintaining a disciplined approach to real-world problems.

Having developed the *Against The Clock* and the *Essentials for Design* series with Prentice Hall/Pearson Education, ATC drew from years of professional experience and instructor feedback to develop *The Professional Portfolio Series*, focusing on the Adobe Creative Suite. These books feature step-by-step explanations, detailed foundational information, and advice and tips from industry professionals that offer practical solutions to technical issues.

Against The Clock works closely with all major software developers to create learning solutions that fulfill both the requirements of instructors and the needs of students. Thousands of graphic arts professionals — designers, illustrators, imaging specialists, prepress experts, and production managers — began their educations with Against The Clock training books. These professionals studied at Baker College, Nossi College of Art, Virginia Tech, Appalachian State University, Keiser College, University of South Carolina, Gress Graphic Arts Institute, Hagerstown Community College, Kean University, Southern Polytechnic State University, Brenau University, and many other educational institutions.

ABOUT THE AUTHOR

Erika Kendra holds a BA in History and a BA in English Literature from the University of Pittsburgh. She began her career in the graphic communications industry as an editor at Graphic Arts Technical Foundation before moving to Los Angeles in 2000. Erika is the author or co-author of more than twenty books about Adobe graphic design software. She has also written several books about graphic design concepts such as color reproduction and preflighting, and dozens of articles for online and print journals in the graphics industry. Working with Against The Clock for more than ten years, Erika was a key partner in developing The Professional Portfolio Series of software training books.

CONTRIBUTING ARTISTS AND EDITORS

A big thank you to the people whose artwork, comments, and expertise contributed to the success of these books:

- **Steve Bird,** Adobe Certified Expert
- **Colleen Bredahl,** United Tribes Technical College
- **Richard Schrand,** International Academy of Design & Technology, Nashville, TN
- **Pam Harris,** University of North Texas at Dallas
- **Debbie Davidson,** Against The Clock, Inc.

Finally, thanks to **Angelina Kendra**, editor, for making sure that we all said what we meant to say.

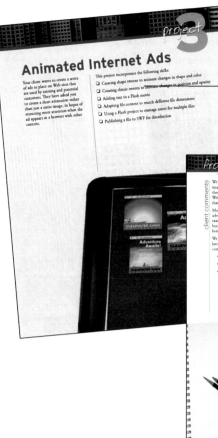

Project Goals

Each project begins with a clear description of the overall concepts that are explained in the project; these goals closely match the different "stages" of the project workflow.

The Project Meeting

Each project includes the client's initial comments, which provide valuable information about the job. The Project Art Director, a vital part of any design workflow, also provides fundamental advice and production requirements.

Project Objectives

Each Project Meeting includes a summary of the specific skills required to complete the project.

Real-World Workflow

Projects are broken into logical lessons or "stages" of the workflow. Brief introductions at the beginning of each stage provide vital foundational material required to complete the task.

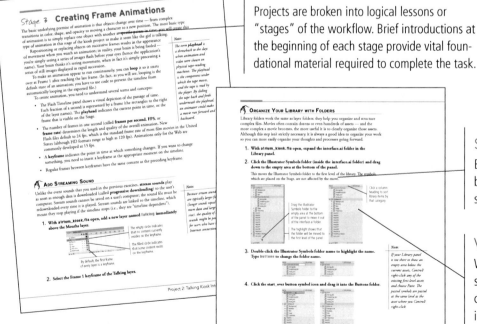

Step-By-Step Exercises

Every stage of the workflow is broken into multiple hands-on, step-by-step exercises.

Visual Explanations

Wherever possible, screen shots are annotated so you can quickly identify important information.

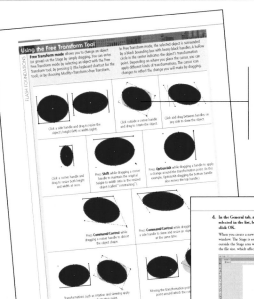

Flash Foundations

Additional functionality, related tools, and underlying graphic design concepts are included throughout the book.

Advice and Warnings

Where appropriate, sidebars provide shortcuts, warnings, or tips about the topic at hand.

Project Review

After completing each project, you can complete these fill-in-the-blank and short-answer questions to test your understanding of the concepts in the project.

Portfolio Builder Projects

Each step-by-step project is accompanied by a freeform project, allowing you to practice skills and creativity, resulting in an extensive and diverse portfolio of work.

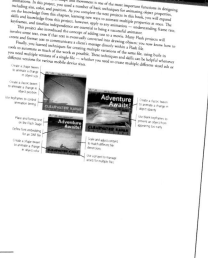

Visual Summary

Using an annotated version of the finished project, you can quickly identify the skills used to complete different aspects of the job.

The Against The Clock *Portfolio Series* teaches graphic design software tools and techniques entirely within the framework of real-world projects; we introduce and explain skills where they would naturally fall into a real project workflow.

The project-based approach in *The Professional Portfolio Series* allows you to get in depth with the software beginning in Project 1 — you don't have to read several chapters of introductory material before you can start creating finished artwork.

Our approach also prevents "topic tedium" — in other words, we don't require you to read pages and pages of information about text (for example); instead, we explain text tools and options as part of a larger project (in this case, as static text in a series of Internet ads and as dynamic text in an interactive game).

Clear, easy-to-read, step-by-step instructions walk you through every phase of each job, from creating a new file to saving the finished piece. Wherever logical, we also offer practical advice and tips about underlying concepts and graphic design practices that will benefit students as they enter the job market.

The projects in this book reflect a range of different types of Flash jobs, from animating creatures in the ocean to programming an interactive Web site interface. When you finish the eight projects in this book (and the accompanying Portfolio Builder exercises), you will have a substantial body of work that should impress any potential employer.

The eight Flash CS6 projects are described briefly here; more detail is provided in the full table of contents (beginning on Page viii).

project 1 — *Corvette Artwork*

- ❏ Setting up the Workspace
- ❏ Drawing in Flash
- ❏ Painting and Coloring Objects

project 2 — *Talking Kiosk Interface*

- ❏ Working with Symbols
- ❏ Working with Sound
- ❏ Creating Frame Animations

project 3 — *Animated Internet Ads*

- ❏ Animating Symbols
- ❏ Working with Text
- ❏ Working with a Flash Project

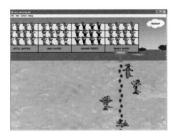

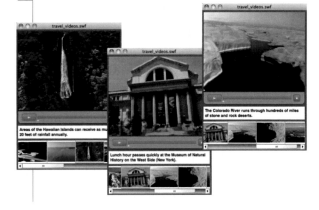

Our goal in this book is to familiarize you with the Flash tool set so you can be more productive and more marketable in your career as a graphic designer.

It is important to keep in mind that Flash is an extremely versatile and powerful application. The sheer volume of available tools, panels, and features can seem intimidating when you first look at the software interface. Most of these tools, however, are fairly simple to use with a bit of background information and a little practice.

Wherever necessary, we explain the underlying concepts and terms that are required for understanding the software. We are confident that these projects provide the practice you need to be able to create sophisticated artwork by the end of the very first project.

Contents

Contents

CONTENTS

PREREQUISITES

The entire Portfolio Series is based on the assumption that you have a basic understanding of how to use your computer. You should know how to use your mouse to point, click, and drag items around the screen. You should be able to resize and arrange windows on your desktop to maximize your available space. You should know how to access drop-down menus, and understand how check boxes and radio buttons work. It also doesn't hurt to have a good understanding of how your operating system organizes files and folders, and how to navigate your way around them. If you're familiar with these fundamental skills, then you know all that's necessary to use the Portfolio Series.

RESOURCE FILES

All of the files you need to complete the projects in this book — except, of course, the Flash application files — are on the Student Files Web page at www.againsttheclock.com. See the inside back cover of this book for access information.

Each archive (ZIP) file is named according to the related project (e.g., **FL6_RF_Project1.zip**). At the beginning of each project, you must download the archive file for that project and expand that archive to access the resource files that you need to complete the exercises. Detailed instructions for this process are included in the Interface chapter.

Files required for the related Portfolio Builder exercises at the end of each project are also available on the Student Files Web page; these archives are also named by project (e.g., **FL6_PB_Project1.zip**).

SYSTEM REQUIREMENTS

As software technology continues to mature, the differences in functionality from one platform to another continue to diminish. The Portfolio Series was designed to work on both Macintosh or Windows computers; where differences exist from one platform to another, we include specific instructions relative to each platform.

One issue that remains different from Macintosh to Windows is the use of different modifier keys (Control, Shift, etc.) to accomplish the same task. When we present key commands or other system-specific instructions, we always follow the same Macintosh/Windows format — Macintosh commands are listed first, then a slash, followed by the Windows command.

Minimum System Requirements for Adobe Flash CS6:

Windows

- Intel® Pentium® 4 or AMD Athlon® 64 processor

- Microsoft® Windows® XP with Service Pack 3 or Windows 7

- 2GB of RAM (3GB recommended)

- 3.5GB of available hard-disk space for installation; additional free space required during installation (cannot install on removable flash storage devices)

- 1024×768 display (1280×800 recommended)

- Java™ Runtime Environment 1.6 (included)

- DVD-ROM drive

- QuickTime 7.6.6 software required for multimedia features

- Some features in Adobe Bridge rely on a DirectX 9– capable graphics card with at least 64MB of VRAM

Mac OS

- Multicore Intel processor

- Mac OS X v10.6 or v10.7

- 2GB of RAM (3GB recommended)

- 4GB of available hard-disk space for installation; additional free space required during installation (cannot install on a volume that uses a case-sensitive file system or on removable flash storage devices)

- 1024×768 display (1280×800 recommended)

- Java Runtime Environment 1.6

- DVD-ROM drive

- QuickTime 7.6.6 software required for multimedia features

The Flash User Interface

Adobe Flash is the industry-standard application for building animations and other interactive content. Mastering the tools and techniques of the application can significantly improve your potential career options.

Typical Flash work ranges from simply moving things around within a space to building fully interactive games and Web sites, complete with sound and video files. Flash is somewhat unique in the communications industry because these different types of work often require different sets of skills — specifically, a combination of both visual creativity and logical programming. Depending on the type of application you're building, you should have a basic understanding of both graphic design and ActionScript code.

Our goal in this book is to teach you how to use the available tools to create various types of work that you might encounter in your professional career. As you complete the projects, you explore the basic drawing techniques, and then move on to more advanced techniques such as adding animation and interactivity — two core functions that make multimedia one of the most popular elements of modern electronic communication.

The simple exercises in this introduction are designed to let you explore the Flash user interface. Whether you are new to the application or upgrading from a previous version, we highly recommend that you follow these steps to click around and become familiar with the basic workspace.

EXPLORE THE FLASH INTERFACE

The user interface (UI) is what you see when you launch Flash. The specific elements you see — including which panels are open and where they ar located— depend on what was done the last time the application was open. The first time you launch the application, you'll see the default workspace settings defined by Adobe. When you relaunch after you or another user has quit, the workspace defaults to the last-used settings — including specific open panels and the position of those panels on your screen.

We designed the following exercise so you can explore various ways of controlling Flash's panels. Because workspace preferences are largely a matter of personal taste, the projects in this book direct you to use specific panels, but you can choose where to place those elements within the interface.

1. **Create a new empty folder named WIP (Work In Progress) on any writable disk (where you plan to save your work).**

2. **Download the FL6_RF_Interface.zip archive from the Student Files Web page.**

3. **Macintosh users: Place the ZIP archive in your WIP folder, then double-click the file icon to expand it.**

Double-click the archive file icon to expand it.

Windows users: Double-click the ZIP archive file to open it. Click the folder inside the archive and drag it into your primary WIP folder.

Open the archive file...

...then drag the Interface folder from the archive to your WIP folder.

The resulting **Interface** folder contains all the files you need to complete this introduction.

4. In Flash, open the Window menu and choose Workspace>Essentials.

Saved workspaces, accessed in the Window>Workspace menu or in the Workspace switcher on the Application bar, provide one-click access to a defined group of tools that might otherwise take multiple clicks each time you need the same toolset.

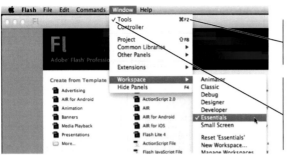

Keyboard shortcuts (if available) are listed on the right side of the menu.

Many menu commands are toggles, which means they are either on or off. The checkmark indicates that an option is toggled on (visible or active).

Note:

When no file is open in Flash, only a few menus are available in the Menu bar. Others become available only when a file is open.

Note:

If a menu command is grayed out, it is unavailable for the current selection.

5. Choose Window>Workspace>Reset 'Essentials'.

Saved workspaces remember the last-used state; calling a workspace again restores the panels exactly as they were the last time you used that workspace. For example, if you close a panel that is part of a saved workspace, the closed panel will not be reopened the next time you call the same workspace. To restore the saved state of the workspace, including opening closed panels or repositioning moved ones, you have to use the Reset option.

Steps 4 and 5 might not do anything, depending on what was done in Flash before you started this exercise. (If you or someone else changes anything and quits the application, those changes are remembered even when Flash is relaunched.) Because we can't be sure what your default settings show, by completing these steps you reset the interface to one of the built-in default workspaces so your screen will match our screen shots.

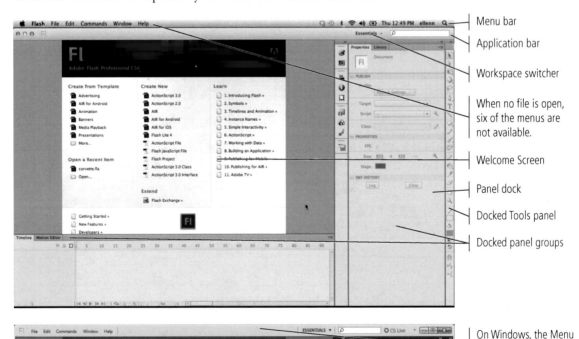

Menu bar

Application bar

Workspace switcher

When no file is open, six of the menus are not available.

Welcome Screen

Panel dock

Docked Tools panel

Docked panel groups

On Windows, the Menu bar and Application bar are the same.

The Welcome Screen appears by default when no file is open, unless you (or someone else) checked the Don't Show Again option. The Welcome Screen provides quick access to recently opened files, and links for creating a variety of new documents. If you don't see the Welcome Screen when no files are open, you can turn this feature back on by choosing Welcome Screen in the On Launch menu of the General pane of the Preferences dialog box. After you quit and relaunch the application, the Welcome Screen reappears.

6. **Move your mouse cursor over the top icon in the leftmost column of the panel dock.**

Panels, whether docked or floating, can be collapsed to icons (called iconic or iconized panels) to save space in the document window. You can see what each icon symbolizes through the tooltip that appears when you hover over a specific icon.

Note:

*The area where the panels are stored is called the **panel dock**.*

7. **Click the left edge of the leftmost column of docked panels. Hold down the mouse button and drag left until the names of the panels are visible.**

When panels are iconized, you can expand the icons to show the panel names. This is useful when you are not yet familiar with the icons. You can also drag the column edge to the right to hide the panel names.

Click the edge of the column and drag left to show the panel names.

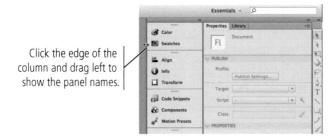

Note:

When we say "click and drag", you should hold down the mouse button while dragging.

8. **Control/right-click the title bar above the docked panel icons. Make sure Auto-Collapse Icon Panels in the contextual menu is checked (toggled on).**

As we explained in the Getting Started section, when commands are different for the Macintosh and Windows operating systems, we include the different commands in the Macintosh/Windows format. In this case, Macintosh users who do not have right-click mouse capability can press the Control key and click to access the contextual menu. You do not have to press Control *and* right-click to access the menus.

If you're using a Macintosh and don't have a mouse with right-click capability, we highly recommend that you purchase one. Two-button mice are inexpensive, they're available in most retail stores, and they save you significant amounts of time when accessing contextual options.

Control/right-clicking a dock title bar opens the dock contextual menu, where you can change the default panel behavior. If you toggle off the Auto-Collapse Icon Panels option (which is active by default), a panel remains open until you intentionally collapse it, or until you open a different panel in the same column of the dock.

This option should be checked.

The Auto-Collapse Icon Panels option is also available in the User Interface pane of the Preferences dialog box, which you can open directly from the dock contextual menu.

Note:

*A **contextual menu** is a menu that offers options related to a specific object. You can access contextual menus for most objects by Control/right-clicking.*

Note:

Throughout this book, we provide instructions for both operating systems using the convention Macintosh/Windows.

9. **Click the Color panel button in the column of iconized panels.**

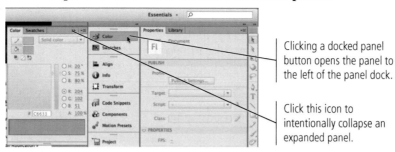

Clicking a docked panel button opens the panel to the left of the panel dock.

Click this icon to intentionally collapse an expanded panel.

Note:

Press F4 to hide all open panels. If any panels are hidden, press F4 again to reshow the hidden panels.

10. **Click away from the expanded panel (in the main workspace area, above or below the Welcome Screen).**

By default, expanded panels close as soon as you click away from them.

11. **Double-click the header above the column of panel icons.**

Double-clicking the dock column header expands iconized panels (or vice versa).

When you first expand the dock column in the Essentials workspace, it occupies nearly half of the available workspace.

Note:

You can create multiple columns of panels in the dock. Each column, technically considered a separate dock, can be expanded or iconized independently of other columns.

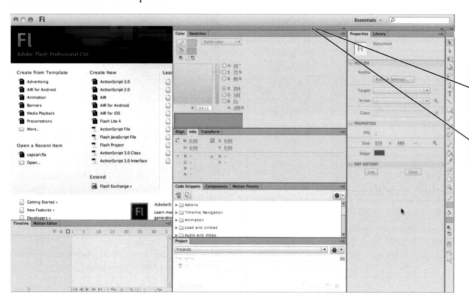

Double-click the header bar over a collapsed dock column to expand that column.

Double-click the header bar over an expanded dock column to collapse that column.

12. **Click the left edge of the expanded dock column and drag right to make the column narrower.**

Columns have a minimum possible width; you should drag right until the column does not get any smaller.

Note:

Click a specific tab to make that panel active in a panel group.

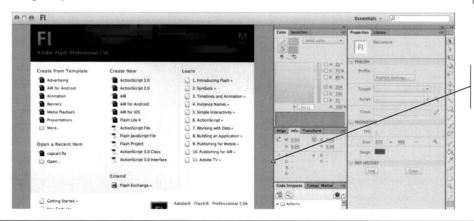

Click the left edge of the expanded dock column to make the panels narrower.

13. Click the Project panel tab and drag left, away from the dock.

Most panels are docked by default, but you can move any panel (or panel group) away from the dock so it will appear as a separate panel (called a **floating panel**). A panel does not need to be active before you move it around in the workspace.

Note:

Floating panel groups can be iconized just like columns in the dock. Double-click the floating panel group title bar to toggle between expanded and iconized mode.

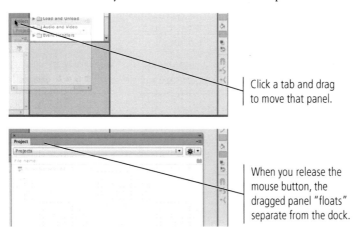

Click a tab and drag to move that panel.

When you release the mouse button, the dragged panel "floats" separate from the dock.

14. Click the Project panel tab again and drag back to the dock until a blue line appears below the docked Properties/Library panel group.

You can move any panel or group to a specific position in the dock by dragging. The blue highlight indicates where the panel will be placed when you release the mouse button.

Note:

You don't need to move a panel out of the dock before placing it in a different position within the dock. We included Step 13 to show you how to float panels and panel groups.

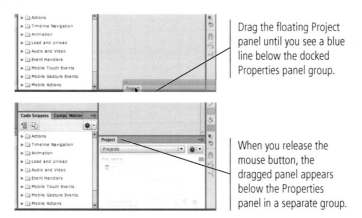

Drag the floating Project panel until you see a blue line below the docked Properties panel group.

When you release the mouse button, the dragged panel appears below the Properties panel in a separate group.

Note:

An individual panel can be dragged to different locations (including into different groups) by dragging the panel's tab. The target location — where the panel will be located when you release the mouse button — is identified by the blue highlight.

15. In the left dock column, click the drop zone behind the Code Snippets/Components/Motion Presets panel tabs, and drag right until a blue line appears between the two panel groups in the middle column.

You can move entire panel groups by dragging from the group's drop zone.

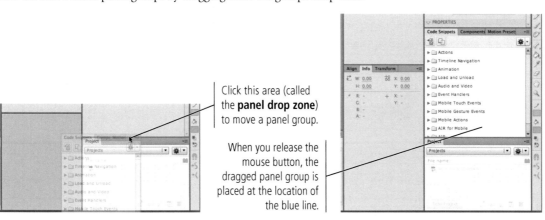

Click this area (called the **panel drop zone**) to move a panel group.

When you release the mouse button, the dragged panel group is placed at the location of the blue line.

16. Click the Code Snippets panel tab and drag down and right until the drop zone behind the Project panel tab turns blue.

When the drop zone of a panel group turns blue, releasing the mouse button will group the moved panel with the existing panel (group).

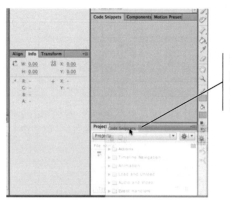

When the drop zone of a panel is blue, releasing the mouse button places the panel into that group.

Note:

Many screen shots in this book show floating panels so we can focus on the most important issue in a particular image. In our production workflow, however, we make heavy use of docked and iconized panels and take full advantage of saved custom workspaces.

17. Double-click the Code Snippets panel tab to collapse the group.

When a group is collapsed but not iconized, only the panel tabs are visible. Clicking a tab in a collapsed panel group expands the group and makes the selected panel active. You can also expand the group by again double-clicking the drop zone.

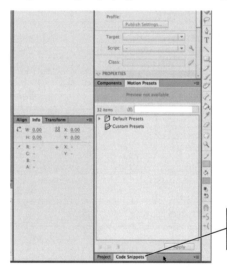

Double-click a panel tab to collapse the group to show only the panel tabs.

Note:

When a file is open, you can access all panels from the Window menu.

If you choose a panel that's open but iconized, the panel expands to the left of its icon.

If you choose a panel that's open in an expanded group, that panel comes to the front of the group.

If you choose a panel that isn't currently open, it opens in the same position as when it was last closed.

18. Place the cursor over the line that separates the Properties and Components/Motion Presets panel groups. When the cursor becomes a double arrow, click and drag down to show as much of the Properties panel as possible.

You can drag the bottom edge of a docked panel group to vertically expand or shrink the panel; other panels in the same column expand or contract to fit the available space.

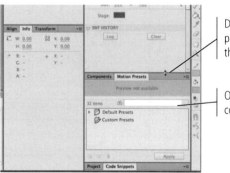

Drag the line between panel groups to change the height of a panel.

Other panels in the same dock column resize accordingly.

19. **Control/right-click the Swatches tab in the left dock column. Choose Close Group from the contextual menu.**

    You can use this menu to close an individual docked panel (Close), or to close an entire panel group (Close Group).

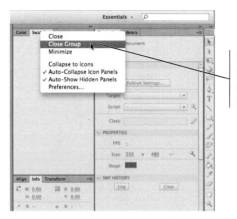

Control/right-click a panel tab or the group drop zone to open the contextual menu.

Note:

If a panel or panel group is floating, you can use the panel or group's close button to put away those panels.

20. **Repeat Step 19 to close the Align/Info/Transform panel group.**

21. **On the right side of the workspace, click the Tools panel title bar and drag away from the dock into the middle of the workspace.**

Click the panel title bar and drag to move it out of the dock.

When docked, the Tools panel appears by default as a single column of tools.

When floating, the Tools panel defaults to a standard rectangular panel with the tools in several rows.

22. **Click the Tools panel title bar and drag left until a blue line appears on the left edge of the workspace.**

    This step docks the Tools panel on the left edge of the workspace.

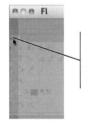

This pop-up "drawer" indicates that you are adding a column to the panel dock (in this case, on the left edge of the workspace).

After docking the Tools panel, the panel might still appear in the same configuration as the floating panel (with multiple rows of tools).

23. **If your Tools panel shows multiple tools in the same row, click the right edge of the Tools panel and drag left until all the tools appear in a single column.**

Click and drag the right edge of the docked Tools panel to show all tools in a single column.

24. **Continue to the next exercise.**

In addition to a wide variety of panels, Flash CS6 includes 30 tools — a large number that indicates the power of the application. You can change the docked Tools panel to more than one column by dragging the right edge of the panel. When the Tools panel is floating, it defaults to show the various tools in rows.

You learn how to use these tools as you complete the projects in this book. For now, you should simply take the opportunity to identify the tools and their location on the Tools panel. The image to the right shows the icon, tool name, and keyboard shortcut (if any) that accesses each tool. Nested tools are shown indented and the names appear in italics.

Keyboard Shortcuts & Nested Tools

When you hover your mouse over a tool, the pop-up tooltip shows the name of the tool and a letter in parentheses. Pressing that letter activates the associated tool (unless you're working with type, in which case pressing a key adds that letter to your text). If you don't see tooltips, check the General pane of the Preferences dialog box; the Show Tooltips check box should be active. (Note: You must have a file open to see the tooltips and access the nested tools.)

When you hover the mouse cursor over the tool, a tooltip shows the name of the tool.

Any tool with an arrow in the bottom-right corner includes related tools nested below it. When you click a tool and hold down the mouse button, the nested tools appear in a pop-up menu. If you choose one of the nested tools, that variation becomes the default choice in the Tools panel.

This arrow means the tool has other nested tools.

Click and hold down the mouse button to show the nested tools.

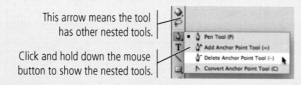

Tool Options

In addition to the basic tool set, the bottom of the Tools panel includes options that control the fill and stroke colors, as well as options that change depending on the selected tool.

Stroke Color
Fill Color
Black and White
Swap Colors
Tool-specific options

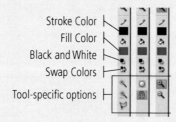

Customizing the Tools Panel

Flash offers a unique feature that makes it easy to customize the Tools panel (Flash>Customize Tools Panel on Macintosh or Edit>Customize Tools Panel on Windows).

If you click a tool icon in the left side of the dialog box, all tools nested in that space display in the Current Selection list. If you add a tool from the Available Tools list, it becomes nested under the selected tool. To remove a tool from the default Tools panel, select the tool name in the Current Selection list and click Remove.

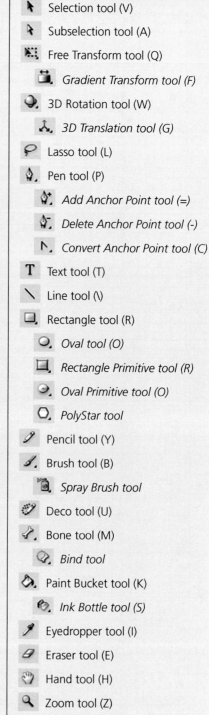

- Selection tool (V)
- Subselection tool (A)
- Free Transform tool (Q)
  - *Gradient Transform tool (F)*
- 3D Rotation tool (W)
  - *3D Translation tool (G)*
- Lasso tool (L)
- Pen tool (P)
  - *Add Anchor Point tool (=)*
  - *Delete Anchor Point tool (-)*
  - *Convert Anchor Point tool (C)*
- Text tool (T)
- Line tool (\)
- Rectangle tool (R)
  - *Oval tool (O)*
  - *Rectangle Primitive tool (R)*
  - *Oval Primitive tool (O)*
  - *PolyStar tool*
- Pencil tool (Y)
- Brush tool (B)
  - *Spray Brush tool*
- Deco tool (U)
- Bone tool (M)
  - *Bind tool*
- Paint Bucket tool (K)
  - *Ink Bottle tool (S)*
- Eyedropper tool (I)
- Eraser tool (E)
- Hand tool (H)
- Zoom tool (Z)

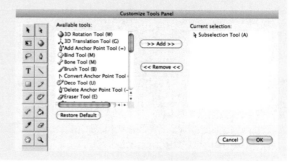

CREATE A SAVED WORKSPACE

By now you should understand that you have extensive control over the appearance of your Flash workspace — you can determine what panels are visible, where and how they appear, and even the size of individual panels and panel groups. Rather than re-establishing every workspace element each time you return to Flash, you can save your custom workspace settings so you can recall them with a single click.

1. **Click the Workspace switcher in the Application/Menu bar and choose New Workspace.**

   Again, keep in mind that we list differing commands in the Macintosh/Windows format. On Macintosh, the Workspace switcher is in the Application bar; on Windows, it's in the Menu bar.

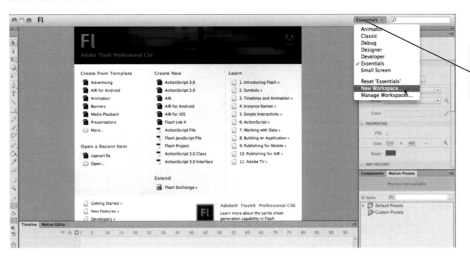

The Workspace switcher shows the name of the last-called workspace.

2. **In the New Workspace dialog box, type Portfolio as the Workspace name and click OK.**

After saving the current workspace, the Workspace switcher shows the name of the new saved workspace.

3. **Click the Library panel tab and drag down until the drop zone of the Components/Motion Presets panel group turns blue. Release the mouse button to add the Library panel to the Components/Motion Presets panel group.**

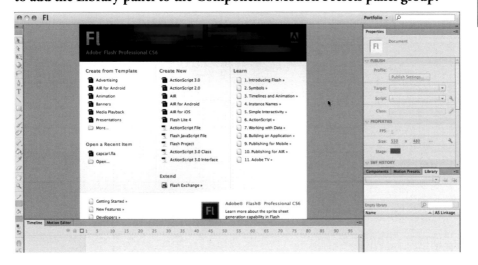

Note:

The Manage Workspaces option in the Workspace switcher opens a dialog box where you can choose a specific user-defined workspace to rename or delete. You can't rename or delete the default workspaces that come with the application.

4. **Click the Workspace switcher and choose Essentials from the list of available workspaces.**

    This restores the original saved version of the workspace because you saved a new workspace based on the changes you made in the previous exercise.

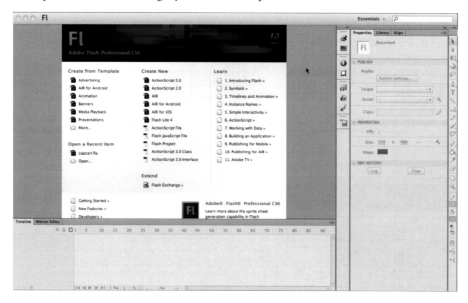

5. **Choose Portfolio from the Workspace switcher.**

    As we explained, calling a saved workspace restores the last-used state of the workspace. Even though you changed the panel arrangement after saving the Portfolio workspace, the adjusted arrangement is restored by calling the saved workspace.

Note:

If you move panels after calling a specific workspace, choosing Reset '[Workspace]' restores the active workspace to its saved settings.

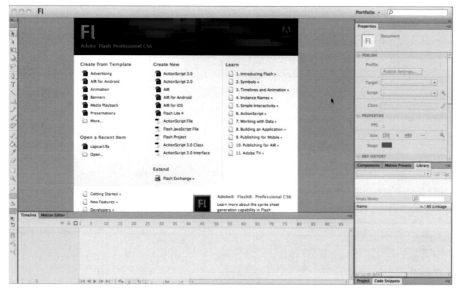

6. **Continue to the next exercise.**

Customizing Flash Behavior

Keyboard Shortcuts and Menus

Different people use Flash for various reasons, sometimes using a specific, limited set of tools to complete only one type of project. In addition to customizing the workspace and the Tools panel, you can also customize the various keyboard shortcuts used to access different commands (Flash>Keyboard Shortcuts on Macintosh or Edit>Keyboard Shortcuts on Windows). Once you have defined custom menus or shortcuts, you can save your choices as a set so you can access the same custom choices again without having to redo the work.

Application Preferences

You can also customize the way many of the program's tools and options function. The left side of the Preferences dialog box (Flash>Preferences on Macintosh or Edit>Preferences on Windows) allows you to display the various sets of preferences available in Flash. As you work your way through the projects in this book, you will learn not only what you can do with these collections of Preferences, but also why and when you might want to use them.

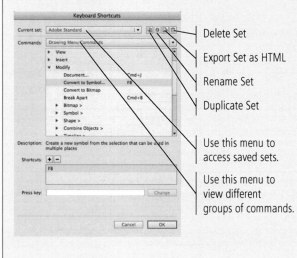

Delete Set

Export Set as HTML

Rename Set

Duplicate Set

Use this menu to access saved sets.

Use this menu to view different groups of commands.

EXPLORE THE FLASH DOCUMENT WINDOW

There is far more to using Flash than arranging panels around the workspace. What you do with those panels — and even which panels you need — depends on the type of work you are doing in a particular file. In this exercise, you open a Flash file and explore the interface elements that you will use to create digital animations.

1. **In Flash, choose File>Open.**

2. **Navigate to your WIP>Interface folder and select `capcarl.fla` in the list of available files.**

    The Open dialog box is a system-standard navigation dialog. Press Shift to select and open multiple contiguous (consecutive) files in the list. Press Command/Control to select and open multiple non-contiguous files.

Note:

Press Command/ Control-O to access the Open dialog box.

3. Click Open.

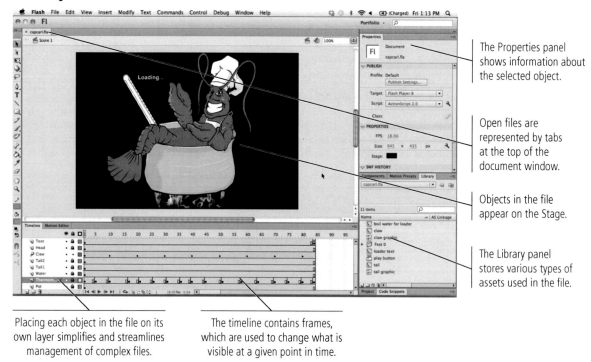

The Properties panel shows information about the selected object.

Open files are represented by tabs at the top of the document window.

Objects in the file appear on the Stage.

The Library panel stores various types of assets used in the file.

Placing each object in the file on its own layer simplifies and streamlines management of complex files.

The timeline contains frames, which are used to change what is visible at a given point in time.

4. Above the top-right corner of the Stage, open the View Percentage menu and choose Fit in Window.

Understanding Auto-Recovery

FLASH FOUNDATIONS

In the General pane of the Preferences dialog box, the Auto-Recovery option is turned on by default. Auto-recovery files are saved every 10 minutes by default; you can use the field to change this interval.

When auto-recovery is turned on, a copy of the active file is saved with the word "RECOVER_" appended to the beginning of the defined file name. (If you haven't saved a file yet, the auto-recovery file is saved in the application's Temp folder.)

If the application or system crashes before you intentionally save and close the file, you will be asked if you want to restore the auto-recovery file when you relaunch Flash. When you intentionally close the active file, the recovery files are deleted.

If auto-recovery is not enabled for a particular file, you will see a warning after the defined interval, asking if you want to enable auto-recovery for that file.

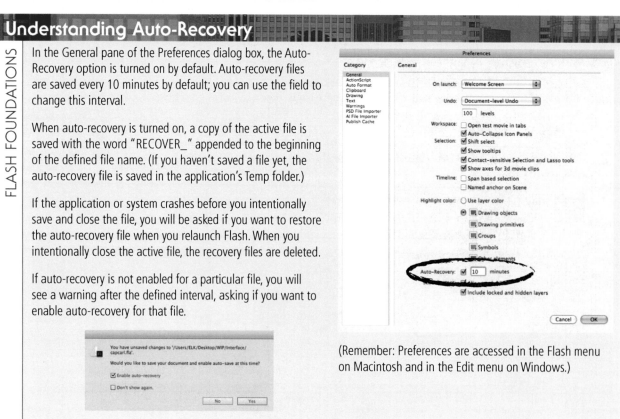

(Remember: Preferences are accessed in the Flash menu on Macintosh and in the Edit menu on Windows.)

5. **Click the bar that separates the Stage area from the Timeline panel. Drag up to enlarge the panel and show all the layers in the file.**

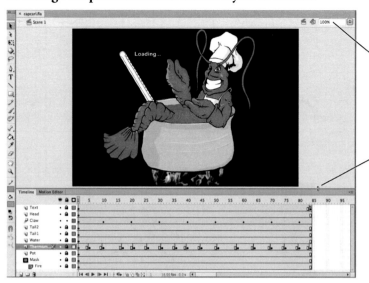

The Fit in Window command enlarges or reduces the view percentage to fill the available space in the document window.

Click here and drag up to show all the layers in the file.

6. **Review the Timeline panel.**

The Timeline panel is perhaps the most important panel in Flash. It represents the passage of time within your animation, and it enables you to control what happens to objects in your file, as well as when and where changes occur.

Note:

Different people prefer larger or smaller view percentages, depending on a number of factors (eyesight, monitor size, and so on). As you complete the projects in this book, you'll see our screen shots zoom in or out as necessary to show you the most relevant part of a particular file. In most cases we do not tell you what specific view percentage to use for a particular exercise, unless it's specifically required for the work being done.

The playhead identifies the point in time that's currently active on the Stage.

The timeline includes frames, which represent fractions of a second in the resulting animation.

Keyframes identify the point in time at which something changes.

Layers can be used to organize elements of complex files.

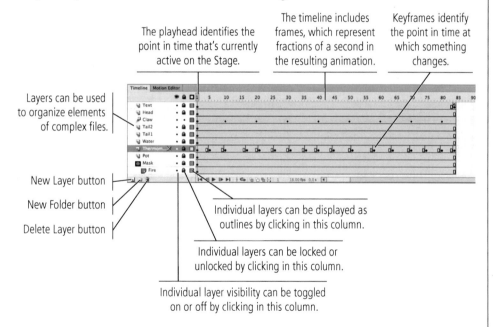

New Layer button

New Folder button

Delete Layer button

Individual layers can be displayed as outlines by clicking in this column.

Individual layers can be locked or unlocked by clicking in this column.

Individual layer visibility can be toggled on or off by clicking in this column.

FLASH FOUNDATIONS

Most Flash projects require some amount of zooming in and out, as well as navigating around the document within its window. As we show you how to complete different stages of the workflow, we usually won't tell you when to change your view percentage because that's largely a matter of personal preference. You should understand the different options for navigating around a Flash file, however, so you can easily and efficiently get to what you want, when you want to get there.

Zoom Tool

You can click with the Zoom tool to increase the view percentage in specific, predefined intervals (the same intervals you see in the View Percentage menu in the top-right corner of the document window). Pressing Option/Alt with the Zoom tool allows you to zoom out in the same predefined percentages. If you drag a marquee with the Zoom tool, you can zoom into a specific location; the area surrounded by the marquee fills the available space in the document window.

Click with the Zoom tool to zoom in.

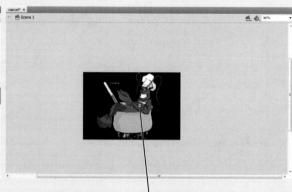

Option/Alt-click with the Zoom tool to zoom out.

Draw a marquee with the Zoom tool…

…to fill the document window with the selected area.

View Menu

The View>Magnification menu also provides options for changing the view percentage, including their associated keyboard shortcuts. (The Zoom In and Zoom Out options step through the same predefined view percentages you see by clicking with the Zoom tool.)

Zoom In	Command/Control-equals (=)
Zoom Out	Command/Control-minus (–)
100%	Command/Control-1
Show Frame	Command/Control-2
Show All	Command/Control-3
400%	Command/Control-4
800%	Command/Control-8

View Percentage Field

In addition to the predefined view percentages in the menu, you can also type a specific percentage in the View Percentage field in the top-right corner of the document window.

Hand Tool

If scroll bars appear in the document window, you can use the Hand tool to drag the file around within the document window. The Hand tool changes the visible area in the document window; it has no effect on the actual content of the image.

When using a different tool other than the Text tool, you can press the Spacebar to temporarily access the Hand tool.

Double-clicking the Hand tool in the Tools panel fits the Stage to the document window.

7. **Choose the Selection tool from the Tools panel and click the lobster's right claw on the Stage.**

Although we will not discuss all 20+ Flash panels here, the Properties panel deserves mention. This important panel is **context sensitive**, which means it provides various options depending on what is selected on the Stage.

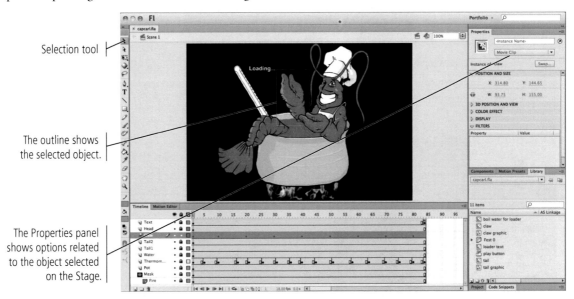

Selection tool

The outline shows the selected object.

The Properties panel shows options related to the object selected on the Stage.

8. **Click the playhead above the timeline frames and drag right.**

This technique of dragging the playhead above the timeline is called **scrubbing**.

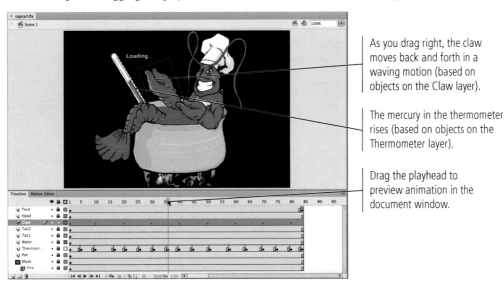

As you drag right, the claw moves back and forth in a waving motion (based on objects on the Claw layer).

The mercury in the thermometer rises (based on objects on the Thermometer layer).

Drag the playhead to preview animation in the document window.

9. **Click the gray area outside of the Stage to deselect everything, then press Return/Enter.**

One final reminder: throughout this book, we list differing commands in the Macintosh/Windows format. On Macintosh, you need to press the Return key; on Windows, press the Enter key. (We will not repeat this explanation every time different commands are required for the different operating systems.)

This keyboard shortcut plays the movie on the Stage from the current location of the playhead. If the playhead is already at the end of the frames, it moves back to Frame 1 and plays the entire movie.

Note:

When a movie is playing on the Stage, press the Escape key to stop the playback.

10. Press Command-Return/Control-Enter.

Instead of simply playing the movie on the Stage, you can preview the file to see what it will look like when exported. This keyboard shortcut, which is the same as choosing Test Movie in the Control menu, exports a SWF file from the existing FLA file so you can see what the final file will look like.

In the Flash Player window, you see some animated elements that did not appear when you played the movie on the Stage. This is a perfect example of why you should test actual animation files rather than relying only on what you see on the Stage. (This is especially true if you did not create the file, as in this case.)

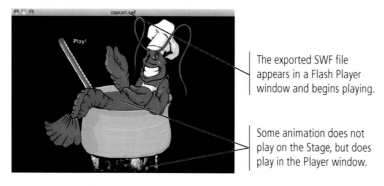

The exported SWF file appears in a Flash Player window and begins playing.

Some animation does not play on the Stage, but does play in the Player window.

11. Close the Player window and return to Flash.

12. Click the Close button on the capcarl document tab. If asked, click Don't Save/No to close the file without saving.

Click here to close the file.

Clicking the Close button on the Application frame closes all open files and quits the application.

Application frame Close button (Macintosh)

Application frame Close button (Windows)

Note:

The Controller panel (Window>Toolbars> Controller) enables you to control movie playback on the Flash Stage, using buttons similar to those on DVD players.

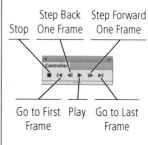

Step Back One Frame   Step Forward One Frame
Stop
Go to First Frame   Play   Go to Last Frame

The Controller panel only affects the Flash Stage. It does not affect files open in the Flash Player window.

Note:

Press Command/ Control-W to close the active file.

Managing Multiple Documents

FLASH FOUNDATIONS

When more than one document is open, each document is represented by a tab at the top of the document window. You can switch from one document to another by clicking a different tab. All open files are also listed at the bottom of the Window menu; you can use these menu options to navigate from one file to another.

If you want to look at more than one file at a time, you can separate a specific document into a floating window by dragging its tab away from the bar.

The active document tab is lighter than the rest.

Each document tab has its own Close button.

The active document is checked.

Dragging the pompano.fla tab away from the tab bar floats the document in its own document window.

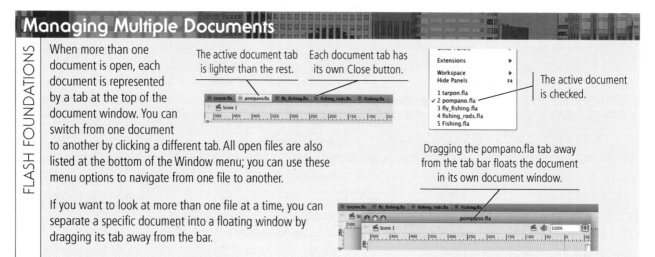

Corvette Artwork

Your client is the main sponsor and event promoter of the annual Car Cruise and Music Festival at the Santa Monica Pier. You have been hired to create digital artwork for a number of elements that will be added to the event's Web site, including a game that he is going to develop with Flash programming functionality. The first step is to build a basic illustration that can later be used for the client's various purposes.

This project incorporates the following skills:

❏ Defining a movie file with appropriate settings for the job you're building

❏ Importing a scanned image provided by the creative director

❏ Using various drawing tools to develop a digital version of a rough sketch

❏ Understanding the different drawing modes that are available

❏ Changing fill and stroke colors using a variety of methods

❏ Adding depth using various shading techniques

Project Meeting

client comments

I'm planning a number of different projects to promote this year's event. The American sportscar area is always a big draw at the event, so I'd like to feature Corvettes on all of the event collateral.

Most of the print work is going to use photos, but I'm planning some other projects where I want to use artwork instead of a picture — animations, Web banners, and I'm even thinking about an interactive game.

I'm a Flash programmer, but not an artist. If you can create the artwork directly in Flash, that would save me some time when I implement my other plans.

One final thing — I don't want it to look like a cartoon. Obviously, I don't want it to be a photo, but I do want it to be realistic looking.

art director comments

I sketched out a Corvette, and scanned the sketch for you. I want you to use that sketch as the basis for your finished piece. Start by tracing that drawing using a bright color that you can see against the gray lines in the sketch.

Once you've finished the basic artwork, try to make it seem more realistic than just a flat color conveys. Add some dimension and depth so it doesn't look too much like a cartoon.

Also, make sure you build the artwork using logical layers. That way, if we need to make changes later, or if the client decides to animate it, the document will be structured properly.

Since Flash produces vector artwork, we can use the artwork for just about anything once it's done. The important thing for now, however, is to create a static file that we can post on our Web site for the client's approval.

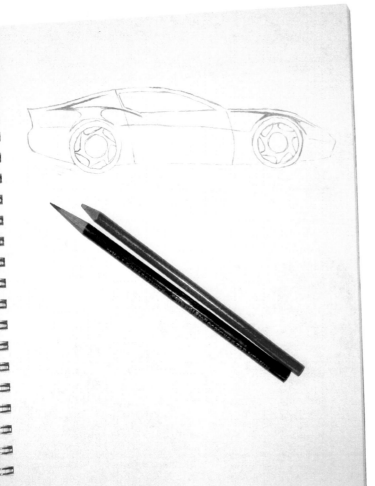

project objectives

To complete this project, you will:

❏ Set up the workspace to match the requirements of the project

❏ Import the scan of the sketched Corvette

❏ Create a set of logical layers to hold the various components that you will create

❏ Understand the difference between object-drawing and merge-drawing modes

❏ Use the basic shape and line tools to develop the components of the drawing

❏ Draw precise Bézier curves with the Pen tool

❏ Use the Pencil tool to draw freehand shapes

❏ Apply color with the Paint Bucket, Ink Bottle, and Brush tools

❏ Use object groups to combine and protect drawing objects

❏ Apply and edit color gradients

❏ Export a static image for online review

Stage 1   Setting up the Workspace

CREATE A NEW DOCUMENT

1. Download **FL6_RF_Project1.zip** from the Student Files Web page.

2. **Expand the ZIP archive in your WIP folder (Macintosh) or copy the archive contents into your WIP folder (Windows).**

   This results in a folder named **Car**, which contains all of the files you need for this project. You should also use this folder to save the files you create in this project.

   If necessary, refer to Page 1 of the Interface chapter for specific information on expanding or accessing the required resource files.

3. **In Flash, choose File>New.**

   When you create a new file, you can define a number of options for the file.

   In the left side of the General tab, you can choose to create a new file for a specific output goal.

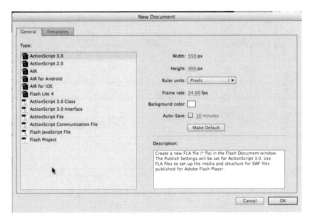

   - The first two options allow you to target a specific version of ActionScript, the code language that is used to create interactivity (ActionScript 3.0 is the current version). You will begin working with ActionScript in Project 4: Gator Race Game.

   - The AIR, AIR for Android, and AIR for iOS options create files that can be distributed as standalone "apps" for (respectively) desktop use, Android tablets/phones, and Apple iPads/iPhones.

   The **Width** and **Height** options define the Stage size, or the area that will be included in the final exported movie. In this project, you are using the Flash Stage as a drawing board (literally). You will adjust the Stage size in the next exercise to meet the needs of the drawing you will create.

   The **Ruler Units** menu defaults to Pixels, which is apropriate for files that will be distributed digitally.

   The **Frame Rate** defines the number of frames that exist in a second (frames per second, or fps). You will learn more about frame rate beginning in Project 2: Talking Kiosk Interface, when you start creating animations.

   The **Background Color** option defines the color of the Stage; this color appears behind all elements that you create or place in the file.

   The **Auto-Save** option can be activated to automatically save a working file at specific intervals. Keep in mind that if you activate this option, you will not be able to use the File>Revert to Saved command to go back to a version older than the set Auto-Save interval.

Note:

You can click the Templates tab to access pre-defined templates in many different styles.

Note:

Although there are slight differences in appearance between the Macintosh (shown here) and Windows dialog boxes, the options are the same.

Note:

Press Command/Control-N to open the New Document dialog box.

4. **In the General tab, make sure the first option — ActionScript 3.0 — is selected in the list, leave all other options at the default settings, and then click OK.**

When you create a new file, the Stage is automatically centered in the document window. The Stage is essentially the workspace, or the digital "page" area; anything outside the Stage area will not be visible in the exported movie (although it will increase the file size, which affects its download time).

Note:

We have arranged our workspace to make the best possible use of space in our screen shots, including two docked and collapsed panels that we will expand when necessary. Feel free to arrange the panels however you prefer; for this project, you will be using the Tools, Properties, Align, Timeline, and Color panels.

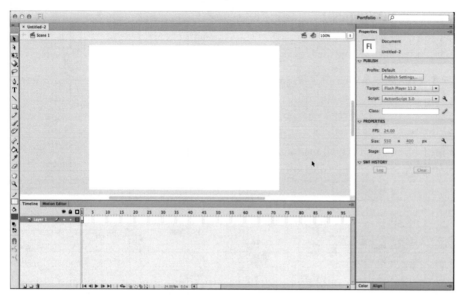

5. **Choose File>Save.**

Because this is a new file that has not yet been saved, choosing the Save command automatically opens the Save dialog box, where you define the file name, save location, and format.

If a file has already been saved at some point, the Save command simply overwrites the previously saved version of the file. If you want to save the new version and maintain the original, you have to use the File>Save As command to define a new name and create a copy of the file.

Note:

Press Command/Control-S to save the active file.

Press Command/Control-Shift-S to use the Save As option.

6. **In the resulting dialog box, navigate to your WIP>Car folder.**

7. **Change the file name to corvette.fla. Make sure Flash CS6 Document is selected in the Format/Save As Type menu and click Save.**

Native Flash files use the FLA extension. By default, the extension is included in the Save As/File Name field. You don't need to type it again if you don't delete the default extension from the field.

If necessary, you can use this menu to save a file to be compatible with the previous version of Flash.

8. **Continue to the next exercise.**

IMPORT A RASTER IMAGE

As we discussed during the production meeting for this assignment, you will base your work on a scanned sketch. Flash can import a range of file formats, including JPG, GIF, PNG, TIF, EPS (Encapsulated PostScript), and native Illustrator and Photoshop files.

1. **With `corvette.fla` open, choose File>Import>Import to Stage.**

   Four options are available when you choose the Import command:

   - **Import to Stage** simply brings an external image onto the current Stage. Imported bitmap images are also added to the file Library panel; if you import a vector graphic that uses no symbols or embedded bitmaps, nothing is added to the Library.

   - **Import to Library** automatically places graphics into the Flash Library panel.

   - **Open External Library** is commonly used to incorporate graphics from one project into another project.

   - **Import Video** allows you to bring digital video into a Flash file.

Note:

Press Command/Control-R to import an object to the Stage.

Raster Images vs. Vector Art

FLASH FOUNDATIONS

There are two kinds of computer images. The first, known as a **raster image** or **bitmap**, is composed of simple pixels. Each pixel has a specific value, and that value determines the color, intensity, and brightness of that pixel. The number of pixels that make up a raster image is fixed at the time of the scan or photograph. If you blow up (enlarge) a raster image, it becomes grainy and loses quality; in fact, you can see the individual pixels in an enlarged raster image.

Flash is a vector-based program. **Vector art** is not based on pixels; instead, it is based on mathematical values. Vector art is essentially a tiny program that tells the printer or monitor to "draw a line from Point A to Point B, bending X degrees coming out of Point A and entering Point B from an angle of Z." Although this might sound confusing, you don't have to write any complicated code to make vectors work properly; you can simply draw with the Flash tools and the vectors are created for you.

Although vector graphics often improve the quality of images in a movie, they can also degrade playback performance because they require greater processing power to perform the necessary mathematical calculations.

An example of the difference between a raster image and vector art is shown below. At 100% size, both the raster image and the type (composed of vector objects) are crisp. If you zoom in closely enough, however, the individual pixels in the photograph start to show, and the image breaks down — but the type remains perfectly crisp and clear.

At actual size, both the type in this image and the photograph appear perfectly crisp and clear.

The detail on the tree breaks down if you look closely enough.

The type object, which is a vector object, remains clear even when enlarged.

2. **Navigate to your WIP>Car folder, select corvette.jpg, and click Open/Import.**

   The imported file is placed on the Stage, aligned to the left edge and centered vertically to the Stage area. As you can see, the placed image does not fit in the default Stage area.

3. **Note the placed image's width and height in the Properties panel.**

   Because the imported image is still selected, the Properties panel shows information about the image — most importantly for this exercise, the image's physical dimensions.

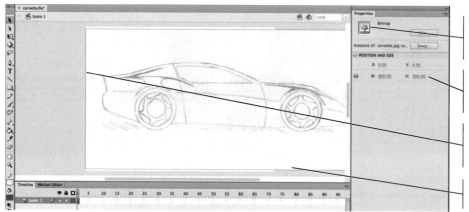

The Properties panel shows that the selected object is a bitmap file.

The object's position and size are clearly listed in the panel.

The placed file is aligned to the left edge and centered vertically on the Stage.

The placed file is wider than the current Stage area.

4. **Save the file and continue to the next exercise.**

 CHANGE DOCUMENT PROPERTIES

The imported artwork is larger than the defined Stage size. Because you're going to trace the imported sketch, you need to make the Stage area large enough to hold the entire sketch (and ultimately, the finished artwork).

1. **With corvette.fla open, choose the Selection tool at the top of the Tools panel.**

2. **Click in the gray area outside the Stage to make sure nothing is selected.**

   When nothing is selected in the file, you can use the Properties panel to access a number of options that affect the entire document — including the file's physical dimensions.

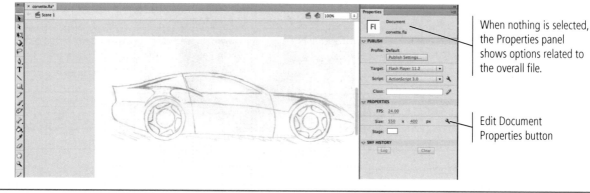

When nothing is selected, the Properties panel shows options related to the overall file.

Edit Document Properties button

3. **In the Properties panel, place the cursor over the left number in the Size setting.**

   When the cursor is over an existing value, you can click and drag left or right to dynamically change the value (called a "scrubby slider").

4. **Click the existing value to highlight the existing field value, then type 850.**

   Clicking an existing value in a panel accesses the actual field; you can type to change the value in the field.

5. **Press Tab to highlight the Height field, then type 450 px, to change the Stage height.**

   Note:

   You can also simply click away from a field to finalize the new value.

6. **Press Return/Enter to finalize the change.**

7. **If you can't see the entire Stage area, use the Hand tool to drag the Stage until you can see the entire Stage. If the entire Stage does not fit into your document window, choose Fit In Window in the View Percentage menu.**

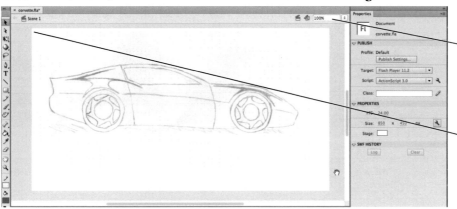

If necessary, use this menu to fit the entire Stage area in the document window.

The placed file's position relative to the Stage is not affected. The top-left corner is in the same spot as before you changed the Stage size.

Using the Document Settings Dialog Box

FLASH FOUNDATIONS

If you click the Edit Document Properties button in the Properties panel (or choose Modify>Document), you can access the same document properties that are available in the Properties panel when nothing is selected in the file.

The Dimensions, Frame Rate, and Background Color options are the same as those available in the Properties panel.

By default, changing the Stage size does not affect objects that are already placed on the Stage. You can check the **Scale Content with Stage** option to resize placed objects when the Stage size changes.

You can also use the **Match** options to automatically change the Stage size to match existing Contents (objects placed on the Stage) or Printer settings.

8. **At the bottom of the Align panel (Window>Align), check the Align To Stage box.**

When this option is active, you can use the Align panel to precisely position one or more objects relative to the defined Stage area in your file. (The Align panel remembers the last-used state of the Align To Stage option, so it might already be checked.)

Note:

Press Command/ Control-K to show or hide the Align panel.

Note:

Remember: all Flash panels can be accessed in the Window menu.

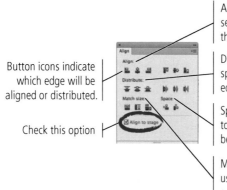

Button icons indicate which edge will be aligned or distributed.

Check this option

Align options move selected objects based on the defined edge or center.

Distribute options average the spacing between the selected edge or center of objects.

Space options can be used to create uniform space between selected objects.

Match Size options can be used to force objects to the same height, width, or both.

9. **Click the sketch with the Selection tool to select it, and then click the Align Horizontal Center and Align Vertical Center buttons in the Align panel.**

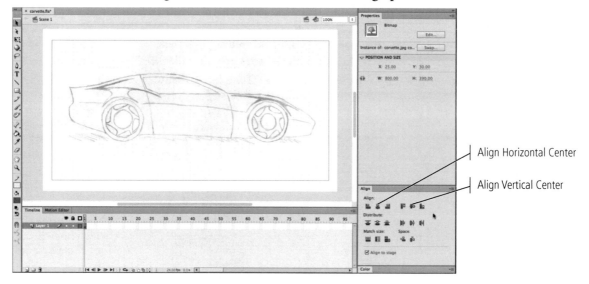

Align Horizontal Center

Align Vertical Center

10. **In the Align panel, turn off the Align to Stage option.**

If you later try to align multiple objects to one another and the Align to Stage option is still checked, the results will not be what you want. Although you can undo the Align process (Edit>Undo Align), it is a good idea to get into the habit of turning off the Align to Stage option as soon as you are finished using it.

11. **Save the file and continue to the next exercise.**

CREATE LAYERS FOR ORGANIZING ARTWORK

The Timeline panel allows you to easily create, arrange, and manage multiple layers in order to better organize your work. Layers can help you organize complex drawings and animations into a logical, easy-to-understand structure, while keeping all the components separate. You will use the timeline layers extensively as you complete the projects in this book.

1. **With corvette.fla open, review the Timeline panel.**

   Layers are listed on the left side of the timeline; the order in which the layers are stacked (from top to bottom) is referred to as the layer **stacking order**. Objects on higher layers in the stack will obscure objects on layers that are lower in the stack. You can easily change the stacking order by dragging a layer to a new position in the stack.

   Every file includes a default layer, named Layer 1. The icons to the right of the layer name provide a number of options, as well as some useful information.

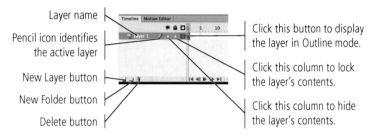

2. **Double-click the name of the layer to highlight the name. Type Sketch to rename the layer, and then press Return/Enter to apply the change.**

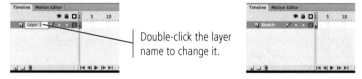

3. **Below the list of layers, click the New Layer button to add another layer on top of the first layer.**

   The new layer is automatically added immediately above the previously selected layer. New layers (and layer folders) are created in the same nesting level as the currently selected layer. In this case there is only one level of layers (there are no layer folders), so all new layers will be placed at the primary level.

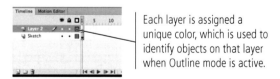

4. **Using the same technique from Step 2, rename the new layer Window and press Return/Enter to apply the change.**

5. **Repeat Steps 3–4 to create four more layers: Body, Wheels, Door, and Accents.**

6. **If you can't see all six layer names, click the bar above the docked Timeline panel and drag up to expand the panel.**

Drag this bar up or down to change the height of the Timeline panel.

Note:

If your Timeline panel is not docked, click the bottom edge of the floating panel and drag down to expand the panel.

7. **If you can't see the full layer names, click the bar to the right of the Outline icons and drag right.**

Drag this bar right to widen the area where layer names are listed.

8. **Lock the Sketch layer by clicking the small dot below the Lock icon.**

Locking the layer prevents you from accidentally deleting or otherwise modifying the sketch, which will be the basis for developing the actual drawing.

Click the dot to lock the layer.

Click the lock icon to unlock the layer.

Note:

If a layer is selected but locked, the pencil is covered with a slash, indicating that you can't change objects on the layer or add objects to the layer.

9. **Save the file and continue to the next stage of the project.**

Drawing Preferences

FLASH FOUNDATIONS

You can control a number of drawing options using the Drawing pane of the Preferences dialog box (accessed in the Flash menu on Macintosh, or in the Edit menu on Windows).

Each time you click with the Pen tool, you add an anchor point; successive anchor points are connected by lines. If the **Show Pen Preview** option is checked, a preview line follows the cursor to show the line that will be created when you add a new anchor point.

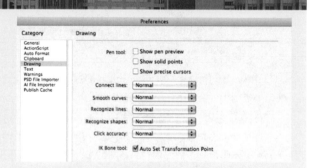

When the **Show Solid Points** check box is active, selecting an anchor point with the Subselection tool shows the selected point as solid instead of hollow.

You can change the Pen tool cursor ( ) to a precise crosshair (×) by activating the **Show Precise Cursors** check box.

The **Connect Lines** options (Must Be Close, Normal, and Can Be Distant) determine how close lines or points must be before Flash automatically connects them.

The **Smooth Curves** options (Off, Rough, Normal, and Smooth) determine how many points will be used to create a line when you're drawing with the Pencil tool with the Smooth or Straighten options turned on. The Rough option results in more points and a jagged line; the Smooth option results in fewer points and a smoother line.

The **Recognize Lines** options (Off, Strict, Normal, and Tolerant) determine how close to straight a line drawn with the Pencil tool must be before Flash recognizes it as straight.

The **Recognize Shapes** options (Off, Strict, Normal, and Tolerant) determine how close to the shape an oval or square drawn with the Pencil tool must be before Flash recognizes it as the actual shape and makes the necessary adjustment.

The **Click Accuracy** options (Strict, Normal, and Tolerant) determine how close to an object you must place the cursor to select the object.

Stage 2 Drawing in Flash

Most drawing objects display two primary attributes: a fill and a stroke. The **fill** is the color contained within a shape, and the **stroke** is the line that defines the outside border of a shape. To change the fill and stroke of an object, use the pop-up Swatches panel that appears when you click the Stroke or Fill swatch at the bottom of the Tools panel.

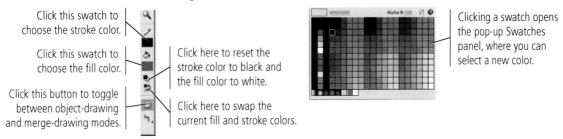

Click this swatch to choose the stroke color.

Click this swatch to choose the fill color.

Click this button to toggle between object-drawing and merge-drawing modes.

Click here to reset the stroke color to black and the fill color to white.

Click here to swap the current fill and stroke colors.

Clicking a swatch opens the pop-up Swatches panel, where you can select a new color.

There are two modes for drawing in Flash: merge drawing and object drawing. When you draw with the basic shape tools, or with the Pen, Line, Pencil, or Brush tool, a toggle near the bottom of the Tools panel determines which type of drawing you will create.

When object-drawing mode is active, the toggle is highlighted.

When object-drawing mode is inactive, the toggle is not highlighted; this is called merge-drawing mode.

All shapes drawn in **merge-drawing mode** exist at the bottom of the stacking order on their layer. In other words, they are always behind other objects — groups, drawing objects, symbol instances, etc. — in the same layer. You cannot rearrange the stacking order of merge-drawing shapes.

When you create an object in merge-drawing mode, the fill and stroke are treated as separate entities. You can individually select each part with the Selection (solid arrow) tool, and you can move or modify each part without affecting the other.

Oval drawn in merge-drawing mode. It has a blue fill and a green stroke.

The fill was selected with the Selection tool and dragged away from the stroke; the tiny cross-hatch pattern indicates the selected area.

You can select a part of a shape by dragging a marquee with the Selection tool.

Drag a selection marquee to select part of a shape.

The selected area can be deleted without affecting the rest of the shape.

When you drag one merge-drawing shape onto another on the same layer, the topmost shape removes whatever portion of the lower shape(s) it covers.

In the image to the right, the second column shows that we moved the orange circle on top of the blue circle. In the third column, you see the result of deselecting the orange circle, and then reselecting it and pulling it away from the blue circle. (This type of destructive interaction does not occur until you deselect the topmost object, which gives you a chance to "finalize" the edit. As long as you still see the crosshatch pattern on the object you're dragging, the underlying objects have not yet been changed.)

This type of interaction does not occur with objects created in object-drawing mode or when you're working with grouped objects. If you select something with the Selection tool and see a bounding box instead of the crosshatch pattern, underlying objects will not be affected.

You can prevent this type of destructive editing by grouping the components of the shape. Simply drag a selection around the entire object and choose Modify>Group. When a standard object is grouped, you see only the group **bounding box** when you select the object on the Stage. (The bounding box marks the outermost boundaries of the shape.)

Bounding box

After grouping the fill and stroke, you can't select the individual components. The bounding box indicates the boundary of the group.

Note:

The color of an object's bounding box is defined based on the Highlight Color options in the General preferences.

Even when merge-drawing shapes are grouped, you can access the individual pixel components by double-clicking them on the Stage. This opens a secondary Stage — called **Edit mode** — where you can access and edit the individual components. The Edit bar above the Stage shows that you are no longer on the main Stage (Scene 1), but editing a group.

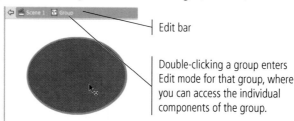

Edit bar

Double-clicking a group enters Edit mode for that group, where you can access the individual components of the group.

Of course, merge-drawing shapes have drawbacks as well; you could easily move or change an object that you don't want to change — such as dragging a fill away from its stroke and not realizing it until 15 steps later when it's too late to easily undo the error.

To solve this problem, you can draw in **object-drawing mode** by toggling on the switch at the bottom of the Tools panel. When you create a shape or line in object-drawing mode, it automatically displays the bounding box as soon as you create it, and you can't select the fill and stroke independently — at least not on the main Stage.

For a shape created in merge-drawing mode, you can use the Selection tool to independently select the shape's fill and stroke.

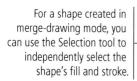  

For a shape created in object-drawing mode, you can use the Selection tool to reveal the shape's bounding box. You can't independently select the object's fill or stroke.

If you want to access the individual components of an object created in object-drawing mode, simply double-click the object. Double-clicking activates Edit mode; the Edit bar shows that you are "inside" (editing) a drawing object.

In Edit mode, you can access the individual drawing shapes that make up an object created in object-drawing mode.

DRAW WITH THE LINE TOOL

The Line tool creates straight lines, which are simply strokes with no fills. You can combine lines with other shapes to develop custom objects. In this exercise, you draw several lines to complete the shape of the car's window.

1. **With corvette.fla open, click the Window layer (in the Timeline panel) to make it the active one.**

   Any objects you draw will be placed on the active layer. It's important to check which layer you're working on whenever you start to create new artwork. Many professionals prefer to lock all but the layer that is being used at a given time, although it isn't necessary if you remember to select the layer you want before you begin drawing.

Note:

You can Control/right-click a layer in the Timeline and choose Lock Others to lock all but the selected layer.

2. **Select the Line tool in the Tools panel. At the bottom of the Tools panel, make sure the object-drawing option is turned off.**

   The Object Drawing toggle is dark gray when active. The Object Drawing toggle is not highlighted when inactive.

3. **Using either the Properties panel or the Tools panel, click the Stroke Color swatch to open the pop-up Swatches panel.**

   When you select a drawing tool, the Properties panel provides a number of options that are relevant to the selected tool. These options remember the last-used settings, so some might already reflect the settings we define in the following steps.

   The Stroke Color swatch in the Properties panel is the same as the Stroke Color swatch in the Tools panel. Changing either swatch reflects the same color in the other one.

   When you use the Line tool, the Fill Color swatch in the Properties panel is always set to None. By definition, a line has no fill; you can't change the fill color of a line. The Fill Color swatch in the Tools panel, on the other hand, retains the last-used fill option. When you switch to a tool that creates filled shapes, the swatch in the Tools panel becomes the default fill color.

Note:

The Line tool does not create filled shapes, so the fill color is not relevant for this exercise.

4. **Choose a bright green color as the stroke color.**

   You can use any color you prefer; we are using green so it is easily visible against the gray lines in the sketch. Later in this project, you will change the stroke and fill colors of the various pieces to create the finished artwork.

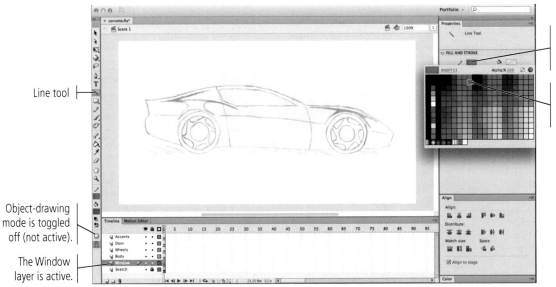

Line tool

Object-drawing mode is toggled off (not active).

The Window layer is active.

Click the swatch to open the Swatches panel.

We are using this green color for the strokes.

5. **In the Properties panel, set the stroke height (thickness) to 2 and choose Solid in the Style menu.**

You can type a specific stroke height in the Stroke Height field, or you can drag the associated slider for a live preview of the size you drag to (between 0.25 and 200 points).

Flash includes a number of built-in line styles, including dashed, dotted, ragged, stippled, and hatched. You can also edit the selected stroke style by clicking the pencil icon to the right of the Style menu.

Note:

The Scale menu defines how strokes are affected by scaling a symbol in which the line exists. (This option only affects strokes that are part of a symbol, which you will learn about starting in Project 2: Talking Kiosk Interface.)

6. **Open the Cap menu and make sure the Round option is active.**

The Cap options define the appearance of a stroke beyond the endpoint of the line.

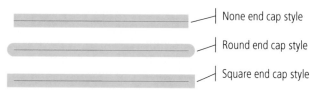

None end cap style

Round end cap style

Square end cap style

Note:

The Hinting check box helps when you draw straight lines that are connected to smooth curves. When this option is not checked, the curves might seem to be disconnected from the straight lines.

7. **Open the Join menu and make sure the Round option is selected.**

The Join options define the appearance of corners where two lines meet. When Miter is selected, you can define a miter limit in the Miter field. (This option is not relevant for Round or Bevel joins.) A miter limit controls when the corner switches from a pointed joint to a beveled joint, as a factor of the stroke height. If you define a miter limit of 2 for a 2-point line, the corner is beveled if the pointed corner extends beyond 4 points (2 × 2).

Miter join results in pointed corners.

Round join applies a circular edge to the corners.

Bevel join cuts off the points of the corners.

8. **Zoom in to the window on the sketch.**

As a general rule, we don't tell you the specific view percentage to use because every user's monitor and personal preferences determine a comfortable working environment. We do, however, tell you where to focus your attention.

9. **Choose View>Snapping>Snap to Objects. If this option is already checked (toggled on), move the mouse cursor away from the menu and click to dismiss the menu.**

As you draw different line segments with the Line tool, you want the ends of individual segments to align with each other. This is much easier when the Snap to Objects option is toggled on.

10. **Click the top-left corner of the window, hold down the mouse button, drag down and right, to the bottom-left corner, and then release the mouse button. When you release the mouse button, do not move the mouse cursor.**

Click here...
...and drag to here.

When you release the mouse button, a line appears with the defined stroke properties.

11. **With the cursor in the same position as when you finished the previous step, click and drag down and right to create the lower-left corner of the window shape (use the following image as a guide).**

Because the Snap to Objects option is toggled on, when you first click you should see a small circle connecting to the end of the line you drew in the previous step. This indicates that the new line is connecting to that open endpoint.

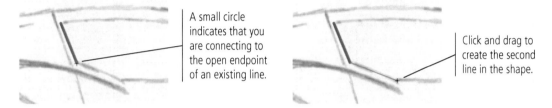

A small circle indicates that you are connecting to the open endpoint of an existing line.

Click and drag to create the second line in the shape.

12. **With the cursor in the exact position as at the end of Step 11, click and drag right to create the bottom edge of the window.**

13. **With the cursor in the exact position as at the end of Step 12, click and drag up and left until you see a small circle over the end of the first line you created, then release the mouse button.**

Again, the circle icon indicates that you are connecting to the open endpoint of an existing line — in this case, you are closing the shape.

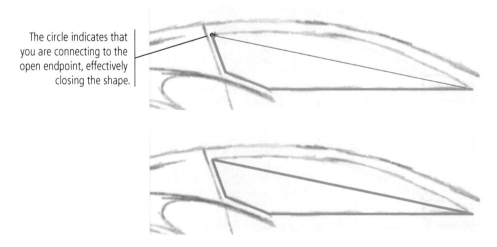

The circle indicates that you are connecting to the open endpoint, effectively closing the shape.

14. **Save the file and continue to the next exercise.**

ADJUST LINES WITH THE SELECTION TOOL

As the name suggests, the Selection tool is used to select objects. It can also be used to push or pull drawing objects into different shapes, almost as if you were pushing a string with your finger. In this exercise you will use this capability to bend the lines you just created into the correct shape for the car's window.

1. **With corvette.fla open, choose the Selection tool in the Tools panel.**

2. **Review the positions of the window corners. If any do not line up with the sketch, move the Selection tool cursor near the corner until a small corner shape appears in the tool icon. Click and drag to reposition the corner point.**

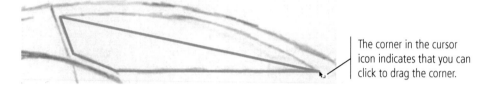

The corner in the cursor icon indicates that you can click to drag the corner.

3. **Move the cursor near the diagonal line that makes up the top of the window shape until you see a curved line in the cursor icon.**

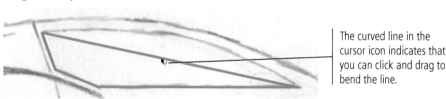

The curved line in the cursor icon indicates that you can click and drag to bend the line.

Note:

If an object is already selected, you have to first deselect it before you can use the Selection tool to bend the lines into different shapes.

4. **Click and drag up to push the line into a curved shape that matches the sketch.**

One unique aspect of drawing with Flash is using the Selection tool to pull or push a line into a different shape. When you release the mouse button, the line is reshaped.

Click and drag up to push the line into a curve that matches the sketch.

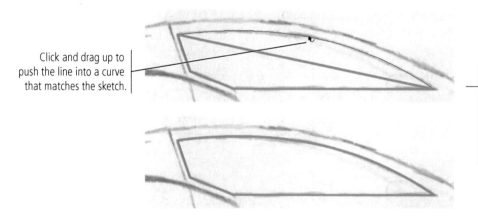

Note:

This type of "bending" works for objects created in merge-drawing or object-drawing mode.

5. **Repeat this process to slightly bend the diagonal line in the bottom-left corner of the window shape.**

6. **Save the file and continue to the next exercise.**

FLASH FOUNDATIONS

Using the Subselection tool. When you click a line with the Subselection tool, you can see the object's anchor points. In the image to the right, both objects were selected by surrounding them with a selection rectangle; you can see all the objects' points and curves.

Paths selected with the Subselection tool show the anchor points.

Subselection tool cursor

Selecting entire objects or single points. As you move the Subselection tool near a stroke, a small, solid black square appears at the bottom right of the icon. If you move the Subselection tool directly over a single anchor point, a hollow square in the icon indicates that clicking will select only that point.

A solid square in the Subselection tool cursor indicates you are selecting the path.

A hollow square cursor indicates you are selecting the anchor point.

Moving points. To move an anchor point, select the point with the Subselection tool and move it to the desired location. You can also **nudge** (move just a little bit at a time) a point by selecting it and then pressing the Up/Down/Left/Right Arrow keys on your keyboard.

Reshaping a curve using its handles. To reshape a curve, first select it with the Subselection tool, and then drag the related handles. (Handles will appear from each smooth anchor point that is connected to the curve.)

As you drag the handle up and out to the right, the shape of the curve changes accordingly.

Converting corner points to curves. To change a corner point into a smooth curve, first select the point you want to convert with the Subselection tool, then press the Option/Alt key and drag the point; handles appear, and the corner becomes a curve. You can then adjust the shape of the curve by adjusting the handles.

First, use the Subselection tool to select the point you want to convert, then Option/Alt-drag to pull out handles.

Converting curves to corner points. If you need to convert a curve to a corner point, place the Pen tool over the curve point, press the Option/Alt key, and click to change the point to a sharp corner. (Alternatively, use the Convert Anchor Point tool.)

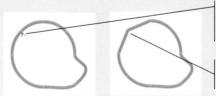

Press Option/Alt to access the Convert Anchor Point option.

Clicking changes a curve point to a corner point.

Adding anchor points. To add an anchor point, first select the path with the Subselection tool. Using the Pen tool, place the cursor near the path (but not over an existing anchor point) until you see a plus sign (+) in the tool cursor. Click the path to add a new point to the line. (Alternatively, use the Add Anchor Point tool.)

A plus sign (+) in the Pen tool cursor means clicking will add a point to the selected line.

Clicking adds a new point to the existing line.

Deleting anchor points. To delete an anchor point, first select the path with the Subselection tool. Using the Pen tool, place the cursor near an existing point until you see a minus sign (-) in the tool cursor. (Alternatively, use the Delete Anchor Point tool.)

A minus sign (-) in the Pen tool cursor means clicking will remove a point from the selected line.

Clicking removes the point from the line.

DRAW WITH THE PEN TOOL

The Pen tool draws curved lines, known as **Bézier curves**, that are shaped based on the position of anchor points and their connected direction handles. This type of drawing creates very precise shapes, but can be difficult to master. In this exercise, you will draw the shape of the car body. While you work through the steps, you should try to get a feeling for how clicking and dragging affects the shape you create. The best way to master the Pen tool is to practice; as you gain more experience using the Pen tool, you will be better able to understand and predict exactly where you should click to create the shape you want.

1. **With corvette.fla open, select the Body layer to make it the active layer. Zoom out so you can see the entire car shape.**

   If the Body layer is locked, click the lock icon for the layer to unlock it.

   Remember, to zoom out you can press Option/Alt and click with the Zoom tool, or press Command/Control-minus (–).

2. **Choose the Pen tool in the Tools panel, and toggle on the Object-Drawing mode option.**

3. **Click the Fill Color swatch in the Tools panel and choose the None swatch in the top-right corner of the pop-up Swatches panel.**

   Because you want to be able to see the sketch through the shapes you create in this stage of the project, you are setting the fill to none.

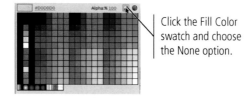

Click the Fill Color swatch and choose the None option.

4. **Using the Properties panel, make sure you are using the same color and size stroke that you used to create the window shape.**

   When you choose a new drawing tool, the object-drawing mode toggle, stroke, and fill properties retain the last-used options. If you took a break from this project after the previous exercise, make sure your drawing options are correct.

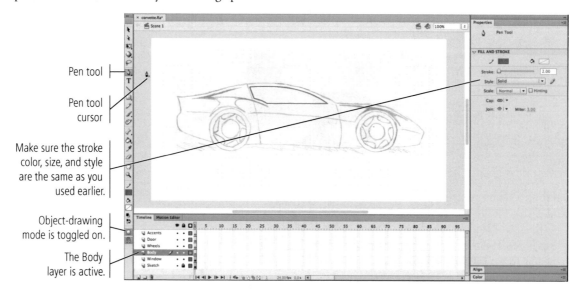

Pen tool

Pen tool cursor

Make sure the stroke color, size, and style are the same as you used earlier.

Object-drawing mode is toggled on.

The Body layer is active.

5. Click once at the top-left corner of the car shape.

Clicking with the Pen tool places an anchor point with no direction handles. (You might have to look very closely to see the anchor point.)

Click here without dragging to place an anchor point with no direction handles.

6. Move to the right until you are above the back edge of the rear wheel. Click and drag to the right to place an anchor point with symmetrical direction handles.

When you click and drag with the Pen tool, you create a smooth anchor point with symmetrical direction handles (the handles on both sides of the point are the same length, exactly opposite one another). The length and angle of the direction handles define the shape of the segment that connects the two anchor points.

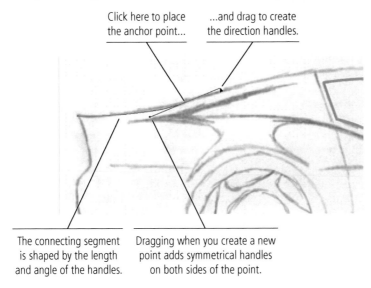

Click here to place the anchor point...

...and drag to create the direction handles.

The connecting segment is shaped by the length and angle of the handles.

Dragging when you create a new point adds symmetrical handles on both sides of the point.

Note:

Pressing Command/ Control while using the Pen tool temporarily switches to the Subselection tool. When you release the key, the Pen tool is again active.

7. Click and drag right again where the back of the door intersects the top of the car.

As you drag, notice how the position of the direction handles affects the shape of the connecting line segment. Experiment with dragging up, down, left, and right to see the effect on the curve shape.

8. Click and drag again where the top of the car meets the hood.

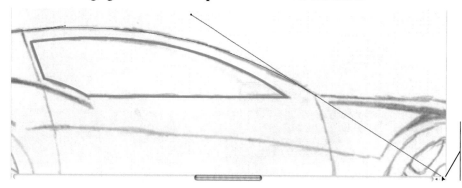

You can drag handles past the edge of the document window if necessary.

9. Place the cursor over the point you just created and click to remove the handle from the outside of the open point.

While you are drawing, clicking a symmetrical point again removes the direction handle from the outside of the point — converting the point to a corner point. The handle on the inside of the point remains because it is required to define the existing curve shape between the converted point and the previous one.

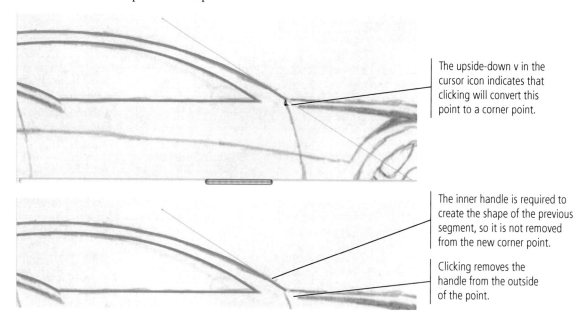

The upside-down v in the cursor icon indicates that clicking will convert this point to a corner point.

The inner handle is required to create the shape of the previous segment, so it is not removed from the new corner point.

Clicking removes the handle from the outside of the point.

10. Click and drag again near the corner of the front bumper.

Do not click the actual corner in the sketch. Because the bumper's edge is actually rounded, it's better to place anchor points on both sides of the curve rather than in the exact center of the curve.

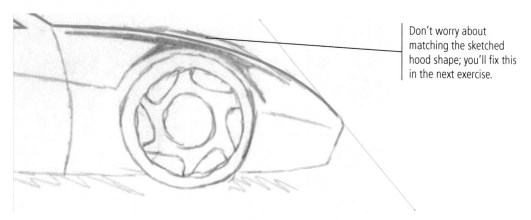

Don't worry about matching the sketched hood shape; you'll fix this in the next exercise.

11. **Click on the front side of the bumper, just below the curve, and drag short handles along the path of the bumper's front edge.**

   As you can see in our example, symmetrical points often result in incorrect paths. You will learn how to fix this type of problem in the next exercise.

The handle from the symmetrical point makes the connecting curve shape incorrect.

12. **Click just before the bottom curve of the bumper shape and drag slightly along the path of the bumper's front edge.**

   The point here is to create a smooth point with handles on both sides, without significantly affecting the shape of the connecting curve. Dragging a very small distance accomplishes this goal.

13. **Click and drag slightly along on the bottom edge of the car, just past the bottom-front corner.**

Short direction handles can be used to create smooth points with very tight connecting curves.

14. **Continue drawing the rest of the car shape by clicking and dragging to place anchor points with handles. For the bottom edge, simply extend the bottom-edge line over the wheel shapes.**

15. **When you get to the end of the shape, place the cursor over the original starting point and click to close the shape.**

   When you hover the cursor over the original starting point, the cursor shows a hollow circle in the icon, indicating that you can click to close the shape.

The circle in the cursor icon indicates that clicking will connect to the open endpoint and close the shape.

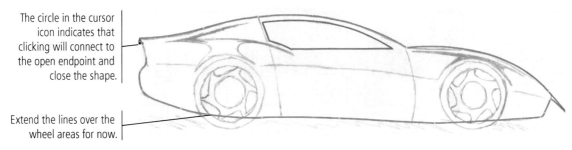

Extend the lines over the wheel areas for now.

16. **Save the file and continue to the next exercise.**

ADJUST BÉZIER CURVE SHAPES

From the previous exercise, you might have guessed that many of the lines that you draw with the Pen tool will not be perfect when you first create them. Fortunately, the Subselection tool makes it easy to edit the position of individual anchor points, as well as the length and angle of the direction handles that define connecting line segments.

Note:

The Selection tool selects entire paths, but the Subselection tool selects individual anchor points. You use the Subselection tool to select and edit anchor points, or adjust curves using the handles that radiate from anchor points.

1. **With corvette.fla open, zoom into the front end of the car.**

2. **Choose the Subselection tool in the Tools panel.**

   You can use the Subselection tool to access and adjust the anchor points that make up a line. You can also use this tool to adjust the direction handles that define the shapes of curves that connect two anchor points.

3. **Click the green line on the hood to select the shape, then click the anchor point on the front edge of the car hood to select it.**

   Clicking a line with the Subselection tool reveals the shape's anchor points.

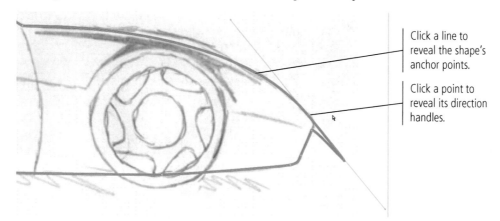

Click a line to reveal the shape's anchor points.

Click a point to reveal its direction handles.

4. **Click the outer direction handle (the one on the bottom) and drag up and left to reshape the curve on the top corner of the bumper.**

   When you drag one direction handle, the point remains smooth but it is no longer symmetrical; changing the length, but not the angle, of a direction handle affects only the segment on the same side of the point as the handle that you change.

   Because this is still a smooth point, however, dragging one handle to a different angle would also affect the point's other handle — affecting the shapes of both segments that are connected to the anchor point.

Note:

If you drag toward or away from the anchor point without changing the angle, you are changing the arc of the segment that is related to the changed handle.

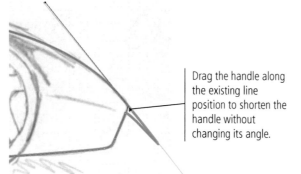

Drag the handle along the existing line position to shorten the handle without changing its angle.

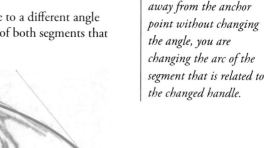

Changing the handle's length without changing its angle affects the shape of the connected segment.

5. **Choose the Add Anchor Point tool (nested under the Pen tool), then click the existing line midway along the hood to add a new anchor point.**

   You can also add a point by moving the Pen tool cursor over an existing line segment.

Add Anchor Point tool

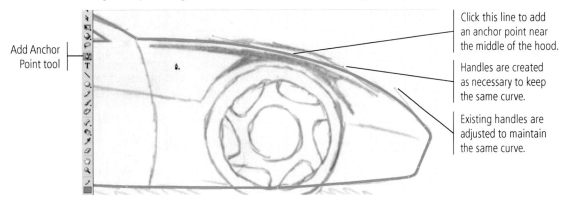

Click this line to add an anchor point near the middle of the hood.

Handles are created as necessary to keep the same curve.

Existing handles are adjusted to maintain the same curve.

6. **Using the Subselection tool, click the new point and drag until the path more closely resembles the line in the sketch.**

   When you move a point, the length and angle of the related handles are not affected. In many cases, you might need to move a point and then adjust the handles to achieve the effect you want.

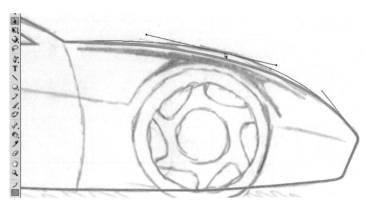

7. **Review the position of each anchor point and line segment in the car shape. Make any necessary adjustments until you are satisfied with your drawing.**

   Keep the following points in mind as you fine-tune your shape:

   • To move a point, click and drag with the Subselection tool.

   • To add an anchor point, click an existing segment with the Pen tool (or use the Add Anchor Point tool that is nested under the Pen tool).

   • To remove a point, click an existing point with the Pen tool (or use the Delete Anchor Point tool that is nested under the Pen tool).

   • To convert a smooth point to a corner point, Option/Alt click the point with the Pen tool (or use the Convert Anchor Point tool).

   • To convert a corner point to a smooth point, Option/Alt click the point with the Pen tool and drag to create symmetrical direction handles for the point (or use the Convert Anchor Point tool).

8. **In the Timeline panel, click the dot under the Eye icon for the Sketch layer. Click the Stage (away from the existing shapes) to deselect everything, then review your progress.**

Any time you are working from a sketch, it can be helpful to hide the sketch so you can better see exactly what you have created.

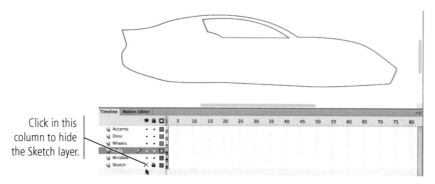

Click in this column to hide the Sketch layer.

9. **Click the red X for the Sketch layer to show that layer again.**

Feel free to repeat this process throughout this project — we won't include this step again.

10. **Save the file and continue to the next exercise.**

USE DESTRUCTIVE EDITING TO CREATE SHAPES

When you work on a merge-drawing shape (in other words, one created without object-drawing mode toggled on), you can use the Selection tool to select specific pieces of objects:

- Clicking a stroke selects the line segment between the nearest two anchor points.
- Clicking a fill selects the entire area of the color you click.
- Double-clicking a merge-drawing shape selects the entire stroke and fill.
- Double-clicking the stroke of a merge-drawing shape selects the entire stroke.
- Clicking and dragging a marquee selects only the pixels inside the marquee area.

This behavior enables a type of **destructive editing** that is unique to Flash drawing tools. You can select and move or delete certain pieces of an object without affecting other pieces of the same object.

1. **With corvette.fla open, click the Body layer icon in the Timeline panel to select the layer.**

When you select a layer, all objects on the layer become selected. In this case, the layer only contains one object — the shape of the car's body. You created this shape in object-drawing mode, so a bounding box identifies the selected object.

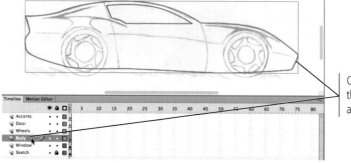

Clicking a layer icon in the Timeline panel selects all objects on that layer.

2. **Using the Selection tool, double-click any line of the car body shape to enter into the drawing object.**

Double-clicking an object on the Stage switches into Edit mode for the object (often referred to as "entering" the object). While in Edit mode for a particular object, you can see but not access other objects on the Stage.

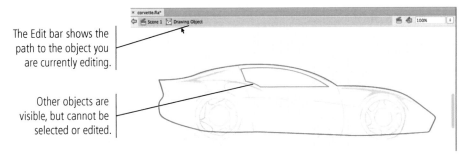

The Edit bar shows the path to the object you are currently editing.

Other objects are visible, but cannot be selected or edited.

3. **Choose the Line tool in the Tools panel. At the bottom of the Tools panel, make sure Object-Drawing mode is toggled on.**

4. **Draw a straight line where the back of the door is indicated in the sketch. Extend the line above and below the car body shape.**

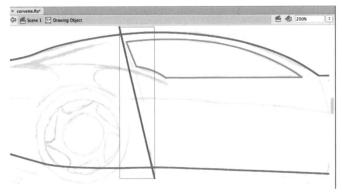

Note:

While drawing, press Command/Control to temporarily access the Selection tool; when you release the modifier key, you return to the last-selected tool.

The easiest way to deselect everything is to press Command/Control, and then click away from any object on the Stage.

5. **Choose the Selection tool. Click away from the line to deselect it, and then use the Selection tool to bend the line into shape.**

6. **Click the line to select it again, and then choose Modify>Break Apart.**

This command converts objects in object-drawing mode to merge-drawing mode so you can use the destructive editing capabilities of the Selection tool.

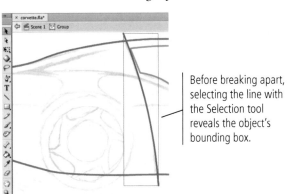

Before breaking apart, selecting the line with the Selection tool reveals the object's bounding box.

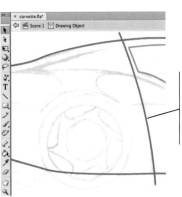

After breaking apart, the line is a merge-drawing shape; you can select the object's pixels with the Selection tool.

7. **Choose the Subselection tool in the Tools panel, and click any of the lines that make up the car body shape.**

   All lines you create in Flash are vectors, which means their position and shape are defined mathematically based on the anchor points that make up the shape. Flash automatically creates whatever anchor points are necessary to reproduce the line you draw — including points wherever two lines overlap, as you can see where the vertical curved line intersects the car body shape.

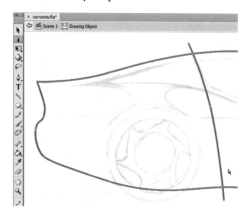

8. **Click away from the lines to deselect everything, then use the Selection tool to click the overhanging line above the car.**

   The selected area displays a small cross-hatch pattern.

9. **Press Shift and click the overhanging line below the car.**

   Pressing Shift allows you to select multiple objects at once; each item you Shift-click is added to the previous selection.

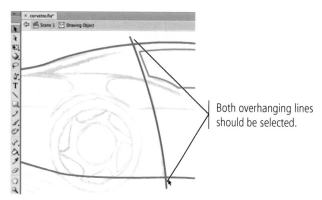

Both overhanging lines should be selected.

10. **Press Delete/Backspace to remove both elements from the drawing.**

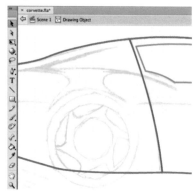

11. Repeat Steps 3–10 to create the line that marks the front of the door.

Later in this project, you need to be able to work with the car body shape as a single piece. If you left the door lines on the Body layer, the car shape would be effectively split into three pieces by the door-edge lines. To solve this potential problem, you are now going to move the door lines onto their own layer so they do not cut the car shape into pieces.

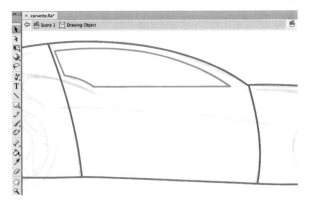

12. Using the Selection tool, click one of the door edges to select it, then Shift-click to select the other.

13. With both lines selected, choose Edit>Cut.

When you cut an object, it is removed from its original position and temporarily stored on the Clipboard — a reserved area of memory that allows you to temporarily hold objects until you need to use them again somewhere else. This is not the same as deleting (as you did in Step 10); when you delete a selection, it is not stored on the Clipboard.

14. In the Edit bar, click Scene 1 to return to the main Stage.

15. Select the Door layer as the target layer and then choose Edit>Paste in Place.

The Paste in Place command puts the cut objects in the exact position as they were cut from, but on the now-selected Door layer instead of the original Body layer.

The lines are pasted in the exact position...

...but on the now-selected Door layer.

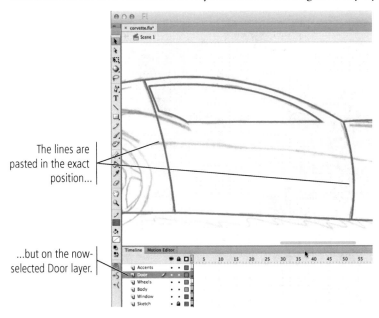

16. Save the file and continue to the next exercise.

DRAW WITH THE OVAL TOOL

Many otherwise complex objects are, in fact, simply a collection of primitive shapes. In the case of this car, you should recognize that the car's wheels are actually made up of three concentric circles. The basic shape tools — including the Oval tool — make it very easy to create this type of object.

1. **With corvette.fla open, select the Wheels layer (in the Timeline panel) as the active layer. Zoom into the rear wheel in the sketch.**

2. **Choose View>Rulers to show the Stage rulers at the top and left edges of the document window.**

3. **Click the horizontal ruler, hold down the mouse button, and drag down to place a horizontal guide in the center of the sketched wheel.**

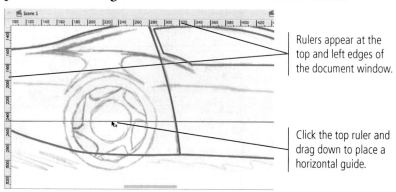

Rulers appear at the top and left edges of the document window.

Click the top ruler and drag down to place a horizontal guide.

FLASH FOUNDATIONS

Using Rulers, Guides, and Grids

Flash rulers, guides, and grids can dramatically improve accuracy and reduce development time. All three can be accessed and controlled in the View menu.

Rulers can be toggled on and off by choosing View>Rulers. Rulers normally measure in pixels, but you can use the General pane of the Preferences dialog box to change this setting to any measurement system you prefer.

The grid is a pattern of horizontal and vertical lines that also aid in the development/alignment process. The View>Grid submenu includes options for toggling the grid visibility (Show Grid), as well as editing the appearance and behavior of the grid (Edit Grid).

Guides are user-defined lines that you place on the Stage to mark a specific location; they are for development purposes only, and do not appear in the final output. The View>Guides submenu includes options for toggling guide visibility (Show Guides), preventing guides from being moved (Lock Guides), modifying the appearance and behavior of guides (Edit Guides), and removing all guides from the file (Clear Guides).

The View>Snapping submenu includes options for making guides (Snap to Guides) and the grid (Snap to Grid) act like magnets when objects are dragged near.

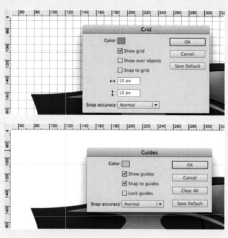

4. **Click the vertical ruler, hold down the mouse button, and drag right to place a vertical guide in the center of the wheel.**

You will perform a number of steps in this and the next exercise that will depend on an exact center point — where these two guides intersect.

Note:

Because this is a hand-drawn sketch, the wheels aren't exactly circular. It's okay if your circle doesn't exactly match the sketch.

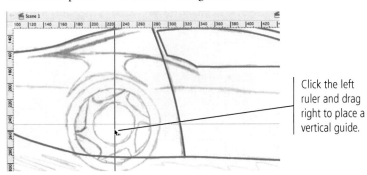

Click the left ruler and drag right to place a vertical guide.

5. **Choose the Oval tool (nested under the Rectangle tool) in the Tools panel. Make sure object-drawing mode is toggled on.**

6. **In the Properties panel, make sure the fill and stroke settings match what you used in earlier exercises. In the Oval Options section, set all three sliders to 0.**

The Oval tool shares many of the same properties as the Pen tool. You can define the fill and stroke color, as well as the stroke style, cap, and join.

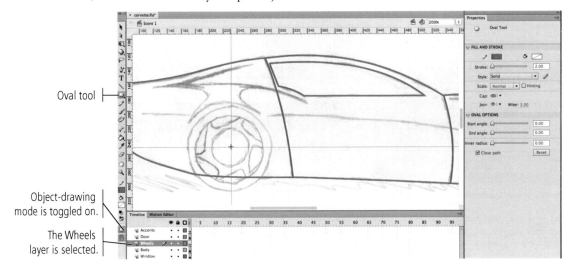

Oval tool

Object-drawing mode is toggled on.

The Wheels layer is selected.

7. **Click at the intersection of the two guides, hold down the mouse button, press Option/Alt-Shift, and drag until the shape approximately matches the outer circle of the sketched wheel.**

You can create a basic shape by simply clicking and dragging; the shape is anchored where you first click. Holding down the Option/Alt key as you drag draws the shape from the center out. Holding down the Shift key while dragging constrains the shape to equal height and width.

Note:

Whenever we say "click and drag", you should hold down the mouse button as you drag. We will not repeat the instruction to hold down the mouse button unless it differs in some way from the normal click-and-drag procedure.

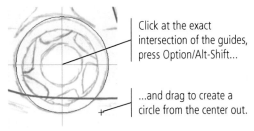

Click at the exact intersection of the guides, press Option/Alt-Shift...

...and drag to create a circle from the center out.

8. **With the new circle selected, choose Edit>Copy.**

   Copying stores the selected object in the Clipboard.

9. **Choose Edit>Paste in Place.**

   Pasting places a copy of the Clipboard contents onto the selected layer. The Paste in Place command puts the copy in the exact position as the original. You can also choose Paste in Center to place the copy in the exact center of the document window.

10. **Choose the Free Transform tool in the Tools panel.**

    When you select an object with the Free Transform tool, the object is bounded by eight transformation handles. You can use these handles to scale, stretch, skew, or rotate the selection. The hollow white circle in the center of the bounding box is the **transformation point**, which identifies the origin point around which the transformation will occur.

Note:

To copy an object, select the object and press Command/Control-C.

To cut an object, select the object and press Command/Control-X.

To paste an object from the Clipboard onto the Stage, press Command/Control-V.

To paste an object in place, press Command/Control-Shift-V.

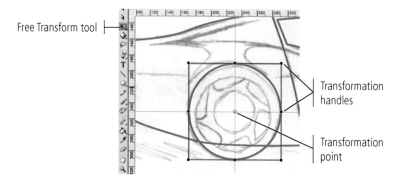

Free Transform tool

Transformation handles

Transformation point

Basic Shape Options

FLASH FOUNDATIONS

When you draw a basic shape, the Properties panel also includes options for defining the shape that will be created.

When you use the Oval tool, you can define the **Start Angle** and **End Angle** to create a shape with a defined wedge removed (such as you might see in a pie chart). You can also use the **Inner Radius** option to remove a circle from inside the circle; the defined value refers to the radius of the inner circle (the part that is removed). The **Close Path** option, which is checked by default, can be unchecked to create an open path rather than a closed shape.

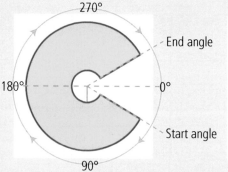

270°

End angle

180° — 0°

Start angle

90°

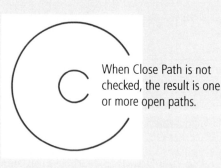

When Close Path is not checked, the result is one or more open paths.

When you use the Rectangle tool, you can easily create a shape with rounded corners. The Options area in the Properties panel defines the **Corner Radius**; by default, all four corners are linked, but you can click the chain icon to define a different radius for each corner. (To understand the concept of corner radius, think of an imaginary circle at the corner of the shape; the radius of that circle is the corner radius, as shown in the following image.)

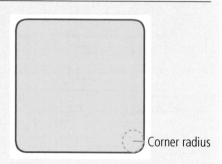

Corner radius

Using the Free Transform Tool

Free Transform mode allows you to change an object (or group) on the Stage by simply dragging. You can enter Free Transform mode by selecting an object with the Free Transform tool, by pressing Q (the keyboard shortcut for the tool), or by choosing Modify>Transform>Free Transform.

In Free Transform mode, the selected object is surrounded by a black bounding box with heavy black handles. A hollow circle in the center indicates the object's transformation point. Depending on where you place the cursor, you can apply different kinds of transformations. The cursor icon changes to reflect the change you will make by dragging.

Click a side handle and drag to resize the object's height (left) or width (right).

Click outside a corner handle and drag to rotate the object.

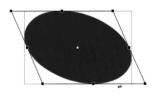

Click and drag between handles on any side to skew the object.

Click a corner handle and drag to resize both height and width at once.

Press **Shift** while dragging a corner handle to maintain the original height-to-width ratio in the resized object (called "constraining").

Press **Option/Alt** while dragging a handle to apply a change around the transformation point. (In this example, Option/Alt-dragging the bottom handle also moves the top handle.)

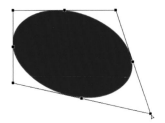

Press **Command/Control** while dragging a corner handle to distort the object shape.

Press **Command/Control** while dragging a side handle to skew and resize an object at the same time.

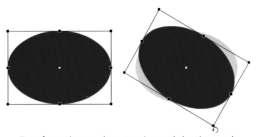

Transformations such as rotation and skewing apply around the transformation point, which defaults to the center of the object.

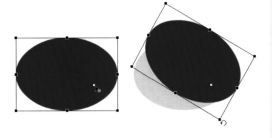

Moving the transformation point (left) changes the fixed point around which the rotation applies (right).

11. **Press Option/Alt-Shift, click the bottom-right bounding box handle, and drag up and left to shrink the circle around the center point.**

Here pressing Option/Alt applies the transformation around the transformation point, which defaults to the center of the selected shape. Adding Shift constrains the transformation so the object's height-to-width ratio remains the same as the original.

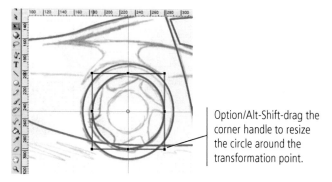

Option/Alt-Shift-drag the corner handle to resize the circle around the transformation point.

12. **Repeat Steps 8–11 to create the third circle in the wheel shape.**

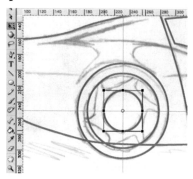

13. **Save the file and continue to the next exercise.**

DRAW WITH THE PENCIL TOOL

The Pencil tool provides another way to develop curved artwork. Drawing with the Pencil tool is easy; simply drag a line or shape, and the stroke — in whatever color is currently set in the Tools panel — follows the cursor.

With the proper equipment — notably a pressure-sensitive drawing tablet — the Pencil tool might very well be the most important tool Flash offers. This notion is particularly true for artists skilled at sketching or drawing with natural media such as pen and ink, pencils, and pastel chalks. Other people find the Pencil tool difficult to use, preferring the Pen tool, Bézier curves, primitive shapes, freeform shapes, and combinations thereof.

1. **With corvette.fla open, make sure the Wheels layer is active. Hide the Body layer so the lines do not interfere with the area where you are working.**

2. **Choose the Pencil tool in the Tools panel. At the bottom of the Tools panel, make sure Object Drawing is toggled on.**

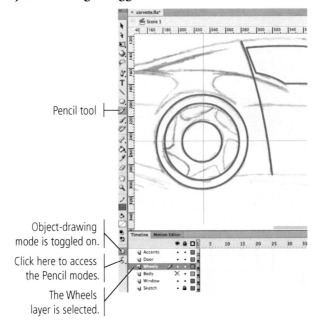

Pencil tool

Object-drawing mode is toggled on.

Click here to access the Pencil modes.

The Wheels layer is selected.

Note:

In the Properties panel, options for the Pencil tool are the same as those for the Pen tool.

3. **Click the mode button in the Tools panel and choose the Smooth option.**

The Pencil tool offers three possible modes:

- Straighten mode favors straight lines and corner points.
- Smooth mode favors curves and smooth anchor points in the resulting paths. This option allows greater tolerance for slightly jerky movement as you drag, which can be very useful if you are using a mouse instead of a drawing tablet.
- Ink mode closely follows the path that you draw, placing a large number of anchor points to capture very small movements of the cursor; this mode often results in a very jagged path, especially if you are using a mouse instead of a drawing tablet.

4. **Click and drag to create one of the inset shapes on the wheel. Make sure you overlap the line ends with the second circle, as shown in the following image.**

When you release the mouse button, the Smooth mode should result in a fairly rounded inset shape.

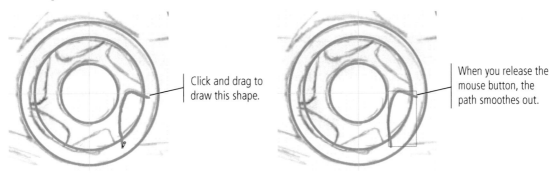

Click and drag to draw this shape.

When you release the mouse button, the path smoothes out.

5. **With the new shape selected, click the Free Transform tool to reveal the transformation handles.**

6. **Click the transformation point and drag to the intersection of the two ruler guides.**

   You are going to copy and rotate this object around the wheel's center point to create the six inset shapes. The ruler guides that you placed mark the center of the wheel, which you used when you Option/Alt dragged from the guide intersection to create the circles. To rotate the inset shape around the same center, you are now moving the first shape's transformation point to the same center point as the overall wheel shape.

   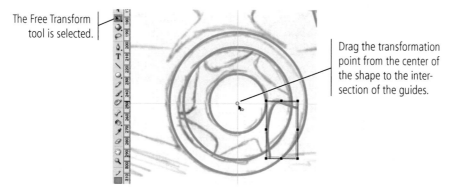

   The Free Transform tool is selected.

   Drag the transformation point from the center of the shape to the intersection of the guides.

7. **With the object selected, choose Edit>Copy.**

8. **Immediately choose Edit>Paste in Place.**

   Because you moved the transformation point before copying and pasting the object, the pasted copy also has the same relocated transformation point.

9. **Move the cursor outside one of the object's corner handles until you see the rotation icon in the tool cursor.**

10. **Click and drag left to rotate the copy, approximating the position of the second inset.**

    As you drag, you should recognize that the object is rotating around the repositioned transformation point; in other words, it's rotating around the wheel center.

    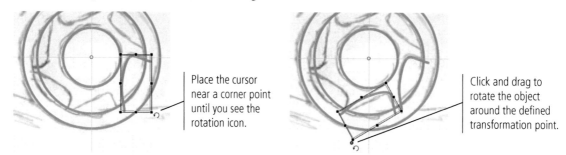

    Place the cursor near a corner point until you see the rotation icon.

    Click and drag to rotate the object around the defined transformation point.

11. **Repeat Steps 7–10 four more times to create the six wheel insets.**

    In this case, don't worry about perfectly matching the sketch. The important point is to have six inset shapes, spaced approximately evenly around the wheel shape.

12. **Choose View>Guides>Clear Guides.**

You no longer need the wheel's center-point reference now that all of the necessary shapes are in place, so you can delete the ruler guides.

Note that this menu also includes options for showing or hiding guides, locking guides, and numerically editing the position of a guide. These options are all useful at some point, depending on the purpose of a particular guide.

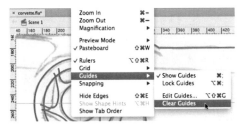

13. **Save the file and continue to the next exercise.**

 WORK WITH AN OBJECT GROUP

You want to be able to access the individual shapes that make up the wheel, but you also want to be able to treat the wheel as a single unit on the main Stage. A **group** is a collection of shapes that is combined so it can be treated as a single object.

1. **With corvette.fla open, hide the Sketch layer to make the wheel shapes more prominent.**

2. **Use the Selection tool to draw a marquee around all the shapes that make up the wheel artwork.**

Because you created these objects in object-drawing mode, you can see a bounding box surrounding each shape. To prevent moving one of these components independently of the others, you can group them into a single unit.

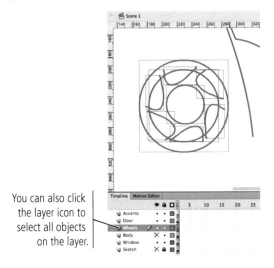

You can also click the layer icon to select all objects on the layer.

Note:

Press Command/Control-A to select all objects (on all layers) that are not locked.

3. Choose Modify>Group.

When you group objects, the bounding box displays around the entire group. By manipulating the bounding box, you can resize, move, and otherwise modify a group as you would a single object. Grouped objects are isolated from other elements on the same layer, so you can place grouped objects on top of one another with no worries of damaging the underlying objects.

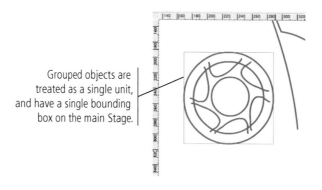

Grouped objects are treated as a single unit, and have a single bounding box on the main Stage.

Note:

You can group objects by pressing Command/Control-G.

To ungroup objects, select the group and choose Modify>Ungroup or press Command/Control-Shift-G.

4. Double-click the group to enter into it (view the group in Edit mode).

Even though the objects are grouped on the main Stage, you can still access the individual components in Edit mode.

Because you drew these shapes in object-drawing mode, you can't access the pixels of the lines. You can't select and delete parts of the lines with the Selection tool, which you need to do.

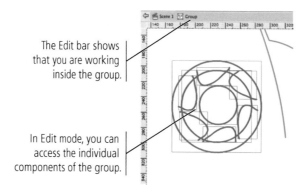

The Edit bar shows that you are working inside the group.

In Edit mode, you can access the individual components of the group.

5. With all objects in the group selected, choose Modify>Break Apart.

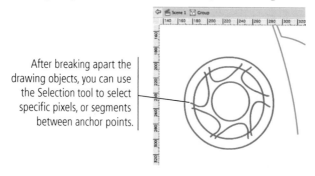

After breaking apart the drawing objects, you can use the Selection tool to select specific pixels, or segments between anchor points.

6. Click away from the lines to deselect everything.

7. Click to select one of the overhanging lines between the outer and middle circles, then press Delete/Backspace.

Remember from the earlier discussion on destructive editing that when you're editing a drawing object, Flash recognizes and separates overlapping elements.

8. **Repeat Step 7 to remove all of the extraneous line segments.**

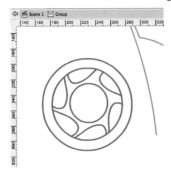

9. **In the Edit bar, click Scene 1 to return to the main Stage.**

   Because these objects started as a group, and you made all the edits inside the group, the selection on the main Stage is still treated a single unit; the bounding box surrounds the outer edge of all the objects in the group.

10. **Show the Body and Sketch layers.**

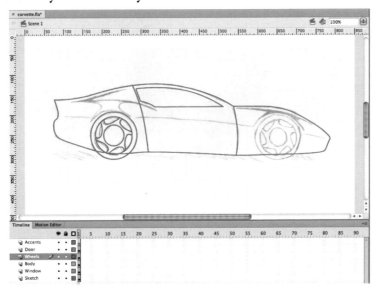

Note:

It's always a good idea to look for ways to avoid duplicating work. You will create the second wheel by duplicating the first wheel after you fill it with color.

11. **Save the file and continue to the next stage of the project.**

Stage 3 Painting and Coloring Objects

You now have all of the primary shapes you need for this project, clearly organized on separate layers. (You will create the front wheel later by duplicating the finished back wheel.) You drew all of these shapes as unfilled objects so you could see the underlying sketch; now you can apply fill colors to the various shapes. You can use a wide variety of methods to paint the car; you will explore these options in this stage of the project.

 APPLY FILL AND STROKE COLORS TO SELECTED SHAPES

The most basic way to change an object's color attributes is simply to select the object and then define the appropriate colors — which you will do in this exercise.

1. **With corvette.fla open, click the Body layer icon in the Timeline panel.**

   Remember, clicking a layer icon selects all objects on that layer.

2. **Click the Fill Color swatch in the Tools panel and choose a dark red color from the pop-up Swatches panel.**

When you change the fill or stroke color when an object is already selected, you affect the relevant attributes of the selected object.

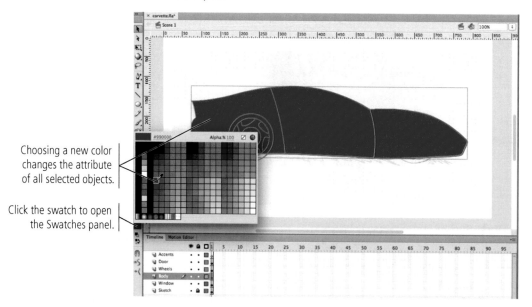

Choosing a new color changes the attribute of all selected objects.

Click the swatch to open the Swatches panel.

3. **With the car shape still selected, click the Stroke Color swatch in the Tools panel and choose the None swatch.**

This removes the green lines which were only necessary for the original drawing process.

4. **Review the layers in the Timeline panel.**

The layer stacking order shows that the window is below the car body. Because the car shape now has a solid fill, you can no longer see the window shape.

5. **Click the Window layer in the Timeline panel and drag up. When a line appears above the Body layer, release the mouse button.**

Rearranging the layer stacking order is as simple as clicking and dragging. When the Window layer is above the Body layer, you can again see the window shape.

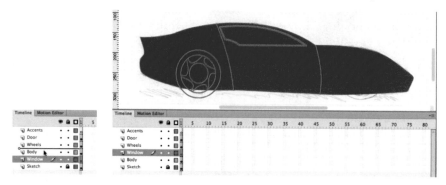

6. **Save the file and continue to the next exercise.**

FILL ARTWORK WITH THE PAINT BUCKET TOOL

The Paint Bucket tool provides an easy way to change objects' fill colors without first selecting the target objects. In this exercise you will use the Paint Bucket tool to change the fill colors of the window and the wheel.

1. **With corvette.fla open, make sure nothing is selected on the Stage.**

   If you change the Fill or Swatch color when an object is selected, the new color will affect the selected object.

2. **Choose the Paint Bucket tool in the Tools panel.**

3. **Use the Fill Color swatch in either the Tools panel or the Properties panel to change the Fill color to Black.**

4. **Click the Gap Size button at the bottom of the Tools panel and choose Close Large Gaps.**

   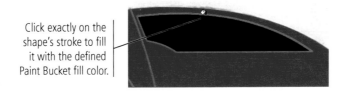

   The Gap Size menu allows Flash to overlook slight openings between line segments, which is a common occurrence when you draw with tools such as the Pencil or Line tool. Because you want all of these objects to be solid fills, you are using the largest setting to allow the application to fill the areas that you have defined.

5. **Click the line that identifies the window shape to fill the shape with the defined fill color.**

   Because the object doesn't yet have a fill color, you have to click exactly on the existing line to fill the closed shape. If you clicked inside the shape, the tool would recognize the fill color of the body shape and change its fill to black.

Note:

If you change the wrong object (on a different layer), press Command/Control-Z to undo the last step, and click again to edit a specific object.

Some users prefer to lock all but the layer currently being edited, to prevent changes to objects other than the one you targeted.

   Click exactly on the shape's stroke to fill it with the defined Paint Bucket fill color.

6. **Using the Selection tool, double-click the wheel group to enter into the group.**

   You have to enter into the group to edit the properties of individual group components.

7. **Choose the Paint Bucket tool again.**

8. **Click between the outer and middle circles of the wheel shape to change the fill color to black.**

   In this case, you are editing only the group. There are no other objects on the group's Stage, so the Paint Bucket tool recognizes the closed area between the two circles and adds the black fill color.

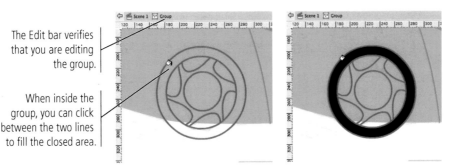

The Edit bar verifies that you are editing the group.

When inside the group, you can click between the two lines to fill the closed area.

9. Click inside the center circle, and then click inside each of the inset shapes to change all of those fills to black (as shown in the image after Step 10).

10. Change the fill color to medium gray, and then click inside the remaining shape (around the wheel insets) to fill that area with gray.

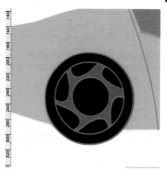

11. Click Scene 1 in the Edit bar to return to the main Stage.

12. Save the file and continue to the next exercise.

CHANGE STROKE ATTRIBUTES WITH THE INK BOTTLE TOOL

The Ink Bottle tool allows you to change the color of strokes without first selecting a specific object. Any stroke you click will be affected by the defined tool settings. In this exercise you will change the bright green strokes that you used during development to something that more accurately suits the project artwork.

1. With **corvette.fla** open, choose Edit>Deselect All to deselect everything.

2. Choose the Ink Bottle tool (under the Paint Bucket tool), change the stroke color to black, and change the stroke height to 1 pt.

   Once you select the Ink Bottle tool, you can use the Properties panel to predetermine the weight, color, stroke style, and other features before you use the tool. The current settings in the Properties panel will be applied to any shape you click with the Ink Bottle tool, and will remain in effect until you change them.

3. Click inside the Window shape to change the object's stroke color to black.

   The stroke instantly changes to the new color; it doesn't matter if you select the layer or not. Since all objects are on unlocked layers, the Ink Bottle tool "knows" which objects it's touching.

Note:

When either selection tool is active, you can also simply click an empty area of the Stage to deselect everything in the file.

Ink Bottle tool

You can click the Ink Bottle tool inside a filled shape to change the object's stroke attributes.

4. **With the Ink Bottle tool still selected, click directly on each line that marks the front and back of the car door.**

   When an object has an existing fill, you can click with the Ink Bottle tool anywhere in the object's fill to change the stroke attributes. If an object has no defined fill, you have to click on the existing stroke to change its attributes.

   To accomplish this step, it might help to zoom into the lines you want to change.

Note:

You don't need to select the objects before clicking with the Ink Bottle tool to change the stroke color.

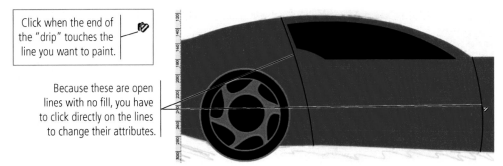

Click when the end of the "drip" touches the line you want to paint.

Because these are open lines with no fill, you have to click directly on the lines to change their attributes.

5. **Zoom in to the top of the rear door line and review the results.**

   Because you removed the stroke from the body shape, the "door" lines now hang a bit past the edge of the body shape. You can fix that by adjusting the end-cap style of the door lines.

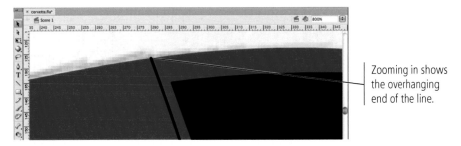

Zooming in shows the overhanging end of the line.

6. **Select the door line with the Selection tool. In the Properties panel, choose None in the Cap menu.**

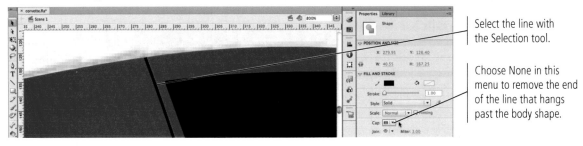

Select the line with the Selection tool.

Choose None in this menu to remove the end of the line that hangs past the body shape.

7. **Repeat Step 6 for the line that marks the front of the door.**

8. **Using the Selection tool, double-click the wheel group to enter into the group.**

9. **Choose the Ink Bottle tool again and change the stroke color to medium gray.**

10. **Click exactly on the edge of the middle circle.**

When you used the destructive editing functionality to create the inset shapes, a number of anchor points were added to the circle shape to create the inset pieces. You need to click the circle edge between and within every inset piece to change the entire stroke to gray.

11. **Continue clicking around the circle until the entire middle circle has a 1-pt gray stroke.**

12. **Change the stroke color to black and then click all of the remaining green strokes in the wheel to change them to black.**

In this case, you have to click the actual strokes. If you clicked the fills, you might inadvertently change the gray stroke of the middle circle back to black.

13. **Click Scene 1 in the Edit bar to return to the main Stage.**

14. **Using the Selection tool, click the wheel group, press Option/Alt, and drag right until the duplicate wheel aligns with the front wheel.**

Pressing Option/Alt while dragging **clones** (makes a copy of) the object you drag.

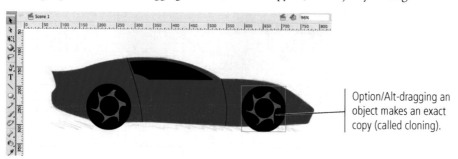

Option/Alt-dragging an object makes an exact copy (called cloning).

Note:

Be careful while making copies of any object. Do not release the Option/Alt key before releasing the mouse. Doing so might move the original object instead of making a copy.

15. **Save the file and continue to the next exercise.**

ADJUST ALPHA TRANSPARENCY

The Alpha property defines the degree of transparency of an object; in other words, it determines how much you can see underlying objects through another object. This can be useful for creating special effects, including the appearance of depth. In this exercise you will adjust alpha values to create the appearance of reflections on the car's surface.

1. **With `corvette.fla` open, click the Outline button for the Body layer so you can see the underlying sketch. Select the Accents layer as the active one.**

   Outline mode allows you to see through filled artwork without changing the objects' fill properties. All objects on outlined layers appear as simple lines that match that layer's defined color swatch; defined fill and stroke attributes are not visible.

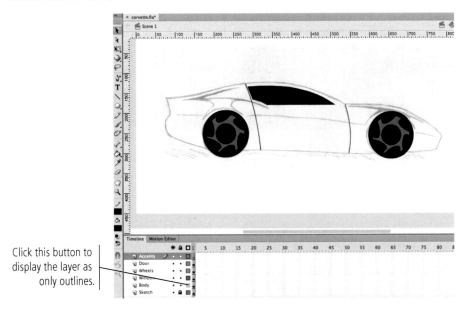

Click this button to display the layer as only outlines.

2. **Use any tool you prefer (with object-drawing mode toggled on) to create the two curved shapes that represent reflections on the car's upper body.**

3. **In the Timeline panel, click the Outline button for the Body layer to restore the layer to the regular view.**

4. **Select both of the accent shapes and apply a white fill and no stroke.**

5. **In the Color panel, click the Fill Color swatch to open the color palette.**

   Remember, all panels can be accessed in the Window menu.

6. **Place the cursor over the Alpha value (in the top-right corner of the palette) until you see a two-headed arrow in the hand icon.**

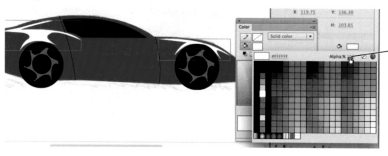

When you see this cursor, you can click and drag left or right to change the associated value (called "scrubbing").

7. **Click and drag left to reduce the Alpha value to 50%.**

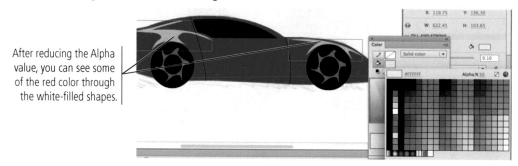

After reducing the Alpha value, you can see some of the red color through the white-filled shapes.

8. **Save the file and continue to the next exercise.**

CONVERT STROKES TO SOFTENED FILLS

The sketch in this project shows three lines that seem to bisect the car, creating an edge that catches the reflection of light. In this exercise, you will create these lines and then adjust them to add depth and texture to the artwork.

1. **With corvette.fla open, click the Outline icon for the Body layer so you can see the underlying sketch. Make sure the Accents layer is the active one.**

2. **Choose the Line tool in the Tools panel and activate object-drawing mode. Change the stroke color to white with a 50% Alpha value, and set the stroke height to 1 pt.**

3. **Draw the three lines on the car that seem to bisect the car horizontally. Use the Selection tool to position and bend the lines as necessary.**

4. **Click the Outline button for the Body layer to restore it to the normal view.**

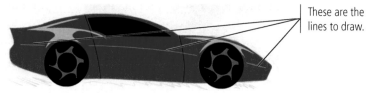

These are the lines to draw.

5. **Using the Selection tool, Shift-click to select all three lines.**

6. **Choose Modify>Shape>Convert Lines to Fills.**

These hard-edged lines appear rather harsh; you are going to modify these lines to more smoothly blend into the underlying shape. To do this, you first need to convert the strokes to filled shapes. The fill of the new shapes adopts the stroke color that was applied to the original line.

7. **With the three lines still selected, choose Modify>Shape>Soften Fill Edges.**

8. **In the resulting dialog box, set the Distance field to 2 px and set the Number of Steps field to 4. Choose the Expand option.**

9. **Click OK to apply the change.**

Each edge of the shape is now extended by 2 pixels, moving in four steps from the original alpha value (50%) to 0%, or entirely transparent.

Before converting to a fill   After converting to a fill   After softening edges

10. **Save the file and continue to the next exercise.**

FLASH FOUNDATIONS

Using the Eyedropper Tool

The Eyedropper tool makes it very easy to choose colors from existing objects on the Stage (called **sampling**). Clicking with the tool cursor pulls the color from the exact pixel where you click.

If you sample color from the fill of an object or from a raster image, the tool automatically switches to the Paint Bucket tool loaded with the sampled color.

If you sample color from an existing object's stroke, the tool automatically switches to the Ink Bottle tool loaded with the sampled color.

Eyedropper tool

Click to sample a fill color from an object or image.

The Paint Bucket tool is automatically loaded with the sampled color.

PAINT WITH THE BRUSH TOOL

The Brush tool, as you might expect, is used to paint areas of color. This tool offers a number of modes that allow you to paint in only specific areas, as you will do in this exercise to create a shadow underneath the car.

1. **With corvette.fla open, select the Body layer as the active one.**

2. **Click away from the artwork on the Stage to deselect everything.**

3. **Choose the Brush tool, change the fill color to a medium gray, and set the Alpha value to 100%.**

4. **At the bottom of the Tools panel, click the Brush Mode menu and choose Paint Behind.**

    The Brush tool has five modes:

    - **Paint Normal** is exactly that: it paints on top of anything already on the Stage.

    - Using the **Paint Fills** mode, you paint over the background and any object fills, but you leave object strokes unaffected.

    - **Paint Behind** allows you to paint only the background; any area with a fill or stroke remains unaffected by the brush stroke.

    - **Paint Selection** applies your brush strokes to currently selected fill areas only.

    - **Paint Inside** applies the brush stroke within the boundaries of the fill where you first click.

Note:

The tool options remember the last-used settings, so some of these might already be set to the options we use in this exercise.

Brush tool

Lock Fill toggle
Brush Mode menu
Brush Size menu
Brush Shape menu

5. **Click the Brush Size button and choose the largest available brush.**

6. **Click the Brush Shape button and choose the full round option.**

7. **Using the sketch as a rough guide, click and drag to paint a shadow shape underneath the car.**

   Don't worry about overlapping the car body shape while you're painting the shadow; although it appears to be on top of (and obscuring) the car shape, the shadow assumes its proper place behind the body when you release the mouse button.

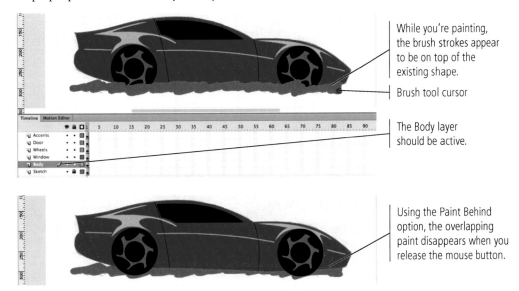

While you're painting, the brush strokes appear to be on top of the existing shape.

Brush tool cursor

The Body layer should be active.

Using the Paint Behind option, the overlapping paint disappears when you release the mouse button.

8. **Using the Selection tool, select the shadow shape and then choose Edit>Cut.**

9. **In the Timeline panel, click the New Layer button. Name the new layer Shadow.**

   As we explained previously, new layers are always added immediately above the previously selected layer.

10. **With the Shadow layer selected, choose Edit>Paste in Place.**

    Even though you couldn't see the entire brush stroke after releasing the mouse button, the shape exactly matches the area that you painted. When moved to a higher layer in the stack, the entire area becomes visible — obscuring some of the car body.

With the Shadow layer above the Body layer, the entire painted shape is visible.

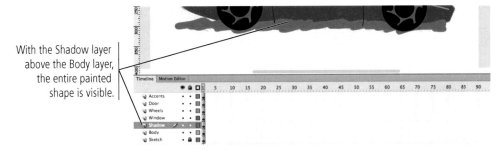

11. **Click the Shadow layer in the Timeline panel and drag it below the Body layer.**

Rearranging the layer stacking order hides the shadow edge behind the car body.

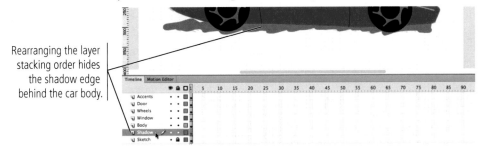

12. **Save the file and continue to the next exercise.**

APPLY LINEAR AND RADIAL GRADIENTS

At this point your artwork is a technically complete piece of artwork; all of the shapes are in place, with the necessary solid fills and strokes to clearly portray a digital Corvette. To add greater depth and dimension, however, you are going to apply gradients to the car's body and window.

1. **With corvette.fla open, click the window shape with the Selection tool.**

2. **Using either the Properties or Tools panel, open the Fill Color palette and choose the white-to-black radial gradient swatch.**

Note:

*Artwork that contains only solid fills is referred to as **flat art**.*

Note:

The terms "gradient" and "blend" are used interchangeably. Flash documentation uses both gradient and blend; both terms mean a smooth transition from one color to another.

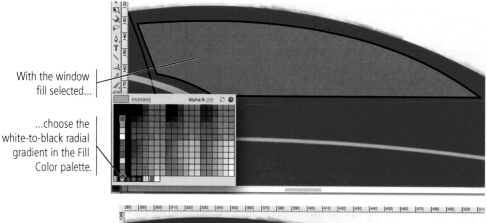

With the window fill selected...

...choose the white-to-black radial gradient in the Fill Color palette.

3. **In the Color panel, click below the gradient ramp near the middle to add a new color stop to the gradient.**

   Clicking below the ramp adds a new stop, which defaults to the color of the location where you first click.

Note:

You can define up to 15 color stops in a single gradient.

Click below the ramp to add a stop to the gradient.

4. **Drag the new stop left, to approximately one-fourth of the way across the ramp.**

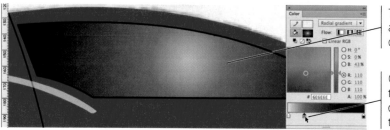

The selected shape adopts the new distribution of color.

Click the stop and drag to move that color to a different position along the gradient.

5. **Click the car body shape with the Selection tool.**

6. **In the Color panel, click the Fill Color icon to activate that attribute.**

7. **In the Color Type menu, choose Linear Gradient.**

The gradient ramp remembers the last-used gradient, so the adjusted white-to-black gradient from the window is applied to the selected shape (the car body).

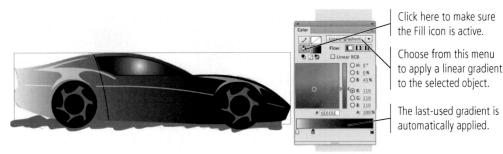

Click here to make sure the Fill icon is active.

Choose from this menu to apply a linear gradient to the selected object.

The last-used gradient is automatically applied.

8. **At the bottom of the Color panel, click the color stop in the middle of the gradient ramp and drag down away from the ramp.**

You can remove any stop from a gradient by simply dragging it away from the ramp.

9. **Double-click the color stop at the left end of the gradient ramp. Choose the dark red color swatch from the pop-up palette.**

Double-click a gradient stop to open the color palette for that stop.

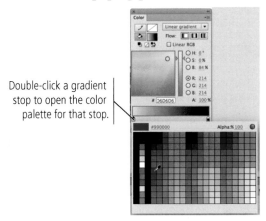

10. **Double-click the right color stop and choose the same dark red color that you applied to the left stop.**

11. **With the right stop still selected, drag the Brightness slider in the Color panel down until the B% option (in the HSB group) shows 10%.**

This is the value you are affecting by dragging the slider.

Drag here to change the stop's brightness value.

12. **Save the file and continue to the next exercise.**

TRANSFORM GRADIENTS

As you saw in the previous exercise, it is fairly easy to apply gradients to any drawing object, and to change the colors that are used in those gradients. Of course, the left-to-right gradient on the car's body does not create the realistic effect that you want. You now need to transform the applied gradients to finish the piece.

1. **With corvette.fla open, choose the Gradient Transform tool (nested under the Free Transform tool).**

2. **Click the window shape to select it.**

   The Gradient Transform tool enables you to adjust the size, direction, and center point of a gradient fill. When you click an object filled with a gradient, a special bounding box shows the gradient-editing handles.

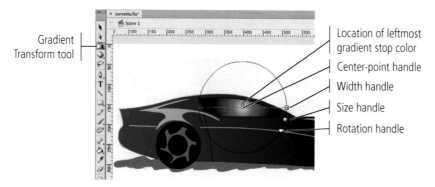

3. **Click the center point of the gradient and drag to the top-left corner of the window shape.**

   Dragging the center-point handle moves the center of the gradient to a new location.

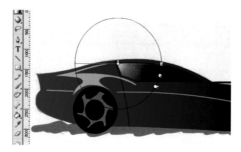

4. **Click the gradient-width handle and drag in about halfway to flatten the gradient.**

   Dragging the width handle of a radial gradient makes the gradient narrower or wider in one direction, essentially creating a skewed effect.

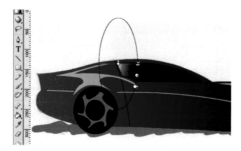

5. **Click the gradient-rotation handle and drag left to change the angle of the gradient.**

   Dragging the rotation handle rotates the gradient around its center point.

6. **Click the gradient-size handle and drag out to enlarge the adjusted gradient.**

   Dragging the size handle makes the overall gradient larger or smaller in both directions (without affecting the gradient's aspect ratio).

   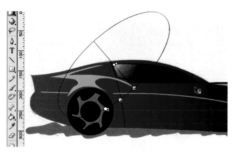

7. **Still using the Gradient Transform tool, click the car body shape to select it and access the gradient transformation handles.**

   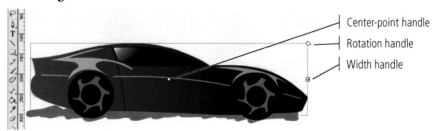

   Center-point handle
   Rotation handle
   Width handle

8. **Click the gradient-rotation handle and drag down and left to rotate the gradient approximately 90°, with the dark part of the gradient toward the bottom edge of the car.**

   Because you did not yet change the gradient width, the rotated gradient transformation handles will be well below the Stage edge. If necessary, zoom out so you can see all the gradient transformation handles.

   The gradient transformation handles are below the Stage edge because you have not yet changed the gradient width.

   Note:

   You can use the rotation handle to reverse a gradient. Simply drag all the way around (180°) until the gradient colors are reversed.

9. **Click the gradient-width handle and drag up until the gradient is just slightly larger than the height of your artwork.**

   For a linear gradient, dragging the width handle makes the gradient narrower or wider.

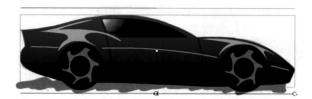

10. **Click the center point of the gradient and drag down, about one-third of the way closer to the bottom of the car body.**

11. **Save the file and continue to the next exercise.**

Locking Bitmap and Gradient Fills

FLASH FOUNDATIONS

You can use a bitmap (raster) image as the fill of an object by choosing Bitmap Fill in the Fill Type menu of the Color panel. All available bitmap files appear as tiles in the lower half of the panel; if none are available, choosing Bitmap Fill in the menu opens a dialog box, where you can import the file you want to use.

Choose Bitmap Fill in this menu.

Click Import to add bitmaps to the file.

Click any available tile to select that bitmap as the fill.

You can use the Paint Bucket tool to fill an existing object with a bitmap, or you can use the Brush tool to paint the bitmap onto the Stage. The **Lock Fill** modifier at the bottom of the Tools panel determines how the fill will be added.

- When toggled on, the selected bitmap image is tiled inside the shape.
- When toggled off, the selected bitmap image is basically pasted to the Stage. The filled area essentially "reveals" the area of the selected bitmap image. (If a bitmap-fill image is too small to fill the Stage, the image is tiled as necessary)

Lock Fill modifier

If you select a bitmap-filled object with the Gradient Transform tool, you can edit the size, skew, rotation, and position of the fill image.

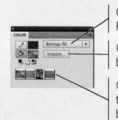

Turned on, the bitmap image is tiled inside the filled shape

Turned off, the shape reveals the relevant area of the bitmap.

Similar Lock Fill options are available when you work with gradients. When this option is toggled on, the gradient is applied relative to the overall Stage; filled areas reveal the relevant area of the gradient. If the option is turned off, the gradient is applied to each object independently.

With the Lock Fill modifier, the same gradient extends across all four squares.

Without the Lock Fill modifier, the gradient is applied individually to each square.

ORGANIZE FINISHED ARTWORK LAYERS

A Flash project can contain dozens of layers representing the various components of the animation or movie. To organize all of these layers into logical groups or categories, you can use layer folders.

1. **With corvette.fla open, select the top layer (Accents) in the Timeline panel.**

2. **Click the New Folder button at the bottom of the Timeline panel.**

New Folder button

3. **Double-click the new folder name. Type Car Parts as the new folder name and press Return/Enter to finalize the new name.**

   You change the name of a layer folder the same way you change the name of a layer; simply double-click the folder name and type in the new name.

4. **Click the Accents layer to select it, then Shift-click the Shadow layer to add it and all contiguous layers to the selection.**

5. **Click any of the selected layers and drag them into the Car Parts folder.**

   The dragged layers now reside inside the layer folder, which makes it easier to treat the entire group of layers as a single object. This will be helpful later when the artwork is used in programming and interactive development.

Note:

You can rename layer folders in the same way you rename layers, and you can reorder layer folders (and their contents) by dragging the folders into new positions in the layer stack.

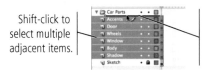

Shift-click to select multiple adjacent items.

The indented line shows that the selected layers will be placed inside the Car Parts folder.

6. **Click the Lock column for the Car Parts folder.**

   By locking the layer folder, you have effectively locked each layer inside the folder (although you can unlock individual layers in the folder by clicking that layer's Lock icon). The same concept is true of the Visibility and Outline buttons for the layer folder.

Note:

Press Shift to select contiguous layers — that is, layers directly on top of or below one another. To select layers that aren't next to each other, press the Command/Control key while clicking the individual (non-contiguous) layers.

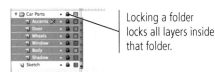

Locking a folder locks all layers inside that folder.

7. **Click the arrow to the left of the Car Parts folder to collapse it.**

   Once collapsed, only the layer folder is visible. This makes it easier to navigate through the Timeline panel, which can grow very large for complex projects.

8. **Select the Sketch layer and click the panel's Delete button.**

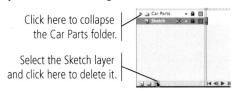

Click here to collapse the Car Parts folder.

Select the Sketch layer and click here to delete it.

Note:

When you use layer folders, you can lock and hide all the contents in the folder by locking and hiding the entire folder.

9. **Save the file and continue to the final exercise.**

EXPORT ARTWORK AS AN IMAGE

Instead of sending the Flash file to your client, you are going to export a static image that can be easily reviewed by anyone with a Web browser. Flash includes the ability to export a number of different static image formats, including JPEG, GIF, Adobe FXG (Flash XML Graphics), PNG, and BMP (Windows only).

1. **With corvette.fla open, choose File>Export>Export Image. Navigate to your WIP>Car folder as the target location.**

2. **Add _drawing to the end of the existing file name (before the file extension).**

   On Windows, the file extension might not be visible in the document window. If this is the case, simply type after the existing file name.

3. **Choose JPEG Image in the Format/Save As Type menu, then click Save.**

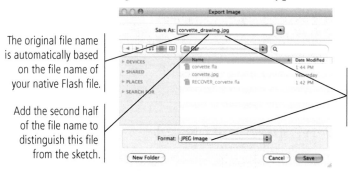

The original file name is automatically based on the file name of your native Flash file.

Add the second half of the file name to distinguish this file from the sketch.

The extension automatically changes to match what you select in the Format/Save As Type menu.

4. **In the resulting Export JPEG dialog box, make sure the Resolution option is set to 72 dpi.**

   You are creating this image for on-screen review, so 72 dpi is fine. If you needed the artwork for high-quality printing, you should use a higher resolution (i.e., 300 dpi).

5. **Choose Full Document Size in the Include menu.**

   If you choose Minimum Image Area, the exported image will be only and exactly as large as the outermost edges of the artwork. The Full Document Size option creates a file that matches the defined Stage size.

6. **Change the Quality value to 100%.**

   Reducing the Quality value will result in a smaller file size, but can significantly decrease the resulting image quality (especially in complex artwork). For review purposes, the highest possible quality setting is a better option.

7. **Click OK to export the JPEG file.**

8. **Close the corvette.fla file.**

Note:

Flash should generally not be used for developing artwork for high-quality commercial print applications.

Note:

You can click the Match Screen button to restore the size and resolution fields to match what you see on your monitor.

Note:

If the Progressive Display option is checked, the image will appear in progressive degrees of detail as it downloads.

1. The _____ can be used to change the size of the Stage.

2. _____ in the Timeline panel can be used to manage the top-to-bottom order of objects on the Stage.

3. When created in _____, the fill and stroke are treated as separate entities; you can individually select each part, and move or modify each part without affecting the other.

4. When you create a shape in _____, it automatically displays the bounding box as soon as you create it; you can't select the fill and stroke independently on the main Stage.

5. A(n) _____ allows multiple objects to be treated as a single entity on the Stage.

6. The _____ tool can be used to select areas of a standard-drawing object, or to bend lines created in either drawing mode.

7. The _____ tool can be used to select individual anchor points and handles that make up a shape.

8. The _____ tool is used to create precise Bézier curves.

9. The _____ tool is used to change the fill color of closed shapes.

10. The _____ tool is used to change the stroke color of any object.

1. Briefly explain how anchor points and handles control the lines in a vector shape.

2. Briefly explain the difference between the Selection tool and the Subselection tool.

3. Briefly explain what is meant by "destructive editing."

Use what you learned in this project to complete the following freeform exercise.
Carefully read the art director and client comments, then create your own design to meet the needs of the project.
Use the space below to sketch ideas; when finished, write a brief explanation of your reasoning behind your final design.

art director comments

Your client is happy with the Corvette, and asked you to create two more illustrations that he can use in different digital projects.

To complete this project, you should:

❑ Search the Internet for photos that you can use as templates for creating the other new illustrations.

❑ Create each illustration in a separate file.

❑ Carefully consider the best approach for each icon and use whichever tool (or tools) are most appropriate to create the icon artwork.

client comments

The cars are a big part of the annual event, but it isn't limited to only cars. Really, the show includes categories for just about anything with a motor. Aside from the American Heavy Metal sports cars, the two most popular parts of the show are Early Antiques and Motorcycles.

I'd like you to create illustrations for both of those categories. For the Early Antiques, the Model T Ford of the 1910s will be a perfect icon for the class. For motorcycles, a 1940s Indian would be ideal for the different projects I want to create.

I really like the Corvette artwork, so I want the two new illustrations to have a similar style. When I look at the three pieces together, I'd like it to be clear that all three were created by the same artist.

project justification

Project Summary

Although Flash is known as an animation program, you can see from this project that the application offers a considerable array of drawing and painting tools. You learned how to import a sketch, control the size of the Flash Stage, and align objects to the Stage. These skills set the groundwork for any type of assignment. You also developed an understanding of layers and how they simplify the construction of complex drawings and artwork.

Flash includes a range of drawing tools — from the basic shape tools, the Line tool for simple lines, and the Pen and Pencil tools for more complex curves. Those tools, combined with the two Flash drawing modes and the unique destructive editing capabilities that are part of the Flash toolset, make it possible to create complex artwork directly on the Stage.

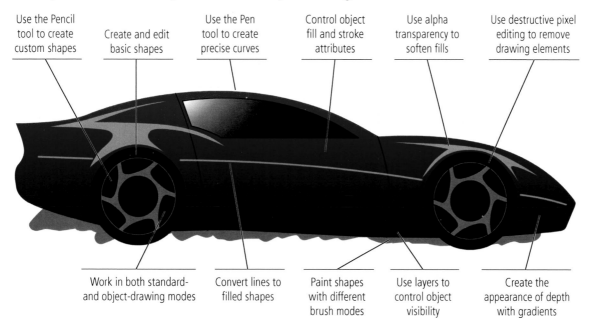

Use the Pencil tool to create custom shapes

Create and edit basic shapes

Use the Pen tool to create precise curves

Control object fill and stroke attributes

Use alpha transparency to soften fills

Use destructive pixel editing to remove drawing elements

Work in both standard- and object-drawing modes

Convert lines to filled shapes

Paint shapes with different brush modes

Use layers to control object visibility

Create the appearance of depth with gradients

Talking Kiosk Interface

You were hired to create an animated introduction for a shopping mall information kiosk. The Flash-based animation must feature a character who offers shoppers assistance in finding various facilities. Each button should also offer audio instructions that explain the link's purpose. As part of the Flash development team, your job is to prepare the interface artwork that will be handed off to the programmer, who will script the interactivity.

This project incorporates the following skills:

❏ Importing and managing artwork from Adobe Illustrator

❏ Using the Library panel to manage a complex file

❏ Building a frame-by-frame animation

❏ Editing various button states

❏ Importing sound files into Flash

❏ Adding event and stream sounds to the Flash timeline

❏ Controlling volume and duration of sound

❏ Applying built-in sound effects

❏ Synchronizing sound to animation

❏ Defining sound compression settings

client comments

Throughout the facility grounds, we are replacing all of the static "You Are Here" maps with interactive kiosks that will help users more quickly find the shops they are looking for.

I'd like the interface to be personal — a person actually talking to the user. We thought about the video route, but I'm convinced an animated character would be better (plus we won't have to pay an actor to use her image).

The interface should provide a link to four different categories of shops: Shoes & Apparel, Home Furnishings, Music & Electronics, and Casual & Fine Dining. We might break it down into more specific categories later, but the important point for now is to get the first version of this thing into use quickly.

art director comments

I already had all of the kiosk components created. I need you to assemble everything in Flash and prepare the various elements for the programmer, who will create all of the necessary code and links.

The artwork was created in Adobe Illustrator. Our illustrator is fairly knowledgeable about Flash requirements, so you should be able to import the artwork without too many problems. He even created the basic appearance of the navigation buttons, so you'll just need to modify those rather than create them from scratch.

When I reviewed the sound files, it seemed like the background music was very loud compared to the spoken intro. You should fix the music so the talking is audible above the background.

The lip-syncing part of the interface requires some careful attention to detail, but overall, it isn't a difficult job. Just take your time and try to make the mouths follow the words.

project objectives

To complete this project, you will:

- ❏ Create symbols from imported Illustrator files
- ❏ Place and manage instances of symbols on the Stage
- ❏ Control timing using keyframes
- ❏ Add visual interactivity to button symbols
- ❏ Import sound files into Flash
- ❏ Add event and stream sounds to a movie
- ❏ Use the start and stop sync methods for button sounds
- ❏ Edit a sound envelope to control volume and duration
- ❏ Swap graphics at precise moments in time
- ❏ Define sound compression settings

Stage 1 Working with Symbols

Although Flash can be used to create extraordinary interactive content, the program can also create extremely large files that take a very long time to download. (The size of a file is often referred to as a file's **weight**.) Users will not wait for more than a few seconds to download a file, so you should always try to keep file weight to a minimum — and that's where symbols come into play.

Symbols are objects that can be used repeatedly without increasing file size. The original symbol resides in the Library panel; **symbol instances** are copies of the symbol that you place onto the Stage. Although a regular graphic object adds to the overall file weight every time you use it on the Stage, a symbol counts only once no matter how many times you use it — which can mean dramatically smaller file sizes.

As another benefit, changes made to the content of an original symbol reflect in every placed instance of that symbol. For example, if you have placed 40 instances of a bird symbol, you can simultaneously change all 40 birds from blue jays to cardinals by changing the primary symbol in the Library panel.

A third benefit of symbols is that you can name placed instances, which means those instances can be targeted and affected by programming — one of the keys to animation and interactive development.

Note:

There are three primary types of symbols — graphic, movie clip, and button — and a number of other types of assets, such as audio and video files. In Flash, all of these assets are automatically stored in the Library panel.

FLASH FOUNDATIONS

The Library Panel in Depth

Assets in Flash are stored in the Library panel. Additional information about each asset is listed on the right side of the panel, including the name by which an asset can be called using ActionScript (AS Linkage), the number of instances in the current file (Use Count), the date the asset was last modified (Date Modified), and the type of asset (Type). To show the additional information, you can either make the panel wider or use the scroll bar at the bottom of the panel. In addition to storing and organizing assets, the Library panel has a number of other uses:

- Each type of asset is identified by a unique icon. Double-clicking a symbol icon enters into Symbol-Editing mode, where you can modify the symbol on its own Stage. Double-clicking a non-symbol icon (sounds, bitmaps, etc.) opens the Properties dialog box for that file.

- You can use the Library menu to switch between the libraries of currently open files.

- The Preview pane shows the selected asset. If the asset includes animation, video, or sound, you can use the Play and Stop buttons to preview the file in the panel. (The Stage background color appears in the Preview pane; if you can't see the Play button, move your mouse over the area of the button to reveal it.)

- If a file has a large number of assets (which is common), you can use the Search field to find assets by name.

- Clicking the Pin button to the right of the Library menu attaches the current library to the open Flash file.

- Clicking the New Library Panel button opens a new version of the Library panel, which allows you to view multiple libraries at one time.

- Clicking the New Symbol button opens the Create New Symbol dialog box, where you can define the name and type of the new symbol you want to create.

- Clicking the New Folder button adds a new folder in the current file's library.

- Clicking the Properties button opens a dialog box that shows information about the selected library asset.

- Clicking Delete removes an asset from the library. Placed instances of that symbol are deleted from the file.

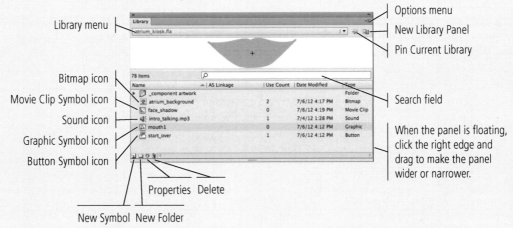

Library menu • Options menu • New Library Panel • Pin Current Library • Search field • Bitmap icon • Movie Clip Symbol icon • Sound icon • Graphic Symbol icon • Button Symbol icon • Properties • Delete • New Symbol • New Folder • When the panel is floating, click the right edge and drag to make the panel wider or narrower.

IMPORT ADOBE ILLUSTRATOR ARTWORK

As you learned in Project 1: Corvette Artwork, you can use the built-in Flash tools to draw complex custom artwork. In many cases, however, your work in Flash will incorporate files that were created in other applications. For example, illustrations and other vector graphics for animation are typically created in Adobe Illustrator. This project incorporates a number of external files, which you need to import into your Flash file.

1. **Download FL6_RF_Project2.zip from the Student Files Web page.**

2. **Expand the ZIP archive in your WIP folder (Macintosh) or copy the archive contents into your WIP folder (Windows).**

   This results in a folder named **Atrium**, which contains the files you need for this project. You should also use this folder to save the files you create in this project.

3. **In Flash, create a new Flash document for ActionScript 3.0 using the default settings in the New Document dialog box.**

   You can choose File>New, or use the Create New ActionScript 3.0 option in the Welcome Screen.

4. **Choose File>Save. Save the file in your WIP>Atrium folder as a Flash document named atrium_kiosk.fla.**

   In this project, you will use the Tools, Properties, Library, Align, and Timeline panels. You should arrange your workspace to best suit your personal preferences.

5. **Review the Properties and Library panels.**

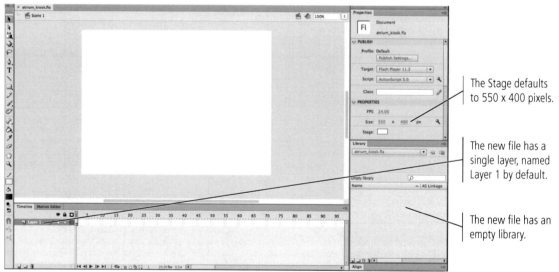

The Stage defaults to 550 x 400 pixels.

The new file has a single layer, named Layer 1 by default.

The new file has an empty library.

6. **Choose File>Import>Import to Stage. Navigate to the file interface.ai in the WIP>Atrium folder and click Open.**

7. **In the top-left section of the resulting dialog box, review the contents of the file that you are importing.**

   The list of imported elements will make more sense if you're familiar with Adobe Illustrator. If you're not, the best choice is usually to import everything and review it carefully once it is in the Flash file.

If the Illustrator file has more than one artboard, access them here.

Layers in the file are the first-level listings.

Groups on the individual layers are listed under the appropriate layers.

Items that make up a group are listed below the appropriate groups.

By default, any item that is visible in the Illustrator file is checked, which means it will be imported.

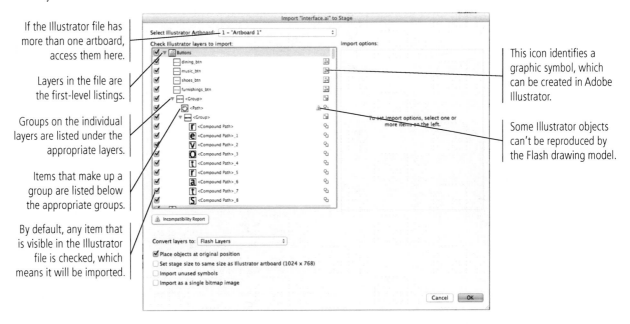

This icon identifies a graphic symbol, which can be created in Adobe Illustrator.

Some Illustrator objects can't be reproduced by the Flash drawing model.

8. **Click the path with the warning icon in the list, and review the information on the right side of the dialog box.**

   Many objects created in Illustrator are fully compatible with Flash drawing capabilities, so they are imported as regular drawing objects. Any objects that don't fit into the Flash drawing model — primarily, ones with some type of applied transparency — are imported in a way that allows Flash to maintain the integrity of the overall artwork. Effects that cannot be reproduced by Flash display a warning icon in the list of elements.

Note:

*Learn more about Adobe Illustrator in the companion book of this series, **Adobe Illustrator CS6: The Professional Portfolio**.*

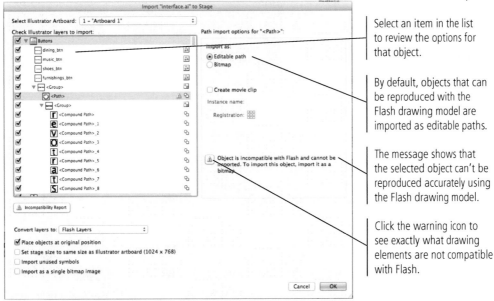

Select an item in the list to review the options for that object.

By default, objects that can be reproduced with the Flash drawing model are imported as editable paths.

The message shows that the selected object can't be reproduced accurately using the Flash drawing model.

Click the warning icon to see exactly what drawing elements are not compatible with Flash.

9. **With the incompatible path selected, choose the Bitmap radio button in the Import As area.**

Importing the object as a bitmap preserves its appearance, but you won't be able to edit the vector paths that make up the shape.

Bitmaps are typically larger files than vector graphics, which means that bitmaps require more time to download. Be careful when you use bitmaps in a Flash movie, especially one that will be downloaded over the Internet; very large bitmap files in the library can still affect the file's overall download time. (Because this project will be placed on a kiosk rather than downloaded over the Internet, file size is not a significant issue.)

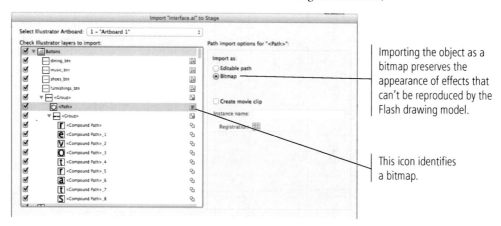

Importing the object as a bitmap preserves the appearance of effects that can't be reproduced by the Flash drawing model.

This icon identifies a bitmap.

10. **Click the arrow to the left of the Buttons layer name to collapse that layer. Review the rest of the items in the list.**

We prepared this Adobe Illustrator file very carefully to meet the requirements of this Flash project, including assigning descriptive names to all layers and groups. Many illustrators will not provide this level of detail in their files.

Illustrator objects that do not directly correlate to Flash drawing objects, such as shapes with applied transparency settings, are automatically converted to movie clip symbols when imported to Flash.

Use these arrows to collapse or expand layers and sublayers.

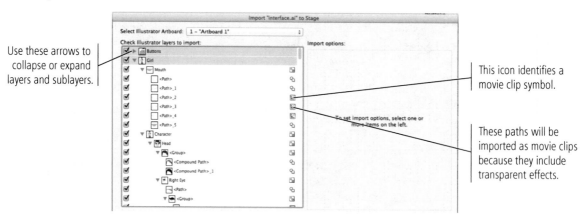

This icon identifies a movie clip symbol.

These paths will be imported as movie clips because they include transparent effects.

11. **At the bottom of the dialog box, choose Flash Layers in the Convert Layers To menu.**

This option maintains the existing layers from the original artwork; each Illustrator layer becomes a layer in the Flash file. This is useful if you aren't sure about what you're importing; you can always change or delete imported layers if you don't need them.

Note:

You will learn about keyframes in Stage 2 of this project.

You can also choose Single Flash Layer, which flattens all objects onto a single layer (named according to the imported file name), or choose Keyframes to add each layer as a keyframe on the default Layer 1.

12. Check the Place Objects at Original Position option.

Using this option places the imported artwork at the same relative position where it appeared on the Illustrator artboard (the "page" or drawing space). For example, if a square is placed 1 inch from the top and left edges of the artboard, it will be placed 1 inch from the top and left edges of the Stage. When this option is not checked, the imported artwork is centered in the document window, regardless of the position of the Flash Stage.

It is important to note that the relative position of imported artwork is maintained, regardless of your selection here. In other words, all imported objects appear in the same place (relative to each other) as in the Illustrator file. The import process treats all objects in the import as a single group for the purposes of the import only; this option simply controls the imported artwork's position relative to the Stage.

13. Check the Set Stage to Same Size as Illustrator Artboard option.

The Illustrator artboard is the area that defines the physical dimensions of the file, just as the Stage defines the physical size of a Flash file. This option shows the dimensions of the imported file's artboard; if you know the Illustrator file was created to the correct dimensions — as it was in this case — you can use this option to automatically change the Stage size to match the imported artwork.

14. Uncheck the Import Unused Symbols option.

Illustrator can be used to create graphic and movie clip symbols (but not buttons), which are stored in a file's Symbols panel. This can include symbols that are not placed in the file, but which might be necessary for the overall project. If you don't know what a file contains, you can check this option to be sure all of the necessary bits are imported; you can always delete unwanted symbols once they have been imported. For this project, we are telling you that all required symbols in the imported artwork are placed on the artboard.

15. Uncheck the Import as a Single Bitmap Image option.

If this option is checked, the entire file is flattened and converted to a bitmap image; you would not be able to access any of the components of the imported artwork.

16. Click OK to import the Illustrator artwork.

17. Fit the Stage in the document window, then click away from all objects on the Stage to deselect them.

All objects are automatically selected after being imported to the Stage. Deselecting them allows you to review the Flash file's properties.

Note:

The options at the bottom of the dialog box remember the last-used settings. Some of the choices we define in Steps 11–15 might already be set on your computer.

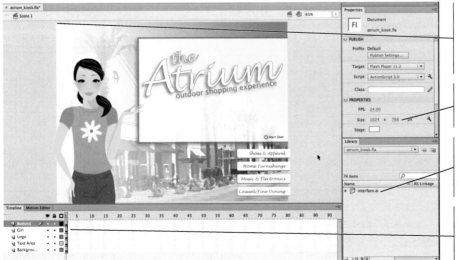

The imported artwork is aligned to the top-left corner of the Stage, matching its relative position on the original Illustrator artboard.

With nothing selected, the Properties panel shows the adjusted Stage size.

A folder (named the same as the imported file) is added to the library, containing all the pieces necessary for the imported artwork.

Five layers are added to the Timeline panel, named the same as the layers in the original Illustrator file.

18. **In the Library panel, click the arrow to the left of the interface.ai folder to expand it, and then click the arrows to expand all but the Girl folder.**

Imported assets are sorted by layer; folder names match the imported layer names to help you understand where different pieces are required. A separate folder for Illustrator Symbols is included.

Note:

The Girl folder includes a long list of paths and groups that were imported as movie clips to preserve transparency effects that were applied in the Illustrator file. We did not include that folder in this instruction simply because the list is so long.

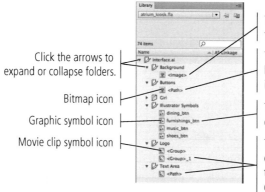

Click the arrows to expand or collapse folders.

Bitmap icon

Graphic symbol icon

Movie clip symbol icon

This bitmap image was placed on the Background layer in Illustrator.

This bitmap image was created by the Import process to solve incompatibility problems.

These graphic symbols were created in the Illustrator file.

These movie clips were created by the Import process to maintain the appearance of transparent effects.

19. **In the Library panel, collapse the subfolders in the interface.ai folder.**

20. **Save the file and continue to the next exercise.**

Illustrator File Import Preferences

FLASH FOUNDATIONS

You can define the default options for importing native Illustrator files in the AI File Importer pane of the Preferences dialog box. (Remember, preferences are accessed in the Flash menu on Macintosh and in the Edit menu on Windows.) Most of these options will become clearer as you work through this project and learn more about the different types of symbols. In some cases, the options refer to specific features in the native application.

- **Show Import Dialog Box** is checked by default. If you uncheck this option, Illustrator files will import using the default options defined here (without showing the Import dialog box).

- If **Exclude Objects Outside Artboard** is checked, objects outside the artboard (Illustrator's "page") are unchecked by default in the Import dialog box.

- If **Import Hidden Layers** is checked, all layers in the file (including ones that are not visible) are listed in the Import dialog box.

- The **Import Text As** options define how text objects in the Illustrator file import into the Flash file. Editable Text, selected by default, imports text objects that you can edit using the Flash Text tool. If you choose Vector Outlines, text objects import as a group of vector shapes; you cannot edit the text in these objects (other than manipulating the vector paths). If you choose Bitmaps, text objects import as raster objects that cannot be edited with either the Text tool or the Subselection tool.

- The **Import Paths As** options determine the default behavior for how vector paths in the Illustrator file are added to the Flash file. Editable Paths means you can use the Flash Subselection tool to manipulate the anchor points and handles on the imported paths; if you select the Bitmaps option, you will not be able to edit the vector paths within Flash.

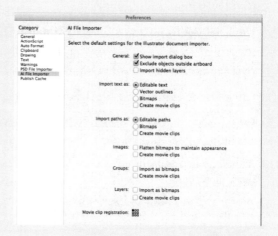

- **Images** placed in an Illustrator file are maintained when the file imports into Flash; if you select an image in the Import dialog box, you can choose to flatten the bitmap file or automatically create a movie clip. You can use the Images preference options to automatically check the related boxes in the Import dialog box.

- **Groups** and **Layers** from the Illustrator file are listed separately in the Import dialog box. You can use the Groups and Layers preference options to automatically check the Import as Bitmap and Create Movie Clip options in the Import dialog box.

- **Movie Clip Registration** defines the default registration option for objects you import as movie clip symbols.

IMPORT FILES TO THE LIBRARY

In addition to importing files to the Flash Stage, you can also import external files directly into the Flash file's Library panel. This option is particularly useful when certain objects aren't going to be placed on the main Stage, or if you don't yet know how you will use a particular object.

1. **With atrium_kiosk.fla open, choose File>Import>Import to Library. Navigate to the file mouths.ai in the WIP>Atrium folder and click Open.**

   When you import an Illustrator file directly to the Library panel, most of the options are the same as for importing to the Stage. The Place Objects at Original Position and Set Stage Size to Same Size as Illustrator Artboard options are not available because they do not apply to files that only exist (for now) in the file's library.

Note:

Flash defaults to the last-used folder, so you might not have to navigate to the folder you used in the previous exercise.

2. **Click OK to import the artwork to the library.**

   The Library panel shows that the resulting object was imported as a graphic symbol. Nothing is added to the Stage or the timeline.

3. **In the Library panel, click the mouths.ai item to select it.**

   The top portion of the panel shows a preview of the selected item.

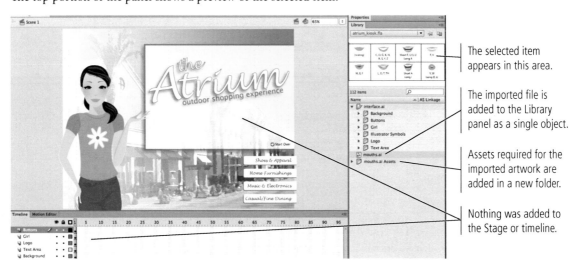

The selected item appears in this area.

The imported file is added to the Library panel as a single object.

Assets required for the imported artwork are added in a new folder.

Nothing was added to the Stage or timeline.

4. **Save the file and continue to the next exercise.**

CONVERT OBJECTS TO SYMBOLS

You now have a number of assets in your file's Library panel. The mouths.ai graphic contains eight groups of graphics — the different mouth shapes that you will use later in this project to synchronize the character to a sound file. For the process to work, you need to separate each mouth shape into a distinct symbol so the correct artwork can be placed at the appropriate point in the file.

1. **With atrium_kiosk.fla open, choose the Selection tool. Double-click the mouths.ai symbol icon to enter into the symbol.**

   Every symbol technically has its own Stage, which is theoretically infinite and separate from the main Stage of the base file. When you double-click the symbol icon in the Library panel, you enter **Symbol-Editing mode** for that symbol; other elements of the base file are not visible on the Stage.

The Edit bar shows that you are now working on the mouths.ai Stage (called **Symbol-Editing mode**).

When you first enter into the symbol, all artwork in the symbol is selected.

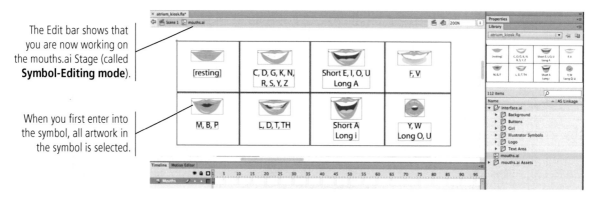

2. **Click away from the artwork to deselect everything, then click the top-left mouth shape to select that group (but not the word "resting").**

   Grouping in the original artwork is maintained in the imported artwork.

3. **Control/right-click the selected artwork and choose Convert to Symbol from the contextual menu.**

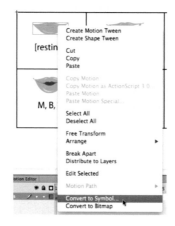

Note:

You can also drag an object onto the Library panel to open the Convert to Symbol dialog box for that object.

4. **In the resulting dialog box, type mouth1 in the Name field and choose Graphic in the Type menu.**

   A graphic symbol is the most basic type of symbol. It is typically used for objects that will simply be placed on the Stage. (A graphic symbol can include animation; you will explore these options in Project 4: Ocean Animation.) The type of animation you create in the third stage of this project — simply swapping one symbol with another at various points in time — is ideally suited to graphic symbols.

5. **Select the center point in the registration proxy icon.**

The registration grid affects the placement of the symbol's registration point, which is the 0,0 point for the symbol. (This will make more sense shortly when you begin editing symbols on their own Stages.)

6. **Click OK to create the new symbol.**

The Properties panel now shows that the selected object is an instance of the mouth1 symbol, which has been added to the Library panel.

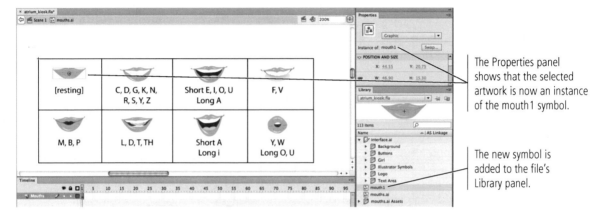

The Properties panel shows that the selected artwork is now an instance of the mouth1 symbol.

The new symbol is added to the file's Library panel.

7. **Click the second mouth shape (to the right) to select it. Control/right-click the group and choose Convert to Symbol.**

The Convert to Symbol dialog box remembers the last-used settings. The Type menu is already set to Graphic, and the center registration point is already selected.

8. **Type mouth2 in the Name field and click OK.**

9. **Repeat Steps 7–8 to convert the rest of the mouth shapes into symbols, working from left to right across the top row and then left to right across the bottom row.**

10. **Click Scene 1 in the Edit bar to return to the main Stage.**

11. **Using the Selection tool, click the mouth shape on the Stage to select it.**

The selected object is a group. It is not an instance of any symbol.

12. **In the Timeline panel, add a new layer immediately above the Girl layer. Name the new layer Mouths.**

    When you selected the mouth shape in Step 11, the layer containing the object (Girl) automatically became the active layer. When you click the New Layer button in this step, the new layer is automatically added above the previously selected layer. The new layer is also automatically selected as the active layer.

13. **Click mouth1 in the Library panel and drag an instance onto the Stage.**

14. **Use the Selection tool to drag the placed instance to the same position as the mouth group on the underlying Girl layer.**

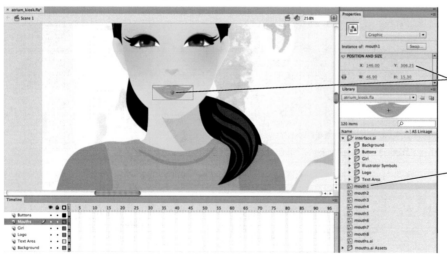

The X and Y fields show the position of the instance's registration point.

Drag the symbol from the Library panel to the Stage to place an instance on the active layer.

> **Note:**
>
> *Don't confuse the symbol registration point (the crosshairs) with the transformation point (the hollow circle).*

15. **Hide the Mouths layer, select the mouth group on the Girl layer and delete it, and then show the Mouths layer again.**

16. **Save the file and continue to the next exercise.**

CREATE A BUTTON SYMBOL

Buttons, one of the three main symbol types in Flash, are interactive assets that change when a user interacts with them. A button symbol has four "states":

- A button's **Up state** (also referred to as the idle or default state) is the basic appearance of a button when a user first loads a file.

- The **Over state** occurs when a mouse pointer rolls over a button. (When a user places a mouse cursor over a rollover area, the cursor often turns into a pointing finger or some other custom shape.)

- The **Down state** occurs when a user clicks a button.

- The **Hit state** defines the size of a rollover area (**hot spot**) of a button.

This file includes five buttons. Four were created as symbols in the Illustrator artwork, and one was imported onto the Stage as a group.

Note:

Buttons can be both animated and idle at the same time; idle simply means that no one has passed over or clicked the button with the mouse pointer.

1. **With `atrium_kiosk.fla` open, use the Selection tool to select the group containing the words "Start Over".**

2. **Control/right-click the selected group and choose Convert to Symbol in the contextual menu.**

3. **In the resulting dialog box, type `start_over` in the name field. Choose Button in the Type menu and choose the center registration point (if it is not already selected).**

4. **Click OK to create the new symbol.**

   Because you created the symbol from objects on the Stage, the Properties panel shows that the selection is automatically an instance of the new symbol.

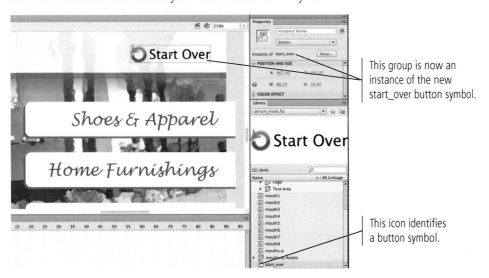

This group is now an instance of the new start_over button symbol.

This icon identifies a button symbol.

5. **Double-click the Start Over button on the Stage to enter into the symbol.**

This method of editing a symbol is called **editing in place**. Other objects on the Stage are still visible, but they are screened back and cannot be accessed.

As we explained earlier, a button is a special type of symbol with four distinct states. Each possible state is represented as a frame in the special Button symbol timeline.

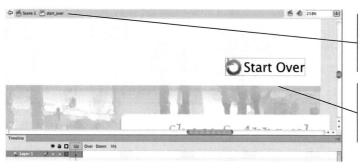

The Edit bar shows that you are editing on the start_over button symbol Stage.

Editing a symbol in place means you can see — but not access — the other objects on the Stage.

6. **In the Timeline panel, Control/right-click the Over frame of Layer 1 and choose Insert Keyframe from the contextual menu.**

A **keyframe** defines a point where something changes. If you want to make something appear different from one frame to the next — whether inside a symbol or on the main Stage — you need to place a keyframe at the point where the change should occur.

Control/right-click the Over frame for Layer 1.

Note:

You can also insert a keyframe by choosing Insert>Timeline> Keyframe, or pressing F6.

7. **Make sure the Over frame is selected in the Timeline panel, then double-click the words "Start Over" in the graphic to enter into that group.**

The contents of the Over frame will appear when the user's mouse moves over the button area. You are going to change the color of the letters in this button.

8. **Double-click any letter in the group to access the individual letters that make up the group.**

Remember, Flash remembers the groupings from the original Illustrator file. Depending on how a file was created, you might have to enter into a number of nested groups before you get to the level you need.

9. **With the individual letter shapes selected, use the Fill Color swatch in the Tools or Properties panel to change the fill color to a medium blue.**

You have to drill down into two levels of grouping to access the individual letter shapes.

You are editing only the Over frame.

10. **Click Scene 1 in the Edit bar to return to the main Stage.**

Even if you have drilled into multiple levels of Symbol-Editing mode, you can return to the main Stage with a single click on the Edit bar. You can also return to any particular nesting level by clicking a specific item (called "breadcrumbs") in the Edit bar.

Note:

You will test the button's functionality in the next exercise.

11. **Save the file and continue to the next exercise.**

DEFINE A HIT FRAME

In the previous exercise, you changed the color of button text in the Over frame. However, there is still a problem — the button currently works only if the mouse pointer touches the icon or one of the letter shapes. If the pointer lies between two letters, for example, the button fails to activate.

All spaces within the button should be active. Moving the pointer close to or on top of the button should trigger the desired action. To resolve the problem, you need to define the Hit frame, which determines where a user can click to activate the button.

1. **With atrium_kiosk.fla open, choose Control>Enable Simple Buttons to toggle that option on.**

   This command allows you to test button states directly on the Flash Stage.

2. **Check the current condition of the Start Over button by positioning the pointer between the two words in the button.**

   There are "dead" areas within the button that don't cause the color change to occur. (You might need to zoom in to verify this problem.)

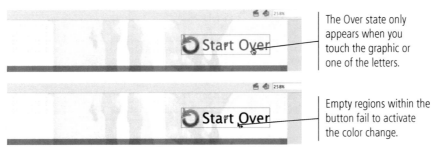

The Over state only appears when you touch the graphic or one of the letters.

Empty regions within the button fail to activate the color change.

Note:

Using a button symbol mimics a four-frame Flash animation; it is the basic concept behind all Flash buttons. You add keyframes, you modify the content of each frame, and then Flash displays the appropriate frame when the user hovers over the object, clicks the object, or — in the case of the Hit frame — approaches the object.

3. **Choose Control>Enable Simple Buttons to toggle that option off.**

   When this option is active, you can't select a button instance on the Stage — which means you can't double-click the button to edit the symbol in place.

4. **Double-click the Start Over button on the Stage to edit the symbol in place.**

5. **Control/right-click the Hit frame and choose Insert Keyframe from the contextual menu.**

   The **Hit frame** defines the live area of the button, or the area where a user can click to activate the button. Objects on this state do not appear in the movie; you only need to define the general shapes.

6. **Choose the Rectangle tool from the Tools panel.**

7. **Change the fill color to some contrasting color, then turn off the Object Drawing option.**

   Because merge-drawing shapes drop to the back of the stacking order, this method allows you to still see the button artwork in front of the "hit" shape.

8. **Draw a rectangle that covers the entire contents of the button.**

   We used a red color that contrasted with the blue text, but any color will work because the Hit frame content doesn't appear on the Stage when you play the movie.

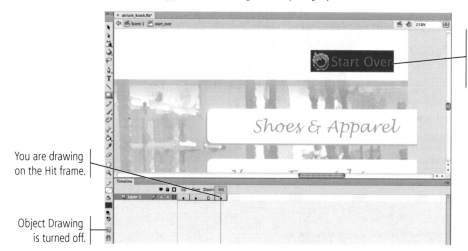

The merge-drawing shape is created behind other objects on the layer.

You are drawing on the Hit frame.

Object Drawing is turned off.

9. **Click Scene 1 in the Edit bar to return to the main Stage.**

10. **Choose Control>Enable Simple Buttons to toggle the option back on, then place the mouse cursor between the words in the button.**

    Now the button works even if you hover over the white areas or between the letters. The Hit frame rectangle determines the live (hit) area of the button.

Now the button works even if you hover between the words.

11. **Save the file and continue to the next exercise.**

FLASH FOUNDATIONS

You can edit symbols in three different ways: in Symbol-Editing mode, in place on the Stage, or in a separate window.

You're already familiar with Symbol-Editing mode, the discrete Stage that appears when you double-click a symbol in the Library panel; the name of the symbol appears to the right of the Scene name in the Edit bar, indicating that you're working on the symbol instead of the main scene. You can also access this option by Control/right-clicking a placed instance on the Stage and choosing Edit.

In basic Symbol-Editing mode, the name of the symbol appears above the Stage in the Edit bar.

When editing a symbol on its own Stage, other placed objects are not visible.

The second option, editing in place, means you can edit a symbol while still seeing other objects on the Stage. If you double-click a symbol on the Stage, you can edit the symbol in the context of the larger movie (as shown by the symbol name in the Edit bar). You can also access this option by Control/right-clicking a placed symbol instance and choosing Edit in Place.

Editing in place is a variant of Symbol-Editing mode, so the name of the symbol appears above the Stage.

When editing a symbol in place, other placed objects remain visible but screened back.

When editing a symbol in place, it is easy to forget that you are actually editing the symbol — including all placed instances of that symbol. Any changes you make will also affect other placed instances, even though those instances are not directly accessible in the document window.

The third option, editing in a separate window, opens the symbol in its own document window. Only the symbol name appears above the Stage; there is no preceding scene. You can access this option by Control/right-clicking an instance on the Stage and choosing Edit in New Window in the contextual menu.

When editing the symbol in a new window, there is no associated scene; only the symbol name appears above the Stage.

The document tabs show that you are editing in a second instance of the file containing the symbol.

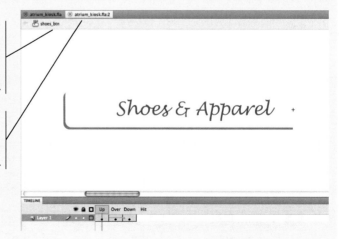

EDIT SYMBOL PROPERTIES

The control buttons, which you imported as graphic symbols from the external file library, are the final pieces of artwork for your Flash movie. Illustrator, however, does not create button symbols; you now need to convert the imported graphic symbols into the necessary button symbols.

1. **With atrium_kiosk.fla open, choose Control>Enable Simple Buttons. Make sure that option is toggled off.**

   Remember, you can't select a button on the Stage if this control is toggled on.

2. **In the Library panel, expand the Illustrator Symbols folder (inside the interface.ai folder) if necessary. Control/right-click the shoes_btn symbol icon in the Library panel and choose Properties from the contextual menu.**

3. **In the Symbol Properties dialog box, choose Button in the Type menu, and then click OK.**

Note:

The Symbol Properties dialog box is nearly the same as the Create New Symbol dialog box; it does not have registration options, because that has already been defined for the symbol. To move the symbol artwork relative to the registration point, you can edit the symbol on its Stage.

4. **Using the Selection tool, click the Shoes & Apparel button on the Stage to select it.**

5. **In the Properties panel, open the top menu and choose Button.**

   Unlike changing the content of a symbol, changes to the symbol type do not reflect in placed instances. When you change the type of a symbol that has already been placed on the Stage, you also have to change the instance type in the Properties panel.

Choose Button in this menu to change the behavior of the placed instance.

The symbol now shows the button icon instead of the original graphic icon.

6. **Repeat Steps 2–5 to convert the remaining three graphic symbols to buttons.**

7. **Save the file and continue to the next exercise.**

EXPLORE THE SYMBOL REGISTRATION POINT

Now that the buttons are symbols rather than graphics, you can define the various states of the buttons. You are going to edit the artwork so the buttons seem to move when the mouse cursor rolls over the hit area.

1. **With atrium_kiosk.fla open, double-click the Shoes & Apparel button instance to edit the symbol in place.**

   The crosshairs in the middle of the symbol artwork identify the **symbol registration point**; all measurements for placed instances begin at this location.

Symbol registration point

When editing the symbol, the X and Y fields show the position of the top-left corner relative to the symbol's registration point.

2. **In the Align panel, make sure the Align To Stage option is active and then click the Align Right Edge button.**

   The right edge of the symbol artwork is now aligned to the symbol registration point. Because you are editing the symbol in place, you can see the effect of the new alignment relative to the overall file artwork. This illustrates that the registration point is fixed, and the artwork is the thing that moves — not the other way around.

Note:

Use the Align panel with the Align To Stage option active to align the placed object to the symbol's registration point.

3. **Click Scene 1 in the Edit bar to return to the main Stage.**

4. **With the Shoes & Apparel button selected, click the current X value in the Properties panel to access the field.**

5. **Type 1034 in the highlighted X field and press Return/Enter to apply the change.**

   As we explained earlier, the symbol registration point is the origin of measurements for placed instances. When you change the X position, you are defining the horizontal location of the symbol registration point for the selected instance.

   The Stage for this file is 1024 pixels wide (as defined by the imported Illustrator artboard); you are placing the right edge of the button 10 pixels past the Stage edge.

On the main Stage, the X and Y fields define the position of the registration point for the instance.

In the next few steps, you will use this position as the basis for changing the object's position when a user moves the cursor over the button (i.e., triggers the Over frame).

6. **Double-click the Shoes & Apparel button again to enter back into the symbol Stage.**

7. **Insert a new keyframe on the button's Over frame. With the Over keyframe selected, click the button artwork to select it.**

    The object must be selected to change its properties. Selecting the frame in the timeline also selects the object on that frame.

8. **In the Properties panel, click the current X value in the Properties panel to access the field.**

Click the field to access the current value.

You should be working on the Over keyframe.

9. **Place the insertion point after the existing value and type `-10` after the existing value. Press Enter to move the selected object.**

    Using mathematical operators makes it easy to move an object a specific distance without manually calculating the change:

    *   Subtract from the X position to move an object left.
    *   Add to the X position to move an object right.
    *   Subtract from the Y position to move an object up.
    *   Add to the Y position to move an object down.

Type **–10** after the current value.

The new X value moves the artwork 10 pixels to the left.

10. **Click Scene 1 in the Edit bar to return to the main Stage.**

11. **Repeat Steps 1–10 for the three remaining buttons.**

12. **Choose Control>Enable Simple Buttons to toggle the option back on. Move your mouse cursor over the buttons to test the Over state functionality.**

13. **Save the file and continue to the next exercise.**

Note:

Because this button artwork includes a solid-filled white rectangle, you don't need to define a separate hit frame. The artwork itself is sufficient to trigger the button.

ORGANIZE YOUR LIBRARY WITH FOLDERS

Library folders work the same as layer folders; they help you organize and structure complex files. Movies often contain dozens or even hundreds of assets — and the more complex a movie becomes, the more useful it is to clearly organize those assets. Although this step isn't strictly necessary, it is always a good idea to organize your work so you can more easily organize your thoughts and processes going forward.

1. **With atrium_kiosk.fla open, expand the interface.ai folder in the Library panel.**

2. **Click the Illustrator Symbols folder (inside the interface.ai folder) and drag down to the empty area at the bottom of the panel.**

   This moves the Illustrator Symbols folder to the first level of the library. The symbols, which are placed on the Stage, are not affected by the move.

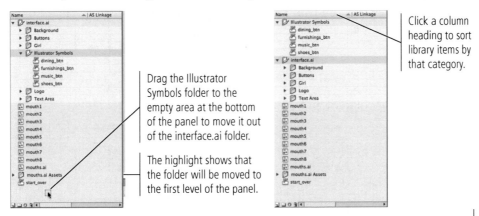

Drag the Illustrator Symbols folder to the empty area at the bottom of the panel to move it out of the interface.ai folder.

The highlight shows that the folder will be moved to the first level of the panel.

Click a column heading to sort library items by that category.

3. **Double-click the Illustrator Symbols folder name to highlight the name. Type buttons to change the folder name.**

4. **Click the start_over button symbol icon and drag it into the Buttons folder.**

5. **Double-click the interface.ai folder name to highlight the name. Type component artwork to change the folder name.**

Note:

If your Library panel is too short to show an empty area below the current assets, Control/ right-click any of the existing first-level assets and choose Paste. The pasted symbols are pasted at the same level as the asset where you Control/ right-click.

6. **Click the mouths.ai Assets folder and drag it into the Component Artwork folder.**

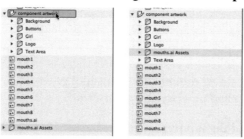

7. **Click the empty area at the bottom of the panel to deselect all assets and folders.**

8. **Click the New Folder button at the bottom of the Library panel. Type mouth graphics as the new folder name.**

   The new folder is added at the main level of the library, alphabetized with other items at the same level. If you didn't deselect in Step 7, the new folder would have been created at the same nesting level as the selected item.

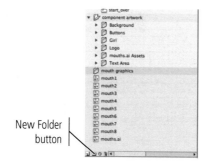

New Folder button

9. **Click the mouth1 symbol to select it, then Shift-click the mouth8 symbol to select it and all files in between.**

   Press Shift to select multiple contiguous items in the panel, or press Command/Control to select multiple, non-contiguous items.

10. **Click the icon of any selected file and drag into the mouth graphics folder.**

All eight of these files are selected.

11. **Click the mouths.ai graphic symbol and click the panel's Delete button.**

    Although you used this artwork to create the individual mouth symbols, this symbol is not used in the file, so it can be safely deleted from the library. If you delete a symbol that is used in the file, the placed instances will also be deleted from the file.

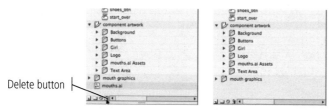

Delete button

12. **Collapse all library folders, then save the file and continue to the next stage of the project.**

Stage 2 Working with Sound

Sound files can be categorized into three basic types: uncompressed, compressed (lossless), and compressed (lossy). **Uncompressed sound files** encode all sounds with the same number of bits per unit of time. In an uncompressed format, two sound files of the same duration — whether a symphony or a simple beep — have the same size (which is typically very large). Such files are commonly used for archiving or other situations where file size is not an issue.

Note:

Compression is the process of removing data to reduce file size.

Lossless compression sound files lose no data during compression; these files are smaller than uncompressed files, but not as small as lossy compression file formats. **Lossy compression sound files** lose some data but retain good sound quality; a large number of these files can be stored in relatively small amounts of space.

Flash handles most major audio formats, including the ones most commonly used today:

- The **MP3** format is the most commonly used audio format. This format compresses a music file in the most efficient manner, so file size is reduced without compromising quality. MP3 playback does require more processing power than other formats because the data has to be decoded every time the file plays.

- The **WAV** format is an uncompressed format with very high quality. This file type can be used in Flash animations for desktop applications, but should be avoided for Web-based movies because the files are huge and take a long time to download.

- The **AIFF** format (Audio Interchange File Format) is common on Macintosh computers. This format is generally uncompressed, so file sizes are large compared to the MP3 format. AIFF files are suitable for applications specifically targeted for Macintosh computers.

- The **Audio** (AU) file format, developed by Sun Microsystems, transmits sound files over the Internet and can be played in Java programs. These files are smaller than AIFF and WAV formats, but the quality of sound is not as good as regular WAV files.

- The **QuickTime** (MOV) format is technically a video format, but it can also include audio.

IMPORT SOUND FILES

In general, there are three methods for incorporating sound into a Flash movie. Sounds in a file's library can be placed directly on the timeline, or you can use code to call a library sound based on a particular event. You can also use code to load and play external sound files (those that don't exist in the Flash library).

In this project, you will use the timeline method to add the various sounds that are needed for the kiosk to function properly. The first step is to import the necessary sound files into the Flash library.

1. **With atrium_kiosk.fla open, choose File>Import>Import to Stage. In the resulting dialog box, select all files in the WIP>Atrium>Audio folder, then click Open.**

2. **Open the Library panel and review the contents.**

Sound files do not exist on the Stage. Even though you chose Import to Stage, the sound files are automatically imported into the file's Library panel.

3. **Click the New Folder button at the bottom of the Library panel. Rename the new folder** `audio`.

4. **Command/Control-click all six imported audio files and drag them into the audio folder.**

5. **Expand the audio folder so you can see the available files.**

6. **Save the file and continue to the next exercise.**

Note:

It's always a good idea to keep your library well organized while developing a file with numerous assets.

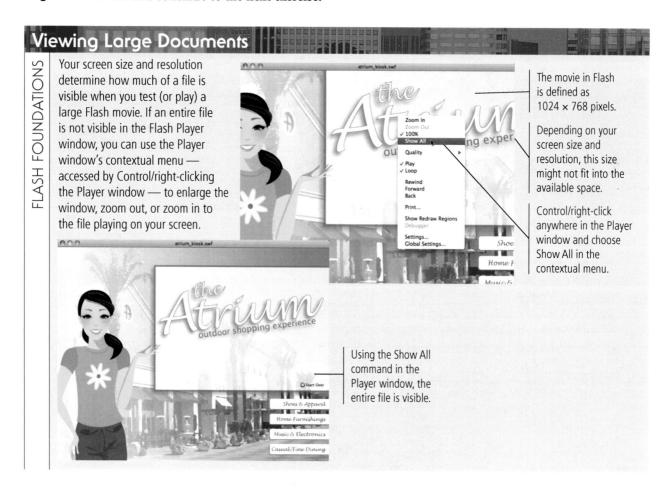

FLASH FOUNDATIONS

Viewing Large Documents

Your screen size and resolution determine how much of a file is visible when you test (or play) a large Flash movie. If an entire file is not visible in the Flash Player window, you can use the Player window's contextual menu — accessed by Control/right-clicking the Player window — to enlarge the window, zoom out, or zoom in to the file playing on your screen.

The movie in Flash is defined as 1024 × 768 pixels.

Depending on your screen size and resolution, this size might not fit into the available space.

Control/right-click anywhere in the Player window and choose Show All in the contextual menu.

Using the Show All command in the Player window, the entire file is visible.

ADD EVENT SOUND

Event sounds are "timeline independent" — they play independently of the movie timeline. They are downloaded completely and stored in the user computer's memory; this means they can be played repeatedly (including continuously looping) without having to redownload the file.

Note:

Because event sounds must be downloaded completely before they can play, they can cause buffering delays in playback.

1. **With the file atrium_kiosk.fla open, click to select the atrium_jazz.mp3 file in the Library panel.**

   This file will be the background music for the entire file. It will play in an infinite loop as long as the kiosk file is open.

2. **Click the Play button in the top-right corner of the Preview area.**

   You can use the Library panel to hear imported sounds before they are used on the Stage.

   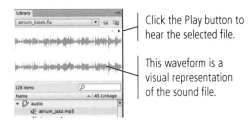

   Click the Play button to hear the selected file.

   This waveform is a visual representation of the sound file.

3. **Select the Frame 1 keyframe on the Background layer.**

4. **In the Sound section of the Properties panel, choose atrium_jazz.mp3 in the Name menu.**

   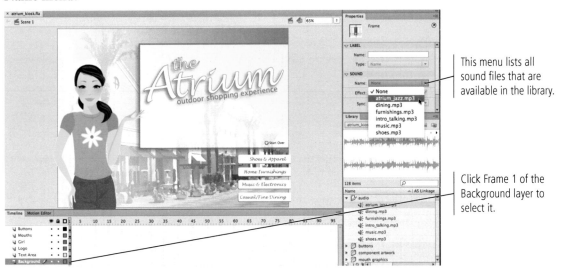

   This menu lists all sound files that are available in the library.

   Click Frame 1 of the Background layer to select it.

5. **In the Sync menu, choose Event.**

   Event sounds default to the Repeat 1 method, which means the sound plays one time. You can change the number in the Repeat field to play the sound a specific number of times.

   Choose Event in this menu.

   A small line, which is actually part of the sound waveform, crosses the selected frame.

6. Choose Control>Test Movie>In Flash Professional.

This command opens the file in a separate Flash Player window. (You will use this command frequently as you develop this and most Flash projects.) Although the sound waveform only appears on Frame 1 of the Background layer, the entire sound plays from start to finish when you test the movie file.

Note:

Press Command-Return/Control-Enter to test the movie in a Flash Player window.

The Test Movie command opens the file in a Flash Player window, showing what the exported movie would look — and sound — like.

7. Close the Player window and return to Flash.

8. With Frame 1 of the Background layer selected, choose Loop in the menu under the Sync menu (in the Sound area of the Properties panel).

Using the Loop method, the event sound plays continuously as long as the movie remains open.

9. Press Command-Return/Control-Enter to test the movie again.

The background sound now plays from start to finish, and then repeats to create a continuous background sound track.

10. Close the Player window and return to Flash.

11. Save the file and continue to the next exercise.

EDIT A SOUND ENVELOPE TO CONTROL VOLUME

Although Flash is not intended to be a sound-editing application, you can apply a limited number of effects to control the volume and length of sounds on the Flash timeline. These options are available in the Effect menu of the Properties panel when a sound is attached to a keyframe.

1. **With atrium_kiosk.fla open, click Frame 1 of the Background layer to select the frame where you attached the sound in the previous exercise.**

2. **In the Properties panel, open the Effect menu and choose Custom to open the Edit Envelope dialog box.**

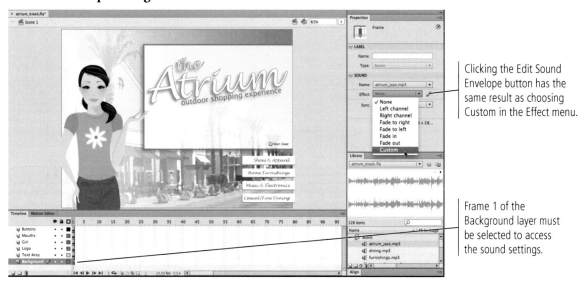

Clicking the Edit Sound Envelope button has the same result as choosing Custom in the Effect menu.

Frame 1 of the Background layer must be selected to access the sound settings.

The Edit Envelope dialog box shows the waveforms for each channel in a sound file. (In many cases, both channels have the same waveform.) You can view the sound waves by seconds or frames, and you can zoom in or out to show various portions of the sound.

The left and right channels refer to sound output systems that have more than one speaker — one on the left and one on the right. You can apply different settings to individual channels to create an effect of aural depth. For example, fading a car sound out for the left channel and in for the right channel helps to reinforce a visual that moves the car from left to right across the Stage.

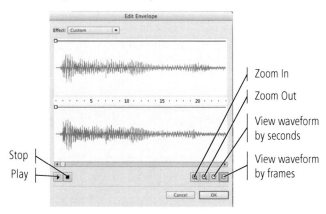

Zoom In

Zoom Out

View waveform by seconds

View waveform by frames

Stop

Play

3. **Click the Frames button to show the sound based on frames (if this isn't already active).**

4. **In the left channel area (the top waveform), click the handle on the left end of the waveform, and drag down to below the existing waveform.**

Click the envelope handle and drag down to this point.

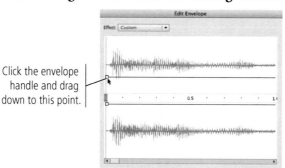

Note:

Click the envelope line to add a new handle to the envelope. Click an existing handle and drag it away from the window to remove a point from the envelope.

Editing Sound Files

FLASH FOUNDATIONS

Applying Built-in Sound Effects

The Effect menu in the Properties panel lists a number of common sound envelope effects built into Flash. These sound effects do not alter the sound in the files; they simply control how the sound data plays.

- **Left Channel** plays only the left channel of the sound.
- **Right Channel** plays only the right channel of the sound.
- **Fade to Right** gradually lowers the sound level of the left channel, and then gradually raises the sound level of the right channel.
- **Fade to Left** gradually lowers the sound level of the right channel, and then gradually raises the sound level of the left channel.

- **Fade In** gradually raises the sound level at the beginning of the sound file.
- **Fade Out** gradually lowers the sound level at the end of the sound file.
- **Custom** opens the Edit Envelope dialog box, where you can define your own sound effects.

Editing Sound Duration

Although it is usually a better idea to edit sound files in a true sound-editing application, you can also use the Edit Envelope dialog box to shorten the duration of a sound file.

When you first open the Edit Envelope dialog box, the Time In Control bar appears at the left side of the waveform. If you use the scroll bar below the waveforms to find the end of the waveform, you will see the Time Out Control bar, which marks the end of the sound. You can drag these control bars to change the starting and ending point of the selected sound.

Warning: Be very careful editing the duration of sound files in Flash. There are known problems with sounds playing from the original starting point and cutting off in the middle, rather than honoring the defined Time In and Time Out points. In many cases, the edited sounds play normally when tested on the Stage in Flash, but the problem is evident when you test your movie in the Flash Player window.

Time In Control bar

Use the scroll bar to find the Time Out Control bar at the end of the waveform.

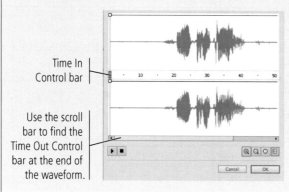

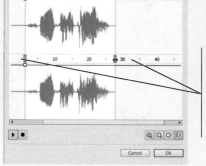

Parts of the waveform to the left of the Time In Control bar or right of the Time Out Control bar will not be included in the movie.

5. **Repeat Step 4 for the right channel (the bottom waveform).**

By lowering the envelope handles, you reduced the volume of the sound file.

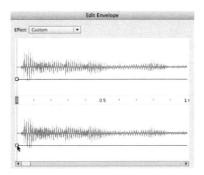

6. **Click OK to close the Edit Envelope dialog box and apply the change.**

7. **Save the file and continue to the next exercise.**

 ### USE THE START AND STOP SYNC METHODS FOR BUTTONS

The four category buttons will link to different screens in the kiosk. Each button needs to trigger a sound that plays when the user's mouse rolls over the button. To achieve this result, you can attach the relevant sound to each button using the same technique you applied in the previous exercise. Because of the four-frame nature of button symbols, however, a few extra steps are required to make the sounds play only when you want them to play.

1. **With `atrium_kiosk.fla` open, choose Control>Enable Simple Buttons to make sure that option is toggled on.**

2. **In the Library panel, double-click the shoes_btn symbol icon to enter Symbol-Editing mode for that symbol.**

Remember, you can edit a symbol by double-clicking the symbol icon. This is especially useful when the Enable Simple Buttons feature is toggled on, because you can't select buttons on the Stage in that mode.

3. **Select the Over frame of Layer 1. In the Properties panel, choose shoes.mp3 in the Sound menu and set the Sync menu to Event.**

You are attaching the sound to the button's Over frame.

4. **Click Scene 1 in the Edit bar to return to the main Stage.**

5. **Move your mouse cursor over the Shoes & Apparel button to hear the attached sound.**

Moving the mouse over the button triggers the Over state, including the attached sound file.

6. **Move your mouse cursor away, and then move back over the Shoes & Apparel button to trigger the sound again.**

   When the mouse re-enters the button area — triggering the Over frame — the message plays again. (Because the sound is very short, this might not be apparent unless you move the mouse back into the button area very quickly.)

7. **Double-click the shoes_btn symbol icon to enter back into the button Stage. Select the Over frame, then change the Sound Sync menu to Start.**

   The Start sync option is similar to the Event method. The difference is that the Start method allows only one instance of the same sound to play at a time; this prevents the overlap problem caused by the Event method.

Apply the Start sync method to the Over frame.

8. **Control/right-click the Down frame and choose Insert Keyframe from the contextual menu.**

9. **In the Properties panel, choose shoes.mp3 in the Sound menu and choose Stop in the Sync menu.**

   The Stop option stops all instances of the selected sound from playing. When a user clicks the Shoes & Apparel button, the sound triggered on the Over frame will stop playing.

Apply the Stop sync method to a keyframe on the Down frame.

10. **Click Scene 1 to return to the main Stage.**

11. **Repeat the same basic process to add the appropriate event sounds to the other three navigation buttons:**
    - Double-click the button symbol icon to enter the symbol's Stage.
    - Select the Over frame and attach the appropriate sound file using the Start sync option.
      - For the Home Furnishings button, use the furnishings.mp3 sound file.
      - For the Music & Electronics button, use the music.mp3 sound file.
      - For the Casual/Fine Dining button, use the dining.mp3 sound file.
    - Add a keyframe to the Down frame.
    - Attach the same sound you used for the Over frame, and apply the Stop sync option.

12. **If you haven't done so already, click Scene 1 in the Edit bar to return to the main Stage.**

13. **Roll your mouse cursor over the four buttons to test all four sounds. Click each to make sure the sounds stop when they're supposed to.**

14. **Save the file and continue to the next stage of the project.**

Stage 3   Creating Frame Animations

The basic underlying premise of animation is that objects change over time — from complex transitions in color, shape, and opacity to moving a character to a new position. The most basic type of animation is to simply replace one object with another at specific points in time; you will create this type of animation in this stage of the kiosk project to make it seem like the girl is talking.

Repositioning or replacing objects on successive frames results in the appearance of movement when you watch an animation; in reality, your brain is being fooled — you're simply seeing a series of images flash before your eyes (hence the application's name). Your brain thinks it's seeing movement, when in fact it's simply processing a series of still images displayed in rapid succession.

To make an animation appear to run continuously, you can **loop** it so it starts over at Frame 1 after reaching the last frame. (In fact, as you will see, looping is the default state of an animation; you have to use code to prevent the timeline from automatically looping in the exported file.)

To create animation, you need to understand several terms and concepts:

- The Flash Timeline panel shows a visual depiction of the passage of time. Each fraction of a second is represented by a frame (the rectangles to the right of the layer names). The **playhead** indicates the current point in time, or the frame that is visible on the Stage.

- The number of frames in one second (called **frames per second**, **FPS**, or **frame rate**) determines the length and quality of the overall animation. New Flash files default to 24 fps, which is the standard frame rate of most film movies in the United States (although HD formats range as high as 120 fps). Animations only for the Web are commonly developed at 15 fps.

- A **keyframe** indicates the point in time at which something changes. If you want to change something, you need to insert a keyframe at the appropriate moment on the timeline.

- Regular frames between keyframes have the same content as the preceding keyframe.

Note:

*The term **playhead** is a throwback to the days when animation and video were shown on physical tape-reading machines. The playhead is the component under which the tape moves, and the tape is read by the player. By sliding the tape back and forth underneath the playhead, an animator could make a movie run forward and backward.*

ADD STREAMING SOUND

Unlike the event sounds that you used in the previous exercises, **stream sounds** play as soon as enough data is downloaded (called **progressive downloading**) to the user's computer. Stream sounds cannot be saved on a user's computer; the sound file must be redownloaded every time it is played. Stream sounds are linked to the timeline, which means they stop playing if the timeline stops (i.e., they are "timeline dependent").

Note:

Because stream sounds are typically larger files (longer sounds equal more data and larger file size), the quality of these sounds might be poor for users who have slow Internet connections.

1. **With atrium_kiosk.fla open, add a new layer named Talking immediately above the Mouths layer.**

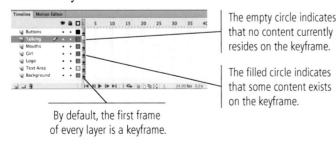

The empty circle indicates that no content currently resides on the keyframe.

The filled circle indicates that some content exists on the keyframe.

By default, the first frame of every layer is a keyframe.

2. **Select the Frame 1 keyframe of the Talking layer.**

3. **In the Properties panel, choose intro_talking.mp3 from the Sound menu and choose Stream in the Sync menu.**

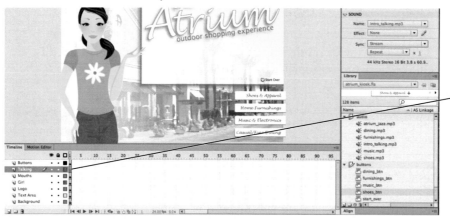

Frame 1 of the Talking layer is selected.

4. **Press Command-Return/Control-Enter to test the movie.**

   The background sound plays as expected, but the intro_talking sound does not. Remember, stream sounds are related to the position of the playhead on the timeline (they are timeline dependent). Because this file currently has only one frame, the playhead has nowhere to move, so the sound file does not play in the Player window.

5. **Close the Player window and return to Flash.**

6. **In the timeline, Control/right-click Frame 95 of the Talking layer and choose Insert Frame from the contextual menu.**

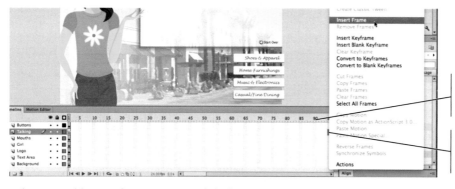

Frame numbers appear in the frame ruler at the top of the timeline.

Control/right-click Frame 95 of the Talking layer to open the contextual menu.

When you add a new frame, you extend the layer's timeline to the point where you place the new frame. The red playhead above the timeline shows the currently active frame.

You can now see the entire waveform of the sound that is attached to Frame 1 of the Talking layer. As you can see, however, none of the graphics are visible on the Stage, because you have not added frames to the other layers. In other words, objects on those layers don't yet exist at Frame 95.

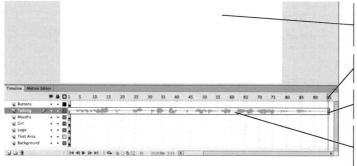

Other layers are not yet extended to Frame 95, so the graphics on those layers are not visible.

The playhead shows the currently active frame.

Adding a frame extends the layer's timeline.

The waveform on the Talking layer, which extends to Frame 91, is now entirely visible.

7. Click Frame 95 of the Buttons layer to select it, then press F5.

This keyboard shortcut inserts a new frame at the selected location on the timeline; it is the same as choosing Insert Frame from the contextual menu.

After adding the new frame to the Buttons layer, objects on that layer are now visible at Frame 95. The other graphics are still not visible because those layers do not yet have frames at Frame 95.

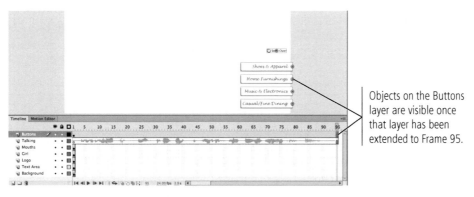

Objects on the Buttons layer are visible once that layer has been extended to Frame 95.

Note:

If you are using a laptop that has system-specific functions assigned to the Function keys, you can either press FN plus the required function key, or use the menu commands (in the Insert>Timeline submenu) to insert frames and keyframes.

You can also insert a frame, keyframe, or blank keyframe by Control/right-clicking a specific frame in the Timeline panel and choosing from the contextual menu.

8. Click Frame 95 of the Mouths layer, then Shift-click Frame 95 of the Background layer.

9. Press F5 to add new frames to all five selected layers.

Because all of the layers now "exist" on Frame 95, all of the kiosk graphics are now visible on the Stage.

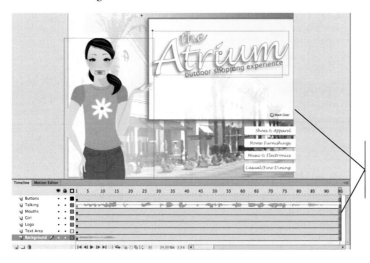

Because all layers now extend to Frame 95, all graphics in the interface are visible on the Stage.

10. Click Frame 1 of any layer to move the playhead to the beginning of the timeline.

The playhead identifies the current point in time on the Flash timeline. If you don't move the playhead back to Frame 1, the background sound will not play.

11. Press Return/Enter to test the movie on the Flash Stage.

You should now hear two sounds: the character talking and the background music. The background music is an event sound (not related to the playhead), so it loops continuously.

Pressing Return/Enter causes the playhead to move, playing the movie directly on the Stage.

When a movie is playing on the Stage, pressing Return/Enter stops the playback but does not stop sounds that are playing.

12. **Press the Escape key (ESC) to stop the background music.**

13. **Press Command-Return/Control-Enter to test the movie in a Player window.**

    The movie plays entirely through, and then starts over again (**loops**) — this is what would happen in the actual exported file.

    To make the timeline play only once, you have to use code to intentionally stop the playhead from looping. This code will be implemented by your developer partner after you are finished creating the lip-syncing animation.

Note:

You can choose Control>Loop Playback to allow the playhead to loop on the Flash Stage.

14. **Close the Flash Player window and return to Flash.**

15. **Save the file and continue to the next exercise.**

PREPARE FOR LIP SYNCING

If you have ever watched cartoons, you have probably seen the results of the time-consuming and painstaking work involved in synchronizing a character's movements to sounds. Realistic lip syncing is an extremely complex art that requires precise attention to detail, as well as in-depth study of behavioral movement. Other projects, such as this one, do not call for the precision and detail required for lifelike animation; rather, they use representative movements to create the effect of a character talking.

1. **Sit or stand in front of a mirror.**

2. **Say the following sentence slowly, paying careful attention to the shape of your mouth for each syllable:**

    Need help? Use the buttons to find exactly what you're looking for.

Note:

To better understand how to sync lip movements to sounds, you should study the different facial movements that are involved in spoken sound (called phonology).

3. **With atrium_kiosk.fla open in Flash, expand the mouth graphics folder in the Library panel.**

    The illustrator for this project created eight different mouth shapes to represent the various "talking" sounds.

4. **Click each mouth symbol in the Library panel and review the shapes.**

    Note that each symbol was created with the registration point at the center.

Symbol		Use for:	Symbol		Use for:
mouth1		Silent, M, B, P	mouth5		M, B, P
mouth2		C, D, G, J, K, N, R, S, Y, Z	mouth6		L, D, T, Th
mouth3		Short E, I, O, U Long A	mouth7		Short A, E Long I
mouth4		F, V	mouth8		Ch, Sh, Qu, W Long O, U

5. **In the timeline layers area, double-click the icon to the left of the Talking layer name to open the Layer Properties dialog box.**

6. **Choose 300% in the Layer Height menu and click OK.**

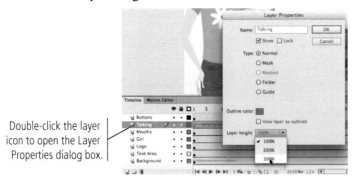

Double-click the layer icon to open the Layer Properties dialog box.

It's easier to sync movement to sound when you can see the variations in the sound file. By enlarging the layer height, you can see the peaks and valleys of the waveform directly on the timeline.

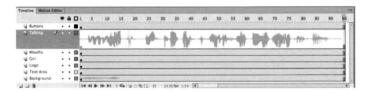

7. **Continue to the next exercise.**

 ## Create Lip Sync Animation

While lip syncing might seem complicated, it's actually quite simple — you show the graphic that supports the sound heard at a particular frame on the timeline. Because the different mouth shapes for this project have already been created, the most difficult part of the process is determining which shape to place at which point on the timeline.

1. **Click Frame 1 above the timeline to reset the playhead to the beginning of the movie.**

2. **Click the playhead and drag quickly to the right.**

   Dragging the playhead, a technique called **scrubbing the timeline**, allows you to manually preview portions of an animation. Because the sound on the Talking layer is a stream sound, you hear the sound as you drag the playhead. The background music — an event sound — is not related to the playhead, so scrubbing the playhead does not play the background music.

   As you drag the playhead from Frame 1, you hear the first sound in the spoken message beginning at Frame 4 (also indicated by the rise in the waveform).

Click the playhead and drag right to find the first spoken sound.

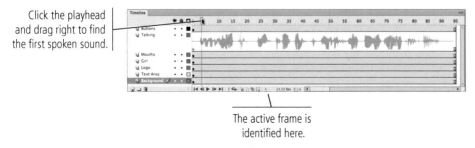

The active frame is identified here.

3. Control/right-click Frame 3 of the Mouths layer and choose Insert Keyframe from the contextual menu.

Remember, a keyframe is the point at which something changes. In this case, you are going to change the mouth shape, so you need to add a keyframe at the appropriate point in time (when the mouth begins to move to make the spoken sound).

Content on the preceding frame is automatically duplicated on the new keyframe.

Although the sound begins at Frame 4, people's mouths usually start moving before actual words are spoken. You are adding a keyframe one frame earlier than the sound to accommodate for this behavior.

4. Click the mouth symbol on the Stage to show its properties in the Properties panel.

The object on the selected keyframe is selected on the Stage.

The Properties panel shows the name of the symbol being used.

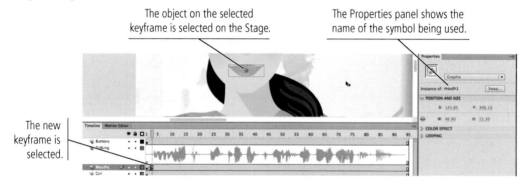

The new keyframe is selected.

5. With the mouth shape on the Frame 3 keyframe selected, click the Swap button in the Properties panel.

Lip syncing requires one primary task: swapping symbols to show the graphics that correlate to the sound at that particular moment.

6. In the Swap Symbol dialog box, choose the mouth2 graphic symbol.

This is the mouth that correlates to the "N" sound at the beginning of the word "Need".

Note:

You can also press the F6 key to add a new keyframe to the selected frame.

Note:

You can use the Insert Blank Keyframe command to add a blank keyframe to the timeline that (as the name suggests) has no content.

7. **Click OK to close the Swap Symbol dialog box.**

The new mouth now appears on Frame 3. The mouth symbol (mouth1) on the previous keyframe will remain visible until the playhead reaches Frame 3.

The new mouth shape appears, starting at Frame 3.

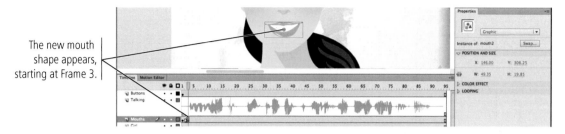

8. **Drag the playhead right to find the next significant change in sound.**

The brief pause between the words "need" and "help" suggests a change in the speaker's mouth position at Frame 8.

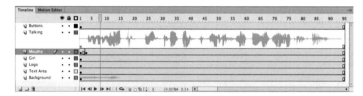

9. **Insert a new keyframe at Frame 8 of the Mouths layer.**

10. **With the mouth on the Frame 8 keyframe selected, click the Swap button in the Properties panel. In the Swap Symbol dialog box, choose the mouth5 symbol and click OK.**

This mouth shape is nearly closed, so it works well for the brief pause between words. It correlates to the short "I" sound, but it also works well as a good transition shape between a wide-open mouth and a closed mouth.

11. **Insert a new keyframe at Frame 10 of the Mouths layer. Open the Swap Symbol dialog box, and replace the mouth shape with the mouth3 symbol.**

This shape correlates to the "short e" sound in "help". (The "h" sound typically blends into the vowel sound.)

12. **Return the playhead to Frame 1 and press Return/Enter to play the movie on the Stage.**

So far you have only three changes in the character's mouth, but you should begin to see how the different symbols appear at the appropriate points in the playback. In general, lip syncing in Flash is a relatively simple process. The hardest parts are determining when to change the graphics in relation to the sound, and deciding which shape best suits the animation at any given point. Once you have determined these two elements, the actual syncing process is simply a matter of swapping symbols.

13. **Press the Escape key to stop the background music playback.**

14. **Applying the same process you used to create the first three mouth changes, continue scrubbing the playhead to identify points of change. Insert keyframes and swap symbols on the Mouths layer at the appropriate locations. In our example, we used the following locations and symbols:**

Frame	Symbol	Frame	Symbol	Frame	Symbol
15	mouth1	43	mouth4	64	mouth3
23	mouth8	46	mouth3	66	mouth2
27	mouth2	48	mouth2	68	mouth8
30	mouth6	50	mouth3	70	mouth2
31	mouth1	52	mouth2	72	mouth6
33	mouth3	55	mouth7	74	mouth2
36	mouth6	57	mouth2	79	mouth4
39	mouth2	60	mouth6	82	mouth8
41	mouth8	62	mouth2	84	mouth2

15. **Return the playhead to Frame 1, then press Return/Enter test the animation.**

By swapping the mouth symbol at various points on the timeline in relation to the sounds on the Talking layer, you now have a character who appears to be talking.

16. **Save the file and continue to the next exercise.**

DEFINE SOUND COMPRESSION SETTINGS

Before you export the final movie file, you should optimize the sounds to produce the smallest possible files while still maintaining the best possible quality. You can define default export settings for all stream sounds and all event sounds, but you can also experiment with different compression settings for individual sound files in the library.

1. **With `atrium_kiosk.fla` open, Control/right-click the atrium_jazz.mp3 file in the Audio folder of the Library panel. Choose Properties from the contextual menu.**

2. **In the resulting Sound Properties dialog box, make sure the Options tab is active at the top of the dialog box.**

3. **Choose MP3 in the Compression menu. If available, uncheck the Use Imported MP3 Quality option.**

   Flash supports five sound compression options:

   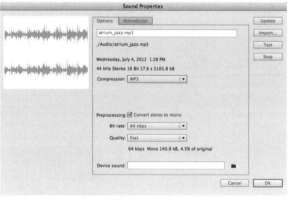

   - **Default.** This option uses the global compression settings (mp3, 16kbps, mono) defined in the Publish Settings dialog box when you export your SWF file. If you select Default, no additional export settings are available.

   - **MP3.** Over the past few years, this format has become a *de facto* standard for audio on the Web. MP3 compression produces small files with very good quality, but it can cause problems for older computers with limited processing power.

   - **ADPCM.** This option converts sounds into binary data. ADPCM encoding is not as efficient as MP3 compression, but it is useful if you need to export a file to be compatible with older versions of Flash (version 3 or lower).

   - **Raw.** This option does not compress the audio data, which results in very large file sizes. This option should only be used for files that will be delivered on the desktop instead of over the Internet.

     ADPCM and Raw use less processing power on each playback than MP3. They are recommended for very short (small) sounds that are played back rapidly. A shooting game in which guns fire many times a second, for example, might benefit from encoding the gun sound in ADPCM or Raw; the cost in file size would probably be less than 1k, and processor performance would be significantly enhanced.

   - **Speech.** This option uses a compression algorithm designed specifically for compressing spoken sounds. Sounds compressed with this option are converted to mono sounds (instead of stereo). Speech-compressed sounds require Flash Player 6 or higher.

Note:

The Preprocessing check box, enabled by default, converts stereo sounds to mono sounds.

4. **Choose 48 kbps in the Bit Rate menu.**

   Depending on the selected compression option, you can also change the bit rate or the sample rate to affect the quality of the exported sound.

   - The **Sample Rate** menu is available for ADPCM, Raw, and Speech compression; lower sample rates decrease file size, but can also decrease sound quality. The 22 kHz setting is recommended for reasonably good quality of most sounds.

   - The **Bit Rate** menu is available for MP3 compression. This option determines the bits per second in the exported sound. Higher bit rates result in better sound quality. Most experts recommend at least 20 kbps for speech, and 48 kbps for reasonably good quality of complex sounds such as music.

Note:

Flash cannot increase the sample rate or bit rate of an imported sound above its original settings.

5. **Choose Best in the Quality menu.**

   Three quality options — in order of file size (from small to large) and quality (from low to high) — are available for MP3 sounds: Fast, Medium, and Best.

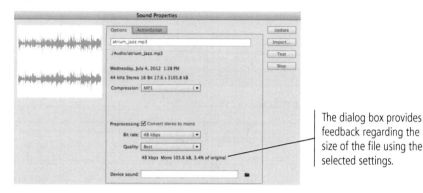

The dialog box provides feedback regarding the size of the file using the selected settings.

6. **Click OK to change the export settings for the selected sound file.**

7. **Open the Sound Properties dialog box for the intro_talking sound file.**

8. **Choose Speech in the Compression menu.**

9. **Choose 11 kHz in the Sample Rate menu, and then click the Test button.**

   When the sound plays, you might notice some popping or hissing noises behind the spoken message.

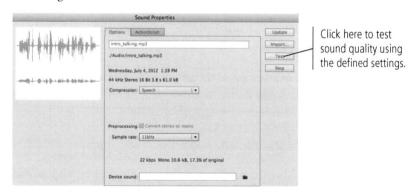

Click here to test sound quality using the defined settings.

Note:

You can change the default sound export settings in the Publish Settings dialog box.

10. **Choose 44 kHz in the Sample Rate menu, and then click the Test button.**

    This sample rate results in much better quality. Because this kiosk will not be downloaded over the Internet, the larger file size is not a problem.

11. **Click OK to apply the new compression settings for this sound file.**

12. **Save the Flash file and close it.**

Project Review

fill in the blank

1. In a Button symbol, the _____ defines the area where a user can click to trigger the button.

2. _____ to edit a symbol in place on the Stage.

3. The _____ marks the location of the defined X and Y values of a placed instance.

4. A(n) _____ defines the point in time when a change occurs.

5. _____ is the number of animation frames that occur in a second.

6. _____ sounds are timeline independent; they must download completely before they play.

7. _____ sounds are timeline dependent; they play as soon as enough of the data has downloaded to the user's computer.

8. The _____ sync method prevents more than one instance of the same sound from playing at the same time.

9. You can use the _____ dialog box to change the length of a specific sound file.

10. Use the _____ option to replace one symbol with another at a specific frame.

short answer

1. Briefly describe at least three uses of the Library panel.

2. Briefly explain the difference between event sounds and stream sounds.

3. Briefly explain the concept of lip syncing, as it relates to symbols and the Flash timeline.

Use what you learned in this project to complete the following freeform exercise.
Carefully read the art director and client comments, then create your own design to meet the needs of the project.
Use the space below to sketch ideas; when finished, write a brief explanation of your reasoning behind your final design.

art director comments

Your client is a company that provides technical support for children's online video games. The owner wants an introduction page for that site similar to the kiosk interface, with a talking character that identifies the options.

To complete this project, you should:

❏ Download the client's supplied files in the **FL6_PB_Project2.zip** archive on the Student Files Web page.

❏ Review the client-supplied sound and artwork files.

❏ Develop a site intro page with a talking robot and two different buttons.

❏ If you use the client's artwork, import the file into Flash and create movie clips as necessary from the different elements.

❏ If you don't use the supplied file, create or find artwork as appropriate.

client comments

We want to build a new introduction page to our video game site. We're using a robot avatar throughout the video game site, and want that character to be featured on the intro page — I even recorded the intro message with a "mechanical" sounding voice. (Feel free to re-record the audio if you want to, as long as the message stays the same.)

I found a robot illustration that I like, but I'm not an artist; I'd be happy to review other artwork if you have a better idea. I also want you to develop some kind of background artwork that makes the piece look like a cohesive user interface.

You need to include two buttons: one that links to online technical support and one that links to a telephone support page.

In the final file, I want the robot to look like it's talking, but I also want the robot to point to the related buttons when the appropriate part of the intro sound plays.

project justification

Project Summary

This project introduced many of the basic concepts and techniques of animating objects in Flash. You learned about frames and keyframes, as well as two different types of symbols that will be used in many Flash projects, both throughout this book and in your professional career. You also learned how to import artwork that was created in Adobe Illustrator — a very common workflow in the graphic design/animation market.

You should understand how frames on the Flash timeline relate to the passage of actual time, and how keyframes are used to make changes at specific points in an animation. You should also understand the basic concept of symbols and instances, including the different ways to edit the conent of a specific symbol. You will build on these skills as you complete the rest of the projects in this book.

This project also showed you how to add audio content to a movie — placing a looped sound in the background of a file, triggering specific sounds with a button's Over state, and even synchronizing graphics to a spoken message.

Import artwork from Adobe Illustrator files

Add a streaming sound as the interface introduction

Create symbols from imported artwork

Synchronize graphics to sound to create a "talking" effect

Add an event sound that loops continuously in the background

Edit a sound envelope to control the sound volume

Attach and control button sounds

Edit button symbols to change different button states

Animated Internet Ads

Your client wants to create a series of ads to place on Web sites that are used by existing and potential customers. They have asked you to create a short animation rather than just a static image, in hopes of attracting more attention when the ad appears in a browser with other content.

This project incorporates the following skills:

❑ Creating shape tweens to animate changes in shape and color

❑ Creating classic tweens to animate changes in position and opacity

❑ Adding text to a Flash movie

❑ Adapting file content to match different file dimensions

❑ Using a Flash project to manage assets for multiple files

❑ Publishing a file to SWF for distribution

Project Meeting

client comments

We've provided you with our logo and an image that we want to use in the ads. Since these are going to be placed into a variety of Web sites, we want some kind of animation that might help catch a user's eye.

Most of the sites where we're planning on advertising use standard ad sizes. I'm not sure exactly which sizes we're going to purchase, but we do like the rectangle and square shapes better than the narrow banners.

We might decide on some of the other options later, but we'd like to get started with three common sizes:

- 300 × 250 pixels
- 336 × 280 pixels
- 250 × 250 pixels

art director comments

Flash includes predefined templates for most of the common ad sizes, so that's the easiest way to start the first file.

I want you to animate different aspects of the client's logo over the course of the animation. The kayaker is ideally suited to move across the stage. He should paddle across the stage while the sun rises. Halfway through, he should pause and wait until the tagline appears, then move the rest of the way across while the image gradually appears in place of the sunrise.

After you create the initial ad, you can use several built-in techniques to repurpose the content for other sizes. You should also take advantage of the Project panel to manage assets that will be used in more than one file, so it will be easier to make universal changes in any of the shared assets.

project objectives

To complete this project, you will:

- ❏ Create a file based on a template
- ❏ Create a shape tween
- ❏ Tween an object's color
- ❏ Create a classic tween
- ❏ Tween an object's opacity
- ❏ Stop the animation timeline
- ❏ Create and control a text object
- ❏ Define font embedding
- ❏ Control object stacking order
- ❏ Create a Flash project
- ❏ Scale content to document properties
- ❏ Edit a shared symbol
- ❏ Publish files to SWF

Stage 1 Animating Symbols

Animation — the true heart of Flash — can be created in a number of different ways. To create the animated ads for this project, you will use **shape tweening** and **classic tweening**.

If you completed Project 2: Talking Kiosk Interface, you already learned a bit about symbols in general, specifically button and graphic symbols. In this project, you work with movie clip symbols. Both graphic symbols and movie clip symbols can include animation, but movie clips offer a number of advantages over graphic symbols.

Movie clips are **timeline independent**; the animation contained in a movie clip requires only a single frame on the timeline where it is placed (called the **parent timeline**). A movie clip timeline might include 500 frames, but the entire animation will play on a single frame of the parent timeline.

An animated graphic symbol, on the other hand, is **timeline dependent**; it requires the same number of frames on the parent timeline that are present inside the symbol's timeline. In other words, a 500-frame animation inside the graphic symbol requires 500 corresponding frames on the timeline where the symbol is placed.

Because movie clip timelines function independently of the parent timeline, you can more easily incorporate animations of different duration onto the same parent timeline.

Note:

Another advantage of movie clips is that placed instances can be named, which means they can be addressed — and controlled — using code. You will learn more about this option beginning in Project 4: Ocean Animation.

Planning a movie

When you start any new project, you should begin by analyzing what you need to accomplish. A bit of advance planning can help you avoid unnecessary rework and frustration — in the project planning phase, you can determine, for example, that an independent movie clip is a better option than animating an object directly on the main timeline.

The ad that you are going to create in this project has the following plan or **storyboard**:

- The entire animation should last four seconds.

- The logotype will change from white to dark blue throughout the entire four-second animation.

- The kayaker will move across the Stage throughout the entire four-second duration, pausing halfway until the client's tagline appears.

- The sun is going to rise while the kayaker moves across the Stage. The sunrise animation should be finished when the kayaker gets halfway across the Stage.

- An image will gradually appear to replace the sunrise.

This information tells you a number of things about what you need to do:

- The finished ad requires four separate animations — the logo changing colors, the moving kayaker, the sunrise, and the image fading in.

- Each animation requires different timing. The sunrise and the image fade-in each occupy only half the time of the moving kayaker.

- The animations also require different starting points. The sunrise and the moving kayaker need to start as soon as the file opens. The image fade-in doesn't start until the sunrise animation is complete.

As you complete this project, you are going to use movie clip symbols and timeline frames to achieve the stated goals.

 **CREATE AN AD FILE**

The final goal of this project is three separate ads that can be placed on Web sites where your client has decided to advertise. Because Internet ads typically use standard sizes, Flash includes those sizes as templates in the New Document dialog box.

In the first stage of this project, you are going to create the initial ad using one of the defined templates. Later you will use Flash's built-in tools to repurpose the existing content into the other required ad sizes.

1. **Download FL6_RF_Project3.zip from the Student Files Web page.**

2. **Expand the ZIP archive in your WIP folder (Macintosh) or copy the archive contents into your WIP folder (Windows).**

   This results in a folder named **Kayaks**, which contains the files you need for this project. You should also use this folder to save the files you create in this project.

3. **Choose File>New. In the New Document dialog box, click the Templates tab to display those options.**

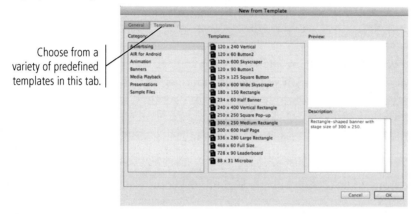

Choose from a variety of predefined templates in this tab.

4. **Select Advertising in the left pane, then select the 300 × 250 Medium Rectangle option in the right pane.**

   Flash includes templates for a number of standard file sizes, including the most common ads that are placed on the Internet.

5. **Click OK to create the new file.**

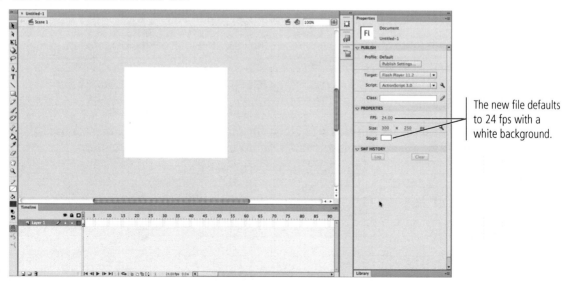

The new file defaults to 24 fps with a white background.

6. **Choose File>Import>Import to Stage. Select `cko_logo.ai` (in the WIP>Kayaks folder) and click Open.**

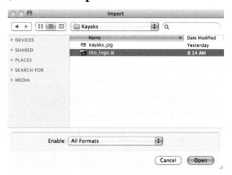

7. **Review the options in the Import to Stage dialog box.**

This Illustrator file was created with four layers, each containing a different element of the logo. As you complete this project, you will use several techniques to animate different parts of the logo artwork.

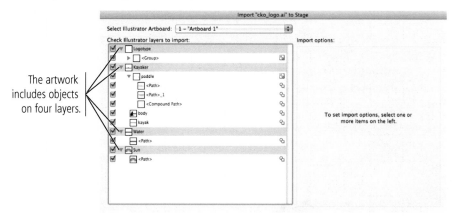

The artwork includes objects on four layers.

8. **At the bottom of the dialog box, choose the following options:**

- **Convert Layers To**   **Flash Layers**
- **Place Objects at Original Position**   **Checked**

  Although you will move the various logo components as you complete this project, you are importing them at their original positions so you can see the logo as it was designed before you make changes.

- **Set Stage Size...**   **Unchecked**

  You created this Flash file at a specific file size to meet specific needs, so you are not converting the Stage to match the imported artwork.

- **Import Unused Symbols**   **Unchecked**

  The Illustrator file does not include any unused symbols. (In a professional environment, it is a good idea to open the Illustrator file and review its contents, if possible, before importing into Flash.)

- **Import as a Single Bitmap Image**   **Unchecked**

  Beginning in the next exercise, you are going to change several of the paths that make up the logo. If you converted the artwork to a single bitmap, you would not be able to access the individual paths that you need to change.

9. **Click OK to import the artwork to the Stage.**

The Flash file now includes four layers, matching the layers in the original Illustrator file. All the imported artwork is automatically selected.

The imported artwork is centered in the document window.

Each separate object has its own bounding box.

Four layers were added to the file.

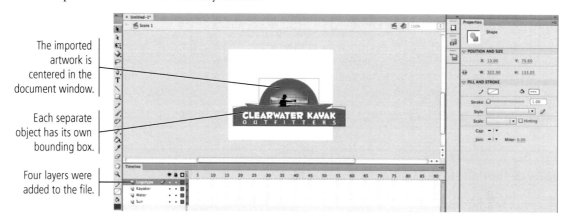

10. **Using the Selection tool, drag the selected artwork until the bottom edge snaps to the bottom edge of the Stage, and the artwork appears to be centered horizontally.**

Each logo component, identified by the various bounding boxes, is an individual object (or group). If you tried to use the Align panel to align the artwork to the bottom and center of the Stage, each individual component would be aligned to the bottom of the Stage; the objects' positions relative to one another would not be maintained.

Note:

Because these objects all reside on different layers, you can't use the Group command to treat them as a single object for positioning purposes.

Use the Selection tool to move all selected objects at one time without changing their positions relative to one another.

11. **Click away from everything on the Stage to deselect everything.**

12. **In the Properties panel, click the Stage swatch and choose Black from the pop-up color panel.**

13. **Click the FPS hot text to access the field, and change the FPS to 15.**

The ads you are creating are only going to be distributed over the Internet; 15 fps is high enough for good-quality display. (Higher frame rate would result in larger file sizes that are unnecessary for this type of file, and could be problematic for users with slower download speeds.)

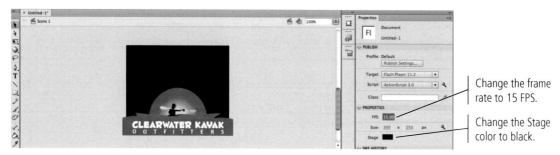

Change the frame rate to 15 FPS.

Change the Stage color to black.

14. **Save the file as `cko_med_rect.fla` in your WIP>Kayaks folder, then continue to the next exercise.**

CREATE A SHAPE TWEEN

A **shape tween** allows you to convert one shape into another over time. You define the starting and ending shape, then Flash creates the in-between frames (hence the name "tween") that create the appearance of continuous movement when the finished animation plays.

You will use this type of tween to create the sunrise animation, as well as change the colors of the logotype.

1. **With `cko_med_rect.fla` open, use the Selection tool to select the sun object on the Stage.**

2. **Open the Color panel.**

   This object was created with a gradient fill. You are going to edit the gradient so the edge of the shape blends smoothly into the Stage background color.

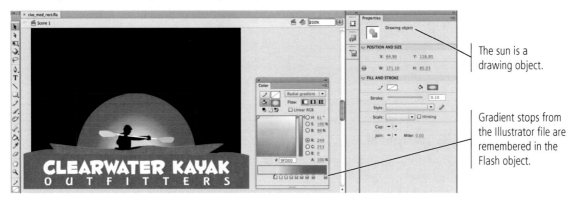

The sun is a drawing object.

Gradient stops from the Illustrator file are remembered in the Flash object.

3. **Click the right gradient stop to select it, then change the Alpha value of the selected stop to 0.**

   Alpha refers to transparency; a value of 0 means something is entirely transparent.

   This step makes the last gradient stop entirely transparent. Colors between the next-to-last and last stop will now transition from entirely opaque (100% Alpha) to entirely transparent (0% Alpha), which allows the object to blend into the background without a harsh edge.

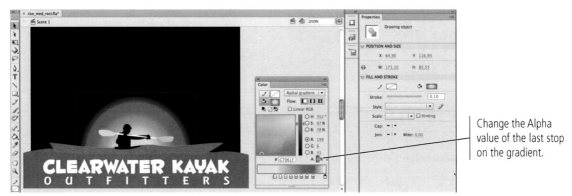

Change the Alpha value of the last stop on the gradient.

4. **Control/right-click the selected object and choose Convert to Symbol.**

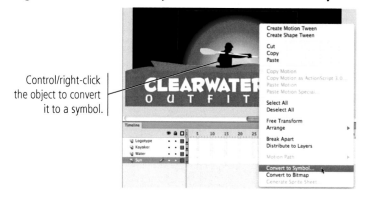

Control/right-click the object to convert it to a symbol.

5. **In the resulting dialog box, type sun_mc in the Name field, choose Movie Clip in the Type menu, and choose the bottom-center registration point.**

You are using the bottom-center registration point because you want the sun object to grow out from that point.

6. **Click OK to create the new symbol.**

When you create a symbol from existing artwork, the original object is automatically converted to an instance of that symbol.

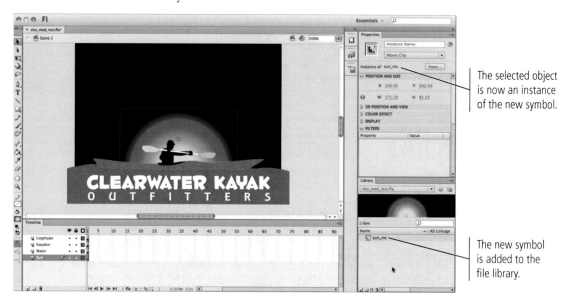

The selected object is now an instance of the new symbol.

The new symbol is added to the file library.

7. **Double-click the instance on the Stage to enter into Symbol-Editing mode.**

You are editing this symbol in place because you need to be able to see the shape's size relative to the Stage on which it is placed.

8. Select Frame 30 in the timeline, then press F6 to add a new keyframe.

The completed ad needs to last four seconds. At 15 fps, the entire ad will require 60 frames; the sunrise should take half that time to complete, so this movie clip needs 30 frames.

When you add a keyframe to a layer, the contents of the previous keyframe are automatically copied to the new keyframe. You can edit the contents on each keyframe independently, without affecting the same contents on other keyframes.

Note:

If you can't use, or don't have, Function keys, you can use the Insert> Timeline submenu, or the frame's contextual menu, to insert a frame, keyframe, or blank keyframe.

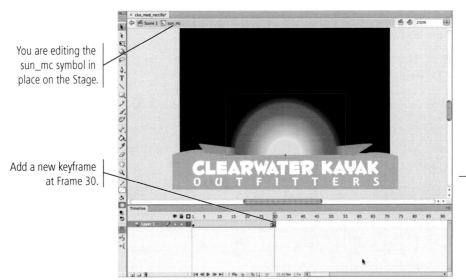

You are editing the sun_mc symbol in place on the Stage.

Add a new keyframe at Frame 30.

Note:

Remember, keyframes are required when an object needs to change in some way at a given point in time.

9. Choose the Free Transform tool.

10. With Frame 30 selected in the timeline and the sun object selected on the Stage, move the transformation point to the object's bottom-center bounding-box handle.

When you create the animation, you want the sun to appear as if it is growing out from the horizon. To accomplish this, you are going to make the sun shape larger, using the bottom-center point as the anchor.

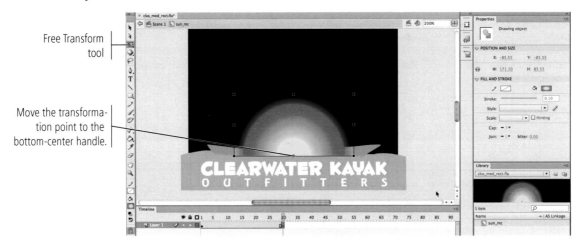

Free Transform tool

Move the transformation point to the bottom-center handle.

11. Zoom out so you can see the area around the Stage on all four sides.

Feel free to zoom in and out as necessary while you complete the projects in this book.

12. **Open the Transform panel. At the top of the panel, make sure the Constrain icon is active.**

   When the Constrain icon appears as two connected chain links, the object's width and height are linked to maintain object's original aspect ratio. If this icon is a broken chain, you can change one value without affecting the other.

13. **Place the cursor over the existing Width value. When you see the scrubby-slider cursor, click and drag right to enlarge the object until none of the black background color is visible.**

   Although you can use the Properties panel to change an object's height and/or width, those changes apply from the top-left corner of the selection instead of the defined transformation point that you set in Step 10. Changes made through the Transform panel respect the defined transformation point, so this is a better option for making this type of change.

Note:

The Transform panel can be used to make precise numerical changes, or simply to monitor the changes you make with the Free Transform tool.

Note:

This technique of dragging to change a property value is called **scrubbing***.*

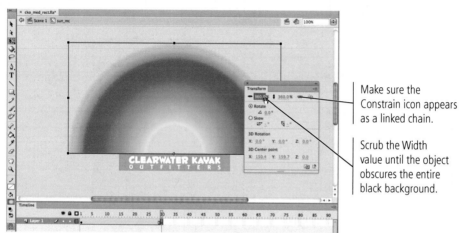

Make sure the Constrain icon appears as a linked chain.

Scrub the Width value until the object obscures the entire black background.

14. **Control/right-click any frame between Frame 1 and Frame 30, and choose Create Shape Tween from the contextual menu.**

   In the timeline, Flash identifies the shape tween with green frames and an arrow between keyframes.

Note:

You can't create a shape tween for a group.

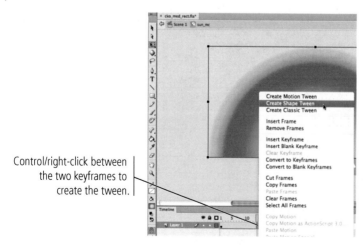

Control/right-click between the two keyframes to create the tween.

15. **Click Frame 1 in the timeline to move the playhead back to the beginning of the animation.**

16. Press Return/Enter to play the animation on the Stage.

Flash defines the object's size at each frame in the tween.

A shape tween is identified by green frames and an arrow.

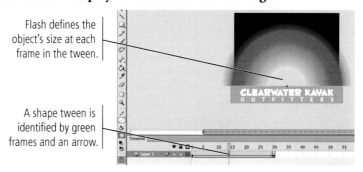

Understanding Transformation Options

The Transform Panel

At times, you might need to apply very specific numeric transformations, such as scaling an object by a specific percentage. Rather than manually calculating new dimensions and defining the new dimensions in the Properties panel, you can use the Transform panel to make this type of change.

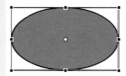

Reset

Constrain

Duplicate Selection and Transform

Remove Transform

When you change a value in the Transform panel, press Return/Enter to apply the change, or click the Duplicate Selection and Transform button to make a copy of the object and apply the change to the copy.

Keep in mind that all transformations made in the Transform panel apply around the defined transformation point.

The Modify>Transform Submenu

The Modify>Transform submenu has a number of valuable options for transforming objects.

Free Transform displays a set of eight bounding box handles, which you can drag horizontally or vertically to scale, stretch, skew, and rotate an object. During a free transform, an object's overall shape is maintained (an oval remains an oval, a square remains a square, and so on).

Distort displays a set of eight bounding box handles. If you drag one of the corner handles, you can "stretch" the object out of its original shape. For example, you can drag one corner of a rectangle to create a polygon with odd angles.

Envelope adds control handles to the object's anchor points, which you can use to warp the shape. You can drag the handles to reshape the connecting curves, and/or drag the anchor points to new positions to create an entirely different shape than the original.

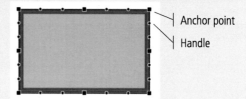

Anchor point

Handle

Scale is a subset of Free Transform. You can see the eight bounding box handles, but you can only scale or stretch the object; you can't rotate or skew it.

Rotate and Skew are also subsets of Free Transform. You can see the eight bounding box handles, but you can only skew or rotate the object; you can't scale or stretch it.

Scale and Rotate opens a dialog box where you can define specific scale percentages or rotation angles. You can also use the **Rotate 90°** (Clockwise and Counterclockwise) options for a selected object.

Flip Horizontal or **Flip Vertical** options allow you to flip objects on either axis.

Even though you can transform objects, Flash remembers the object's original size and shape. You can remove any transformation — except envelope distortion — from drawing objects and symbol instances using the **Remove Transform** option. (You can't remove a transformation from a merge-drawing object after you have deselected the object.)

To remove an envelope distortion, you have to choose Modify>Combine Objects>Delete Envelope.

17. Click Scene 1 in the Edit bar to return to the main Stage.

Remember, pressing Return/Enter plays the *current* timeline on the Stage. Because the main timeline has only one frame, this command would have no effect. Testing a movie on the Stage does not initiate the timeline of movie clips that are placed on the Stage.

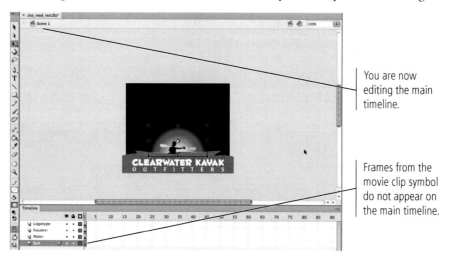

You are now editing the main timeline.

Frames from the movie clip symbol do not appear on the main timeline.

18. Save the file and continue to the next exercise.

 ## TWEEN AN OBJECT'S COLOR

Changing an object from one color to another is a common animation task. This is simply accomplished using a shape tween, using the same method you used to change the size of the sun symbol.

1. With `cko_med_rect.fla` open, use the Selection tool to select the logotype on the Stage.

2. Control/right-click the selected group and choose Convert to Symbol.

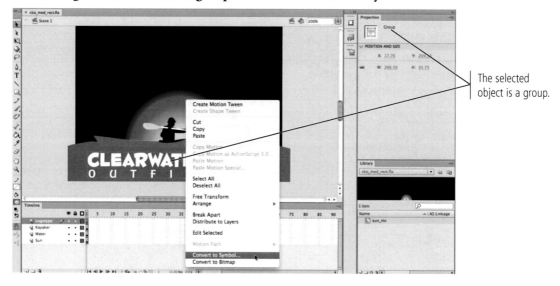

The selected object is a group.

3. Type `logo_mc` in the Name field and choose Movie Clip in the Type menu.

4. **Click OK to create the new symbol.**

   In this animation, you are simply going to change the color of the text. Because nothing is moving, the symbol registration point does not matter; you can leave it at the default location.

5. **Double-click the symbol instance on the Stage to enter into it.**

6. **With the logotype selected on the Stage, choose Modify>Ungroup.**

   You can't create a shape tween with a group, so you first have to ungroup the letters. After ungrouping, you can see that each letter is a separate drawing object.

Note:

You chould also choose Modify>Break Apart (or press Command/Control-B) to accomplish the same general effect.

You are editing the logo_mc symbol in place on the Stage.

7. **Click Frame 60 in the timeline to select it, then press F6 to add a new keyframe.**

   When you add or select a frame on the timeline, the frame becomes the active selection.

   This is deceptive because the objects' bounding boxes are still visible, suggesting that they are selected — even though they are not.

8. **Click the filled area of any of the drawing objects on the Stage to make them the active selection.**

   To edit an object's properties, you first have to remember to intentionally reselect the object(s) on the Stage.

Note:

If you aren't sure what is actually selected, look at the top of the Properties panel.

After adding the keyframe, the Properties panel shows that the frame is selected.

The bounding boxes of objects on the selected frame are visible.

Click any filled area inside any bounding box to make the drawing objects the active selection.

Creating and Controlling Shape Tweens

In addition to changing object properties in a shape tween, you can also create a **shape tween** to change one shape into another over time (as the name suggests). A shape tween requires two existing keyframes — one with the starting shape and one with the ending shape.

If you Control/right-click between two keyframes, choosing Create Shape Tween generates the shape tween; the tween frames automatically change the shape of the object as necessary to convert Shape A into Shape B. The following illustrations show a simple shape tween that changes a blue square into a green circle.

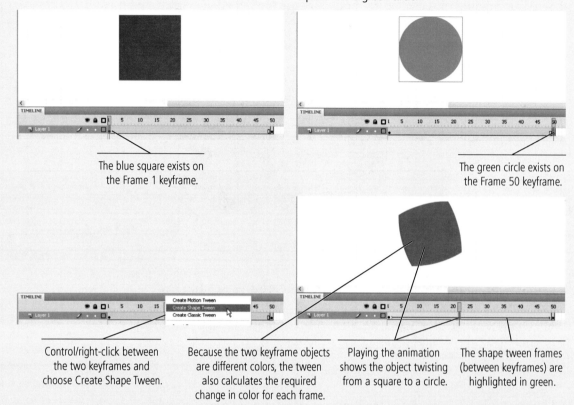

The blue square exists on the Frame 1 keyframe.

The green circle exists on the Frame 50 keyframe.

Control/right-click between the two keyframes and choose Create Shape Tween.

Because the two keyframe objects are different colors, the tween also calculates the required change in color for each frame.

Playing the animation shows the object twisting from a square to a circle.

The shape tween frames (between keyframes) are highlighted in green.

In many cases, shape tweens don't work exactly as expected. In the previous examples, the shape tween animation twists the square shape over time into the circle shape. If you want the square corners to simply move without twisting (for example), you can use **shape hinting** to control the change that occurs in a shape tween.

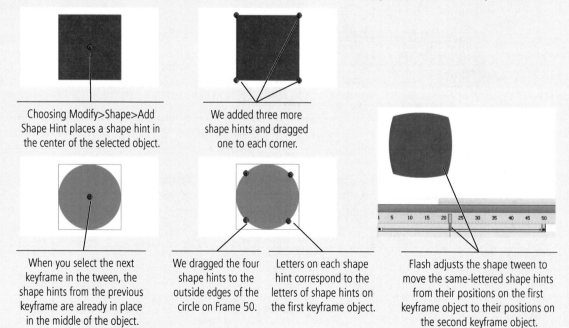

Choosing Modify>Shape>Add Shape Hint places a shape hint in the center of the selected object.

We added three more shape hints and dragged one to each corner.

When you select the next keyframe in the tween, the shape hints from the previous keyframe are already in place in the middle of the object.

We dragged the four shape hints to the outside edges of the circle on Frame 50.

Letters on each shape hint correspond to the letters of shape hints on the first keyframe object.

Flash adjusts the shape tween to move the same-lettered shape hints from their positions on the first keyframe object to their positions on the second keyframe object.

9. **Using the Color panel, change the objects' fill color to #000033.**

After ungrouping, click any of the active objects to select them.

Use the Color panel to change the fill color to #000033 with 100% Alpha.

10. **Control/right-click any frame between Frame 1 and Frame 60, and choose Create Shape Tween from the contextual menu.**

    Although you did not change the objects' shapes, a shape tween is still an appropriate method for changing an object's color over time.

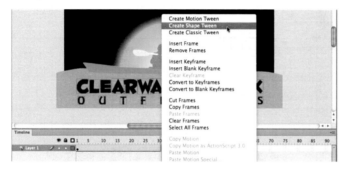

11. **Move the playhead to Frame 1, then press Return/Enter to test the animation on the Stage.**

12. **Click Scene 1 in the Edit bar to return to the main Stage.**

    Again, there is only one frame on the main timeline, so there is nothing to play. Remember, a movie clip symbol operates independently of the main timeline; you can't view the movie clip's animation directly on the Flash stage.

13. **Save the file and continue to the next exercise.**

CREATE A CLASSIC TWEEN

As you can probably guess, there is much more to animation than changing an object's shape or color. In this exercise, you will create very simple classic tweens that move the kayaker across the Stage, and use keyframes to control the movement's timing.

1. **With `cko_med_rect.fla` open, click the Kayaker layer in the timeline to select all objects on the layer.**

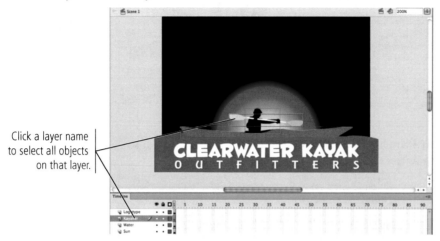

Click a layer name to select all objects on that layer.

2. **Control/right-click the selected objects and choose Convert to Symbol.**

3. **In the resulting dialog box, type `kayaker_mc` in the Name field. Choose Movie Clip in the Type menu and choose the right-center registration point. Click OK to create the new symbol.**

Extending the Length of the Timeline

FLASH FOUNDATIONS

If you create animations with very long timelines, you might need to extend the timeline beyond what is available by default. To accomplish this, scroll the timeline all the way to the right, and then add a regular frame to the layer near the end of the visible timeline.

When you add a frame after the last visible frame, the timeline scroll bar moves to the middle of the panel, indicating that more frames are now available in the timeline. You can then scroll again to the new end of the timeline and add another frame, which again extends the length of the available timeline. Continue this process until you have the number of frames you need.

The default timeline only goes up to a certain frame number.

After adding a regular frame near the end of the default timeline, the scroll bar indicates that more frames are available.

Continue adding regular frames and scrolling the timeline until you have the number of frames you need.

4. **With the new symbol instance selected on the Stage, change the X value in the Properties panel to 0.**

Remember, the X and Y properties define the position of a symbol's registration point. Because you chose the right-center registration point, changing the X value to 0 moves the kayaker entirely off the Stage.

You are editing the symbol instance on the main timeline.

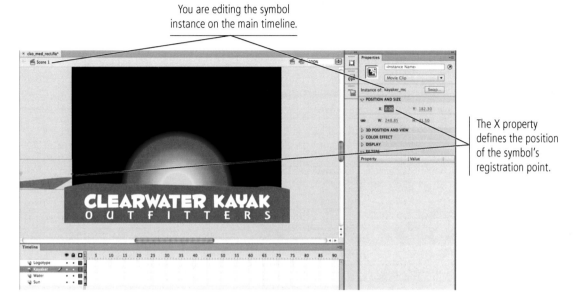

The X property defines the position of the symbol's registration point.

5. **Click Frame 30 of the Kayaker layer and press F6 to add a new keyframe.**

6. **Click the kayaker symbol instance to select it, then scrub the X property in the Property panel until the instance is approximately centered on the Stage (as shown here).**

Other objects are not visible because those layers don't yet exist on Frame 30.

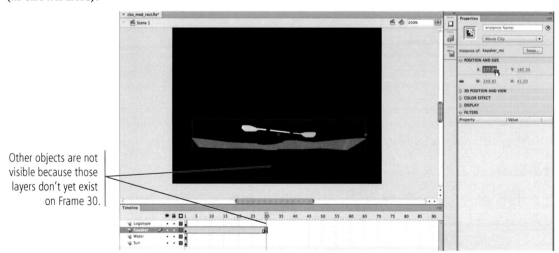

7. Control/right-click any frame between Frame 1 and 30 of the Kayaker layer and choose Create Classic Tween from the contextual menu.

A classic tween is one method for creating motion in a Flash animation. Like the shape tweens you created already, you have to define the starting and ending keyframes, then create the tween; Flash automatically determines the instance's position on the in-between frames.

In Project 4: Ocean Animation, you will learn about Motion Tweens, which provide far more control over numerous aspects of a tween. You should understand what a classic tween is, though, so you can recognize one if you find one in a file — especially files created in older (pre-CS4) versions of the software.

Note:

This is called a "classic" tween because this technique was available in previous versions of the application that did not include the Flash Motion Editor.

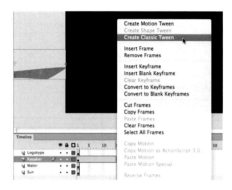

In the timeline, Flash identifies the classic tween with blue frames and an arrow between keyframes.

A classic tween is identified by blue frames and an arrow between keyframes.

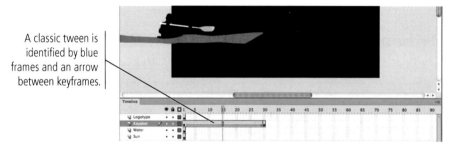

8. Select Frame 38 on the Kayaker layer, then press F6 to add another keyframe.

Remember: when you add a new keyframe, Flash duplicates the content on the previous keyframe. By adding this keyframe at Frame 38, you are holding the symbol instance in place for approximately half a second.

9. Select Frame 60 of the Kayaker layer and add a new keyframe.

10. Select the symbol instance on the Stage, then change the X property in the Properties panel until the instance is entirely past the right edge of the Stage.

Note:

We say "approximately half a second" because the frame rate in this file is 15 fps, which is not equally divisible by 2. Because you can't have half a frame, you are using slightly more than half a second for the pause in animation.

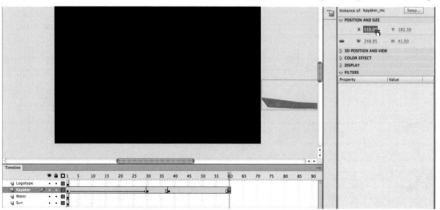

11. Control/right-click any frame between Frame 38 and 60 of the Kayaker layer and choose Create Classic Tween from the contextual menu.

The new tween occupies the entire range of frames between the keyframes that you defined on Frames 38 and 60.

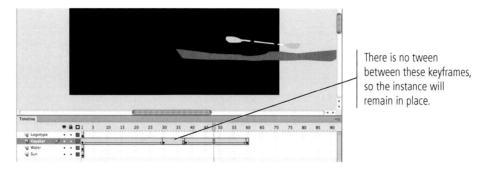

There is no tween between these keyframes, so the instance will remain in place.

12. Select Frame 60 of the Logotype layer. Press Command/Control, then click Frame 60 of the Water and Sun layers to add them to the active selection.

You can press Shift to select contiguous frames, or press Command/Control to select non-contiguous layers.

Although two of these layers contain animated movie clips, those movie clips' timelines do not transfer to the main movie timeline. They will not exist in the main movie beyond Frame 1 unless you extend those layers' timelines on the main Stage.

Command/Control-click to select non-contiguous frames.

13. Press F5 to add regular frames to the three selected layers.

This extends all three layers to Frame 60, so their content will be visible throughout the entire animation.

14. Click Frame 1 to select it, then press Return/Enter to test the animation on the timeline.

Remember, you can't see the sunrise and logotype animations because those are created inside the individual symbols. To see all three animations together, you have to test the movie in the Flash Player window. (You will do this after you create the final required animation in the next exercise.)

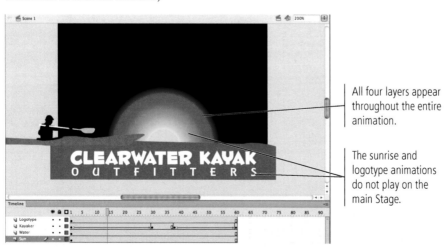

All four layers appear throughout the entire animation.

The sunrise and logotype animations do not play on the main Stage.

15. Save the file and continue to the next exercise.

TWEEN AN OBJECT'S OPACITY

The last required animation for this movie is an image that fades in after the sun finishes rising. In this exercise you will create a new layer on the timeline and use a blank keyframe to prevent the image from appearing too early.

1. **With `cko_med_rect.fla` open, create a new layer named `Photo` directly above the Sun layer.**

2. **Control/right-click Frame 30 on the new layer and choose Insert Blank Keyframe.**

   You are inserting a blank keyframe before placing the image onto the Stage at Frame 30, so the preceding frames (1–29) will remain blank — preventing the image from appearing until halfway through the movie.

3. **With the blank keyframe on Frame 30 selected, choose File>Import>Import to Stage.**

4. **Choose the file `kayaks.jpg` in your WIP>Kayaks folder, then click Open.**

5. **Align the image to the top of the Stage, centered horizontally.**

   Unlike the imported Illustrator artwork, you can use the Align panel to move the image into the correct position.

6. **Choose the Free Transform tool, then move the image's transformation point to the top-center bounding-box handle.**

   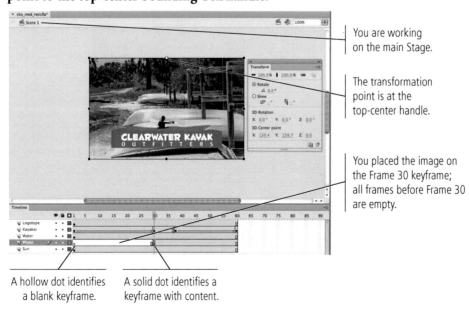

   You are working on the main Stage.

   The transformation point is at the top-center handle.

   You placed the image on the Frame 30 keyframe; all frames before Frame 30 are empty.

   A hollow dot identifies a blank keyframe.

   A solid dot identifies a keyframe with content.

7. **Using the Transform panel, reduce the image scale (proportionally) until the bottom edge is just hidden by the blue shape that makes up the water.**

   You can't animate a bitmap object; to cause this image to appear gradually over time, you first need to convert it to a symbol.

   Because you moved the transformation point, the top-center point of the image remains in place when you scale it.

8. **Control/right-click the image and choose Convert to Symbol. In the resulting dialog box, type** `photo_mc` **in the Name field, and choose Movie Clip in the type menu. Click OK to create the new symbol.**

   You are not animating the instance's position, so the registration point doesn't matter.

   Symbols have a number of properties that are not available for bitmap or drawing objects. In this case you are going to edit and animate the Color Effect property to change the alpha (transparency) value of the symbol over time.

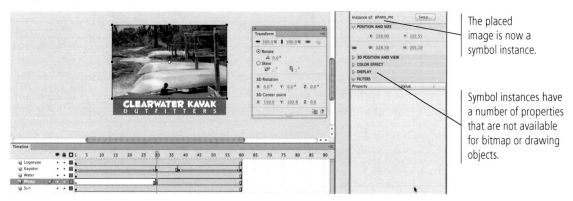

   The placed image is now a symbol instance.

   Symbol instances have a number of properties that are not available for bitmap or drawing objects.

9. **In the Properties panel, expand the Color Effect options, then choose Alpha in the Style menu. In the secondary Alpha field that appears, change the Alpha value to 0.**

   Remember, the Alpha value controls an object's opacity; a value of 0 means the object is not visible.

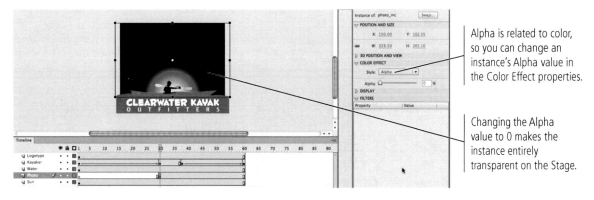

   Alpha is related to color, so you can change an instance's Alpha value in the Color Effect properties.

   Changing the Alpha value to 0 makes the instance entirely transparent on the Stage.

10. **Add a new keyframe to Frame 60 of the Photo layer.**

Controlling Frame and Tween Properties

With one or more frames selected in the timeline, you can use the Properties panel to control a number of options.

- **Name field.** Frames are sequentially numbered by default; you can also assign text-based labels to specific frames so you can more easily remember where in the timeline a specific action occurs. For example, "Arc_Tween" is more descriptive than "Frame 54."
- **Type menu.** If you assign a frame label, you can use this menu to treat the label as a regular name, as an anchor that can be targeted in a hyperlink, or as a comment.
- **Sound options.** The Properties panel includes several options for controlling sound files. You used these options in Project 2: Talking Kiosk Interface.

Classic and Shape Tween Properties

When a classic or shape tween is applied to selected frames, you can use the Properties panel to control a number of options related to the tween itself.

- **Ease.** This option is used to change the speed of movement as an object moves through a tween. You can set negative or positive values to increase or decrease speed at a consistent rate, or you can click the Edit button to define custom Ease values with different speeds at different points in the tween.
- **Rotate menu.** By default, an object simply moves from one place to another through the tween frames. You can choose Auto in the Rotate menu to rotate the object one time over the length of the tween; the object will turn in the direction that requires the least change. You can also exert greater control by choosing CW (clockwise) or CCW (counterclockwise), and then defining a specific number of times the object should rotate throughout the tween.
- **Snap option.** When this option is checked, the registration point of the tweened object attaches to the motion path.
- **Sync option.** This option synchronizes the number of frames in a tween within a graphic symbol to match the number of frames on the timeline where the graphic symbol is placed. With this option checked, replacing the animated symbol in the first frame of the tween replaces the symbol in all tweened frames.
- **Orient to Path.** If you are using a motion path, you can check this option to orient the baseline (bottom edge) of an object to the motion path.

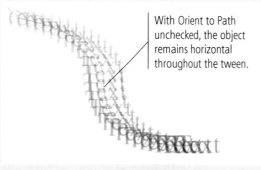

With Orient to Path unchecked, the object remains horizontal throughout the tween.

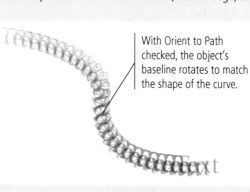

With Orient to Path checked, the object's baseline rotates to match the shape of the curve.

- **Scale.** This option specifies custom ease settings for the scale of an animated object. If you modified the size of the object on the Stage, select Scale to tween the size of the selected item.
- **Blend menu.** When you apply a shape tween, you can use the Blend menu to control how shapes distort as they change from one shape to another. Distributive, the default option, creates smoother shapes throughout the transition. The Angular option can produce better results if the starting and ending shapes have sharp corners that you want to preserve throughout the transition.

11. Click the symbol instance's transformation point to select the instance on the Stage.

The symbol instance is not visible because its Alpha value is currently 0; you can still use the transformation point to select it.

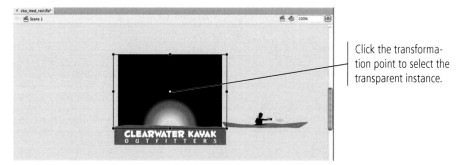

Click the transformation point to select the transparent instance.

12. In the Properties panel, change the Alpha value to 100%.

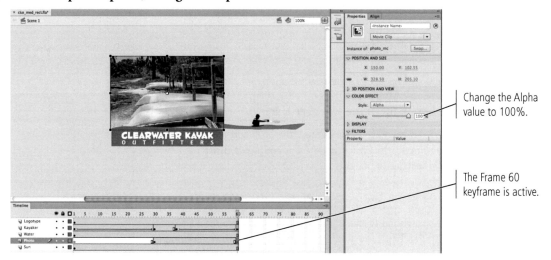

Change the Alpha value to 100%.

The Frame 60 keyframe is active.

13. Control/right-click anywhere between Frames 30 and 60 in the Photo layer and choose Create Classic Tween.

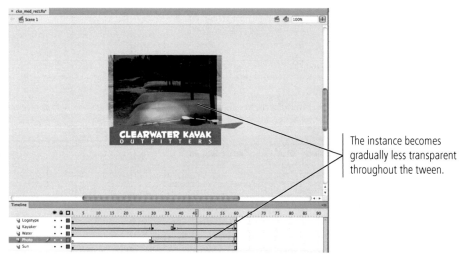

The instance becomes gradually less transparent throughout the tween.

14. Save the file and continue to the next exercise.

STOP THE ANIMATION TIMELINE

Remember, movie clips each have their own timelines, which are independent of other movie clips and of the main movie timeline. As soon as the playhead reaches the end of the timeline in each symbol, it automatically returns to the beginning and plays again (called **looping**). To prevent this, you have to add the Stop command to the timeline.

1. **With `cko_med_rect.fla` open, press Command-Return/Control-Enter to test the movie in the Flash Player window.**

   As you can see, the four animations in your movie play repeatedly. You should also notice that the sun rises twice in the time it takes the logotype to change colors and the kayaker to move out of the movie area.

All four animations in the movie loop repeatedly.

2. **Close the Flash Player window and return to Flash.**

3. **In the timeline, select Frame 60 of the Kayaker layer.**

   It really doesn't matter what layer you selected because the Stop command applies to the entire timeline. Any animation on this timeline — the main movie timeline — will be stopped when the playhead reaches Frame 60.

4. **Open the Code Snippets panel and expand the Timeline Navigation folder.**

   The Code Snippets panel is intended to make it easier for non-programmers to add a certain level of interactivity to a Flash movie. Different types of common commands are available, grouped into logical sets or folders. Each item includes a plain-English name.

Expand the various folders to find the available commands.

Click these icons to find more information about the selected snippet.

You can find more information about a snippet by clicking the icons to the right of the selected item in the panel.

Clicking the ⓘ icon shows a description of the selected snippet.

Clicking the {} icon shows the code that will be added by the selected snippet.

5. Double-click Stop at this Frame in the Code Snippets panel.

The Actions panel opens and shows the code that is required to stop the timeline from playing more than once.

A new Actions layer is added at the top of the layer stack on the main timeline. A small "a" in the selected keyframe (Frame 60) indicates that code exists on that frame.

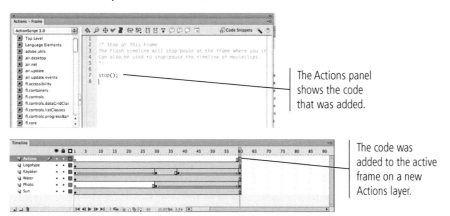

The Actions panel shows the code that was added.

The code was added to the active frame on a new Actions layer.

Note:

Although we are not going to go into any depth about code until Project 4: Ocean Animation, this is a fundamental requirement for many Flash animations. The Code Snippets panel makes it easy to add the necessary code without any programming knowledge.

6. Press Command-Return/Control-Enter to test the animation in the Flash Player window.

The kayaker and photo animations play once and stop; however, the two movie clip animations continue to loop. It is important to note that the stop command does not stop movie clip animations that are placed on the current timeline.

Remember, movie clip timelines are independent of other movie clips and of the main timeline; you have to add the stop command to each symbol timeline to prevent them from looping.

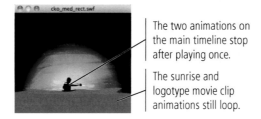

The two animations on the main timeline stop after playing once.

The sunrise and logotype movie clip animations still loop.

7. Close the Flash Player window and return to Flash.

8. In the Library panel, double-click the sun_mc icon to enter into the symbol.

9. Select Frame 30 in the timeline, then double-click Stop at this Frame in the Code Snippets panel.

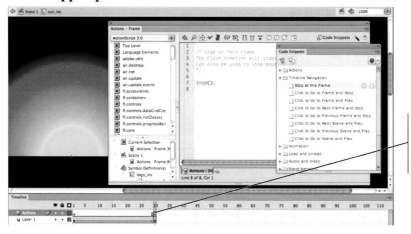

You are adding the Stop command to Frame 30 of the sun_mc symbol.

10. **In the Library panel, double-click the logo_mc symbol icon to enter into that symbol.**

When you use the Library panel to enter into a symbol, you don't need to return to the main movie Stage before entering into a different symbol. You can simply double-click the symbol icons to navigate from one symbol to another.

Double-click the symbol icon to edit that symbol without navigating back to the main Stage.

11. **Select Frame 60 in the timeline, then double-click Stop at this Frame in the Code Snippets panel.**

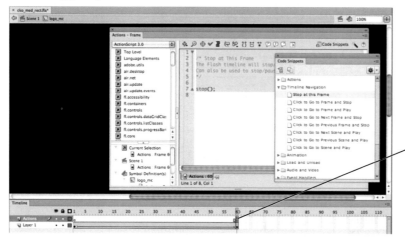

You are adding the Stop command to Frame 60 of the logo_mc symbol.

12. **Click Scene 1 in the Edit bar to return to the main Stage.**

13. **Press Command-Return/Control-Enter to test the movie in Flash Player.**

All four animations now play once and stop.

14. **Close the Flash Player window and return to Flash.**

15. **Save the file and continue to the next stage of the project.**

Stage 2 Working with Text

One of the more frustrating aspects of Web design is working with text elements; this is because the appearance of type is sometimes dependent on the available fonts on a user's computer. Flash movies are not subject to this limitation because used fonts can be embedded in the exported SWF file; you can use any font in a Flash file, and it will appear exactly as expected in the movie as long as it is embedded.

To do this, you can use the Properties panel Text Engine menu to create a Classic Text object. In this case, you create one of three specific types of text object:

- **Static text** is placed, kerned, aligned, and manually edited with the Text tool. This type of text does not change.

If you create a Classic Text object, you can also define the type of text.

Text tool

- **Dynamic text** is basically an area into which information can be read; this type of text object can be named and addressed by code, which means the content can be changed as the result of a certain action.

- **Input text** is a field in which users can type to submit information (as in an online form).

Flash CS6 also supports Text Layout Framework (TLF) text, which makes it very easy to add text to a movie by clicking and dragging with the Text tool. You have a wide range of formatting options that define the behavior of the frame, as well as the appearance of text within the frame. Using TLF text, you don't need to specify whether the object is static or dynamic. Rather, you can determine whether text in the area is Read-Only, Selectable, or Editable.

If you create a TLF Text object, you can define the behavior of the text.

Text tool

CREATE A NEW TEXT OBJECT

Your client wants a very simple text message added to the top of the ad. The message shouldn't appear until halfway through the animation, so you will again use blank keyframes to prevent the message from appearing until it should.

1. **With `cko_med_rect.fla` open in Flash, choose View>Magnification>Fit in Window.**

2. **Add a new layer named Text above the Logotype layer. Select the Text layer as the active layer.**

3. **Control/right-click Frame 37 on the Text layer and choose Insert Blank Keyframe in the contextual menu.**

   You are going to add the text object to this new blank keyframe. On Frames 1–36, the layer remains empty.

4. **Choose the Text tool in the Tools panel.**

5. **At the top of the Properties panel, choose TLF Text in the Text Engine menu and choose Read Only in the Text Type menu.**

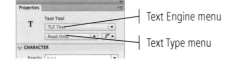

Text Engine menu

Text Type menu

6. **In the Character section of the Properties panel:**

   - **Change the Family menu to a sans-serif font such as Arial or Helvetica.**

   - **Choose a Bold, Black, or Heavy variation of the selected font in the Style menu.**

   - **Click the current Size link to access the field. Type 24 in the field and press Return/Enter.**

   - **Make sure the Leading menu is set to % and change the associated value to 90.**

   - **Click the Color swatch to open the Swatches panel. Choose white (#FFFFFF) as the text color, and make sure the Alpha value is 100%.**

   - **Choose Readability in the Anti-Alias menu.**

7. **In the Paragraph section of the Properties panel, click the Align to End option.**

   Any formatting you define before clicking with the Text tool will be applied in the new text area.

8. **Click near the top-left corner of the Stage and drag to create a rectangular text area in the top half of the Stage.**

   To include text in a movie, you first have to create an area to hold the text characters. As you drag, notice that dragging affects only the width of the rectangle. The area's minimum height is determined by the currently defined text formatting options.

   When you release the mouse button, the text area appears as a white box. When the insertion point is flashing, you can type in the box to add text.

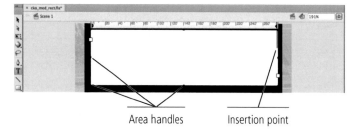

   Area handles   Insertion point

9. **Type Adventure Awaits!.**

The text appears in the text area you just created, using the character and paragraph formatting options that you already defined. (Because you chose white as the text color, the text appears gray in the area so you can see it as you type.)

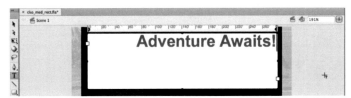

10. **Choose the Selection tool, and then expand the Position and Size options in the Properties panel. Make sure the Constrain icon to the left of the W and H fields is broken, then change the properties of the selected text area to:**

    X: 10   Y: 10
    W: 280   H: 50

When the Selection tool is active, you can drag the area handles but you can't change the text content in the area.

Break this icon to change the W and H fields independently (changing the object's aspect ratio).

Using the Position and Size properties, you can define specific numeric parameters for the area. You can also reposition a text area by clicking inside the area with the Selection tool and dragging to a new location, or resize the area by dragging any of the handles that appear on the outside edges of the object.

11. **In the Character section of the Properties panel, change the Size value to 40.**

When a text area is selected, changes to character and paragraph formatting apply to all text inside the area. You can change the formatting of only certain characters by highlighting the target characters with the Text tool before making changes in the Properties panel.

Overset text icon

The red plus sign icon (called the **overset text icon**) on the bottom-right corner of the area indicates that more text exists than will fit into the defined space. You can either change the size of the area to accommodate all of the text, or click the overset text icon to "load" the cursor with the rest of the text; once loaded with the overset text, clicking again on the Stage creates a second linked text area to contain the overset text.

12. **Click the bottom-center handle of the text area and drag down until all of the text is visible.**

When you drag a text area handle with the Selection tool, text formatting is not affected.

Click the handle and drag down
to change the area height.

13. **Click away from the text area to deselect it.**

14. **Save the file and continue to the next exercise.**

Controlling Text Properties

FLASH FOUNDATIONS

A number of text formatting options are available in the Properties panel when the Text tool is active and TLF text is selected in the Text Engine menu. (If you create a Classic Text object, many of these options are not available.)

Character Options

- **Family and Style menus.** As with any computer application, the font defines the appearance of the text. The entire list of fonts installed on your machine appears in the Family menu. The Style menu lists available variants of the selected font (Bold, Italic, etc.).

- **Embed button.** This button allows you to embed selected fonts or specific characters to ensure that text will appear on users' computers in the font you define.

- **Size.** The size of the text (in points) can be changed in the Font Size field or selected from the menu.

- **Leading.** This defines the distance between the bottom of one line of text and the bottom of the next line of text. You can define leading based on a percentage of the applied type size, or using a specific value.

- **Color.** This swatch changes the fill color of selected characters.

- **Letter Spacing (Tracking) field.** This option is used to increase or decrease the spacing between a selected range of characters.

- **Highlight.** This swatch defines the color immediately behind the selected text.

- **Auto Kern check box.** You can uncheck this option to prevent Flash from applying a font's default kerning values in static text.

- **Anti-aliasing menu.** This menu determines the level of anti-aliasing applied (if any).

- **Rotation menu.** This option changes the orientation of text within the area.

- **Type Style buttons.** These buttons change the style of text to (from left) underlined, strikethrough, superscript (such as the TH in 4th), and subscript (the 2 in H_2O).

Paragraph Options

- **Paragraph Alignment buttons.** Text can be aligned to the start (left), center, or end (right) of its containing area, or it can be justified so all lines fill the available horizontal space. If you use the Justify options on the right, you can also determine how the last line in the paragraph is formatted (start, centered, end, or forced to fill the horizontal space).

- **Margins.** These options define how far text is placed from the left and right edges of the containing area.

- **Indent.** This option defines how far the first line of the paragraph is indented from the left edge of the remaining text in the same paragraph.

- **Spacing.** These options define how much space is added above or below each paragraph.

- **Text Justify.** When using justified paragraph alignment, this menu determines how extra space is distributed: between individual letters or only between words.

FLASH FOUNDATIONS

When an existing text area is selected, other options become available for controlling the selected text.

Advanced Character Options

- **Link.** If you want selected text to act as a hyperlink, you can type the destination URL in this field.

- **Target.** This menu defines where the hyperlink target will open: in the same browser window (_self), in a new blank browser window (_blank), in the parent container of the SWF file (_parent), or in the main browser window that contains the frame with the link (_top).

- **Case.** This menu can be used to change text to all UPPER CASE, all lower case, or SMALL CAPS.

- **Digit Case.** This option can be used to apply Lining or Old Style character glyphs in place of regular numbers (if those options exist in the applied font).

- **Digit Width.** This option controls the horizontal spacing of number characters. Tabular forces all digits to occupy the same amount of space. Proportional allows each number to occupy only the space that is required for the digit shape.

- **Ligatures.** This option can be used to control the replacement of select character pairs (for example, fl or ff) with a single-character representation of those letters.

- **Break.** This menu can be used to force each selected character onto a new line (All), allow text to break naturally where necessary (Any), or prevent selected text from breaking across multiple lines (No Break).

- **Baseline Shift.** This option raises selected characters away from the invisible line on which the bottoms of characters rest.

- **Locale.** This menu defines the language that is used for selected text.

Container and Flow Options

- **Behavior.** This menu can force text to only a Single Line of text, in which all text appears on one line regardless of return characters; display Multiline wrapped text in which text will reflow in the container if you change its width; or Multiline No Wrap text, in which new lines require pressing Return/Enter as you type.

- **Max Chars.** If you define the text type as Editable, you can use this value to limit how much text can be typed in an area.

- **Columns.** You can use this option to create multiple columns in a single text area. If you use multiple columns, you can also define the gutter width, or the space between columns.

- **Padding.** These fields define how far text is placed from the edge of the containing area (sometimes called text inset, especially in applications such as Adobe InDesign).

- **Border Color** and **Background Color.** These swatches can be used to apply different colors to the text area.

- **1st Line Offset.** This menu can be used to adjust the position of the first line of text in the area, either by a specific distance or based on attributes of the applied font and formatting (Ascent and Line Height).

DEFINE FONT EMBEDDING

When you use text in a movie, the fonts you use must be available to other users who open your file. If not, the users' systems will substitute some font that is available — which can significantly change the appearance of your movie. To solve this potential problem, you can embed fonts into your movies so the required fonts are always available.

1. **With cko_med_rect.fla open, select the Text tool.**

2. **Click inside the heading text area to place the insertion point and reveal the formatting for that area.**

3. **In the Character section of the Properties panel, click the Embed button.**

   This button opens the Font Embedding dialog box. The currently applied font is automatically selected in the Family and Style menus.

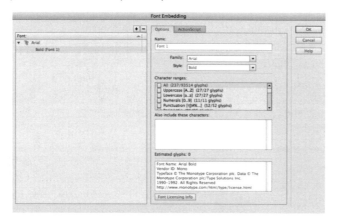

Note:

Keep in mind that embedding fonts adds to the resulting file size and increases the download time of Flash movies. The font outline (embedded character) is stored as part of the SWF file, so repeatedly using the same character of the same font in the movie does not increase the file size (the same theory as using instances of symbols). Keeping the number of fonts in a movie to a minimum ensures a faster-loading movie.

4. **Change the Name field to Heading Font.**

5. **In the Character Ranges list, check the Uppercase, Lowercase, and Punctuation options.**

   Remember, embedding characters from fonts increases the resulting file size. You know only letters and punctuation were used in this document, so you can limit the embedded characters to only these ranges rather than embedding every possible character of the font.

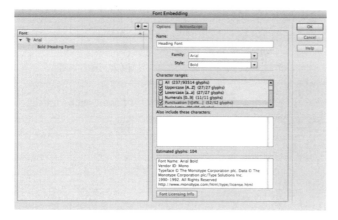

Note:

The item on the left side of the dialog box won't reflect the new name until you click away from the Name field.

6. **Click OK to close the dialog box.**

7. **Review the Library panel.**

   When you embed a font into the file, it is added to the file's library.

The embedded font is added to the file's library.

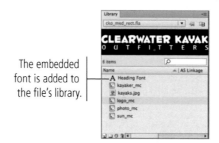

8. **Press Command-Return/Control-Enter to test the file in the Flash Player.**

9. **Read the resulting warning, then click OK.**

When you use TLF text in a file, it requires an external file called a Runtime Shared Library (RSL). Flash warns you about the issue when you test a file containing TLF text.

Note:

In the case of TLF text, the RSL describes a number of classes that apply to TLF text. Although the RSL relates to ActionScript coding methods, the RSL is still required even if you don't use code in the file.

An RSL stores assets that are common to multiple files. Other SWF files can all refer to the external RSL file, so the common assets don't need to be separately embedded in each SWF file. The RSL is downloaded once, and multiple SWFs can then refer to those shared assets without the need to constantly re-download the same assets. This allows SWF files to be smaller and more efficient.

When the SWF file runs, it first looks for the RSL on the local machine, then on the Adobe.com Web site, then in its own directory on the host server. In many cases, the download process for RSLs is virtually transparent. However, on devices that don't already have the RSL available (cached), users might be prompted to allow the SWF file Internet access to download the file from Adobe's site.

You can avoid this issue by embedding the RSL directly into your SWF file, which means it will automatically be available as soon as the SWF file is run. This does increase the file size, though, so you should carefully weigh your options and requirements before embedding the RSL into each SWF file.

Note:

You will learn how to embed an the RSL in the final exercise of this project.

10. **Click OK to dismiss the warning message.**

When the movie ends, you might notice a problem — the white text is difficult to read when the image is entirely opaque. You will solve this problem in the next exercise.

11. **Close the Flash Player window and return to Flash.**

12. **Save the file and continue to the next exercise.**

CONTROL OBJECT STACKING ORDER

If you completed Project 1: Corvette Artwork, you learned how to use layer stacking order to determine how various objects affect others on the Stage. It is not uncommon for designers to use individual layers for every single object in the file — which makes things much easier to find as long as you use descriptive layer names.

You should also understand, however, that stacking order applies to multiple objects that are created on the same layer. Drawing objects and symbols exist from bottom to top in the order they were created *on a single layer*. It is easy to create something in the wrong order, but fortunately, Flash makes it relatively easy to rearrange the stack.

Note:

Merge-drawing shapes are always created at the back of the stacking order, behind any other objects on the layer (symbol instances, object-drawing shapes, type areas, etc.).

1. **With `cko_med_rect.fla` open, select Frame 37 of the Text layer.**

2. **Choose the Rectangle tool, and make sure Object-Drawing mode is active.**

3. **In the Color panel, change the Stroke color to None. Open the Fill color palette and choose the white-to-black linear gradient swatch.**

4. **Click the left gradient stop to select it, then change the Alpha value to 0%.**

5. **Click the right gradient stop to select it, and drag it to the 50% point along the gradient. Change the stop color to #000033, and change the Alpha value to 50%.**

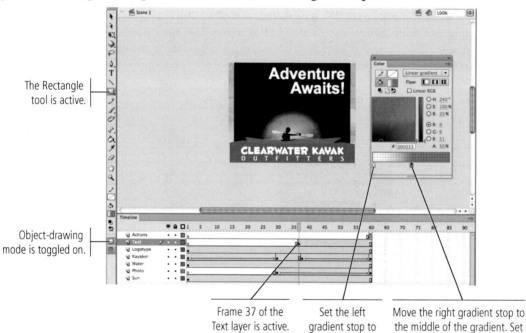

The Rectangle tool is active.

Object-drawing mode is toggled on.

Frame 37 of the Text layer is active.

Set the left gradient stop to 0% Alpha.

Move the right gradient stop to the middle of the gradient. Set it to #000033, 50% Alpha.

6. **With Frame 37 of the Text layer selected, click and drag to create an object that fills the top half of the Stage.**

The gradient defaults to fill the object from left to right.

The gradient goes from entirely transparent to 50% transparent (the Alpha value of each stop on the gradient).

7. **Choose the Gradient Transform tool in the Tools panel, then click the gradient-filled rectangle to reveal the gradient-editing handles.**

Gradient Transform tool

Rotation handle

Width handle

8. **Rotate the gradient 90° counterclockwise, then use the gradient width handle to make the gradient the same size as the object's height.**

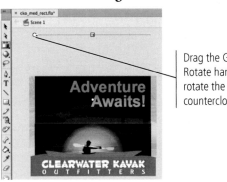

Drag the Gradient Rotate handle to rotate the gradient 90° counterclockwise...

... then drag the Gradient Width handle down until the gradient width is the same as the shape height.

9. **Using the Selection tool, select the gradient-filled rectangle.**

As you can see, the gradient-filled shape is currently on top of the text object. You want the text to appear on top of the gradient, so you need to rearrange the object stacking order.

10. **Choose Modify>Arrange>Send to Back.**

Options in the Modify>Arrange submenu control the stacking order of objects on the active layer. These options have no effect on the relative stacking of objects on different layers.

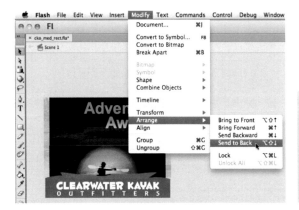

By placing the gradient-filled rectangle behind the text object but in front of the underlying layers, the white text will be more easily visible when the background photo is entirely visible at the end of the animation.

11. **Save the file and continue to the next stage of the project.**

Stage 3 Working with a Flash Project

The first ad required for this project is now complete. However, the entire job calls for three versions of the same ad, using different standard ad sizes but the same content in each ad. Rather than simply creating a new file and then copying the existing content into it, you can use the Project panel to define a new project, which will allow you to share assets across multiple files.

CREATE A FLASH PROJECT

The Flash Project panel provides an easy interface through which you can share assets across multiple Flash files. This is useful whenever you need to make multiple versions of the same movie, whether it is a series of ads that will be placed on the Internet, or an application interface that you need to optimize for multiple different devices.

1. **With `cko_med_rect.fla` open, open the Project panel (Window>Project).**

2. **Open the Projects menu and choose New Project.**

3. **In the resulting dialog box, type `Kayak Ads` in the Project Name field.**

4. **Click the Folder icon to the right of the Root Folder field to open a navigation dialog box.**

   You can't drag this navigation box away from the panel; if necessary enlarge the Project panel until you can see the dialog box.

   The root folder is the folder on your computer that contains (or will contain) the files that are part of the project.

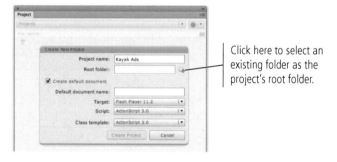

Click here to select an existing folder as the project's root folder.

5. **Macintosh: In the Choose a Folder dialog box, navigate to your WIP>Kayaks folder and click Choose.**

   Windows: In the Browse for Folder dialog box, navigate to your WIP>Kayaks folder and click OK.

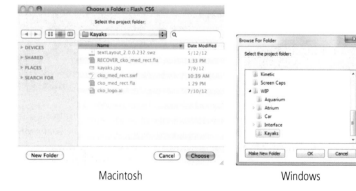

Macintosh   Windows

6. **In the Create New Project dialog box, uncheck the Create Default Document option.**

   When Create Default Document is checked, Flash automatically generates a new file that will be placed in the defined root folder. In this case, you already have a file that you want to include in the project, and you will create other files based on the existing ad; you don't need to check this option.

   Uncheck this option.

7. **Click Create Project.**

   When the process is complete, all Flash files in the WIP>Kayaks folder appear in the Projects panel. The tab for the open document now shows that the open file is part of the Kayak Ads project file. The AuthortimeSharedAssets.fla file stores assets that you want to share across multiple files.

   This file is now part of the Kayak Ads project.

   Note:

   The SWF file is the result of testing the file in the Flash Player window. You will overwrite it later when you publish your files.

8. **Open the Library panel for the active file.**

   When a file is part of a Flash Project, the Library panel includes an option to share specific symbols.

9. **Check the Link option for all four symbols in the file.**

   After you check an item, you might notice a brief "Saving" message. When you opt to share a symbol, that symbol is copied into the AuthortimeSharedAssets.fla file; any changes you make to the symbol are actually made inside that authortime file, and then propagated out to placed instances in all files within the project.

   You do not need to share all symbols in every project file. In this case, however, you do want all pieces of the ad to be the same in all versions, so you are sharing all four symbols.

   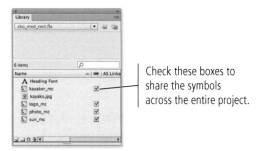

   Check these boxes to share the symbols across the entire project.

10. **Save the file, then continue to the next exercise.**

SCALE CONTENT TO DOCUMENT PROPERTIES

To complete the entire assignment, you need two more versions of the ad: one is a slightly larger rectangle size (336 × 280 px), and one is a 250 × 250-pixel square.

You could use the New Document dialog box to create the file using the built-in templates, then copy and paste all of the necessary content from one file to another. However, that process is time-consuming and introduces the potential for error since you need to copy the entire timeline as well as the objects on the Stage.

In this exercise, you will use the existing file as the basis for the other two files.

1. **With `cko_med_rect.fla` open, make sure nothing is selected on the Stage and the Selection tool is active.**

2. **In the Properties panel, click the Edit Document Properties button.**

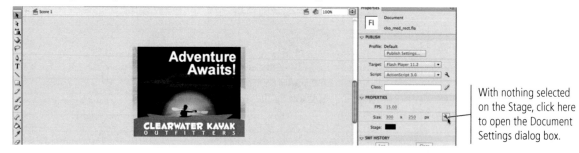

With nothing selected on the Stage, click here to open the Document Settings dialog box.

3. **In the resulting dialog box, change the Width dimension to 336 px and change the Height dimension to 280 px. Check the Scale Content with Stage option, then click OK.**

Make sure this option is checked.

Because the new file dimensions have the same width-to-height aspect ratio (6:5), the objects in the new file are easily scaled up, and require no further manipulation to function properly. By using the existing file as the basis of the new one, the Scale Content with Stage option made the entire process possible in only a few clicks.

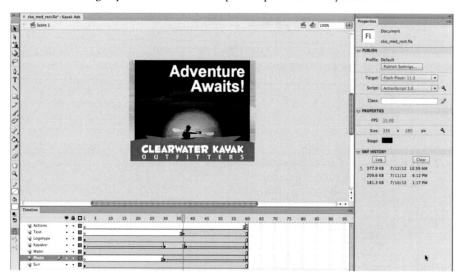

4. **Save the file as `cko_lg_rect.fla` in your WIP>Kayaks folder.**

Because you saved the file in the defined root folder of the Kayak Ads project, the new file should now appear in the Project panel as part of the project file.

5. **If you don't see the new file in the Project panel, open the panel Options menu and choose Refresh.**

Click here to access the panel Options menu.

6. **Save and close the Flash file, then continue to the next exercise.**

MANUALLY ADJUST CONTENT TO DOCUMENT PROPERTIES

In the last exercise, you created a new file by adjusting the document properties of an existing file, allowing Flash to scale objects as necessary to meet the new document size. Because the new file size had the same aspect ratio, you did not need to make any further changes other than saving the file with a new name.

In many cases, the new file size will not have the same aspect ratio as the original. In this situation, you will have to make some manual adjustments to keep the content in the same general position as in the original.

1. **In the Project panel, double-click `cko_med_rect.fla` to open that file.**

You can use the Project panel to open any file in the active project.

2. **With `cko_med_rect.fla` open, deselect everything on the Stage and then click the Edit Document Properties button in the Properties panel.**

3. **In the resulting dialog box, change the Width and Height dimensions to 250 px. Check the Scale Content with Stage option, then click OK.**

In this case, the new file does not use the same aspect ratio as the original. As you can see, Flash is not able to interpret the necessary positions of all elements in the file. Although this is a good start, you still need to make some adjustments manually.

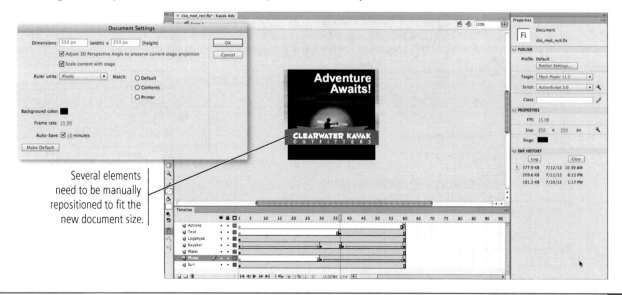

Several elements need to be manually repositioned to fit the new document size.

4. **Save the file as `cko_square.fla` in your WIP>Kayaks folder.**

   By immediately saving the new file with a different name, you avoid accidentally changing something in the wrong file.

5. **Click Frame 1 in the timeline to move the playhead to the beginning of the movie.**

6. **In the Timeline panel, Command/Control-click to select the Logotype, Kayaker, Water, and Sun layers.**

   You need to move the content on all four of these layers down to the bottom of the adjusted Stage. You don't want to move the Photo layer content because you want it to remain attached to the top of the Stage.

7. **Using the Selection tool, click any of the selected objects on the Stage, press Shift, and drag down until the water aligns with the bottom of the Stage.**

   By moving the content on all four layers at once, you maintain the same relative positions between the selected objects.

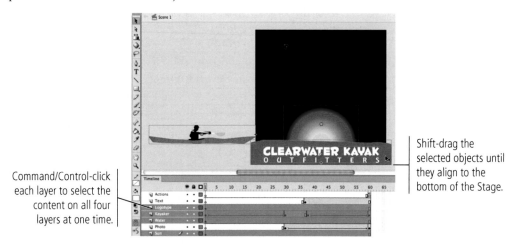

Command/Control-click each layer to select the content on all four layers at one time.

Shift-drag the selected objects until they align to the bottom of the Stage.

8. **Press Return/Enter to play the movie on the Stage.**

   You should notice the kayaker symbol instance moves up and away from the water as the playhead progresses. If you review the timeline, remember that this layer has four separate keyframes, each of which define a specific position in a classic tween. To keep the kayaker paddling straight across the Stage, you have to change the instance's position on each keyframe in the layer.

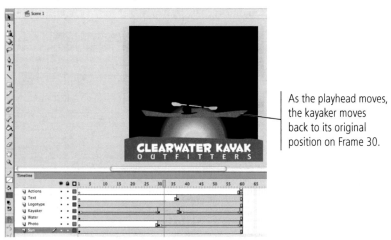

As the playhead moves, the kayaker moves back to its original position on Frame 30.

9. **Return the playhead to Frame 1 on the timeline, then click the kayaker symbol instance to select it.**

10. **In the Properties panel, note the instance's Y position.**

In our example, the Y position is 193.90. If yours is different, you should use the exact value from your file in the following steps. The point is to make the instance's Y (vertical) position consistent in all keyframes.

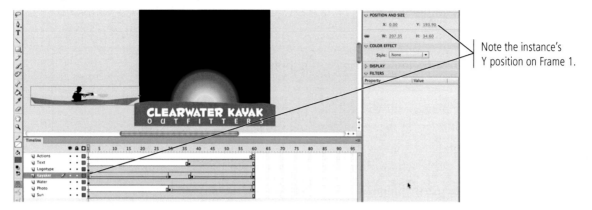

Note the instance's Y position on Frame 1.

11. **Click Frame 30 of the Kayaker layer, then click the symbol instance on the Stage to select it.**

12. **In the Properties panel, change the Y position to the same value you noted in Step 10.**

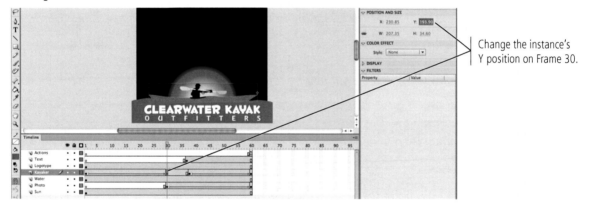

Change the instance's Y position on Frame 30.

13. **Repeat Steps 11–12 for the two remaining keyframes on the Kayaker layer.**

When you get to Frame 60, you should notice another problem — the photo no longer fills the background area. (This was harder to see when the image was entirely or semitransparent). You now need to adjust the photo to fill the space.

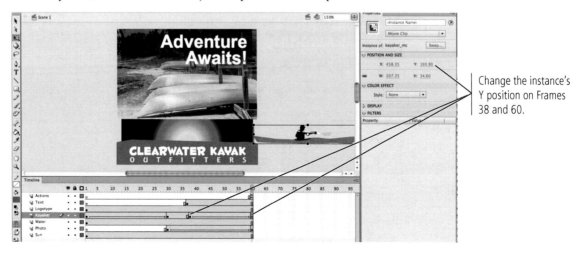

Change the instance's Y position on Frames 38 and 60.

14. Select Frame 30 on the Photo layer, then select the symbol instance on the Stage.

The first keyframe (Frame 1) on the Photo layer is blank; the image doesn't exist on that frame, so you don't need to edit that keyframe.

Remember, this image is actually an instance of a symbol, which you created from the original bitmap image.

15. Choose the Free Transform tool, then move the transformation point of the instance to the top center handle.

16. Using the Transform panel, scale the instance proportionally until the bottom edge is just hidden by the top edge of the water shape.

Be careful when you scale bitmap images, especially making them larger. Bitmap images have a fixed resolution, which means enlarging them can significantly reduce the quality.

If you remember the first part of this project, you actually reduced the image before creating the symbol instance. Enlarging this particular instance above 100% still keeps the image smaller than the original bitmap's physical dimensions, so you should not see any significant lack of quality in this case.

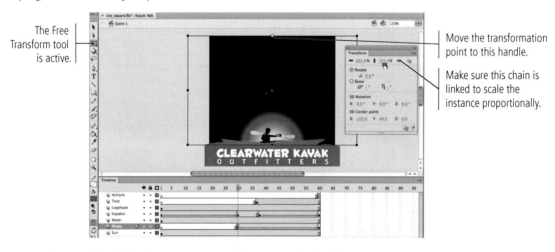

The Free Transform tool is active.

Move the transformation point to this handle.

Make sure this chain is linked to scale the instance proportionally.

17. Repeat Steps 14–16 for the instance on the Frame 60 keyframe.

You should apply the exact same transformation on Frame 60 as you did on Frame 30. Rather than scrubbing the values in the Transform panel, you can click one of the existing values to enter the field, constrain the two dimensions, and then type the exact same value that you applied in Step 16.

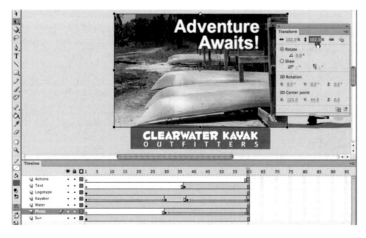

18. Save the file and continue to the next exercise.

EDIT A SHARED SYMBOL

As we explained earlier, changes in shared symbols propagate to all instances in the project's files — one of the benefits to using the Project panel. In this exercise, you will edit a shared symbol in the project file so the change is applied to every instance in any file in the project.

1. **With `cko_square.fla` open, double-click the kayaker_mc symbol icon in the Library panel to edit that symbol.**

   The symbol Stage reflects the defined background color of the open file; as you can see, it is difficult to see the black shape of the kayaker's body against the black background. You are going to change the shape's color to be more easily visible against the background.

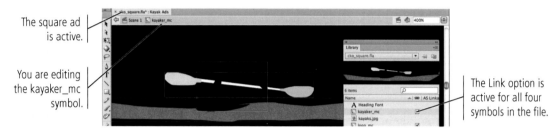

The square ad is active.

You are editing the kayaker_mc symbol.

The Link option is active for all four symbols in the file.

2. **Using the Selection tool, click away from all objects on the Stage to deselect them, then click the shape of the kayaker's body to select it.**

3. **Using the Properties or Color panel, change the Fill and Stroke colors of the selected drawing object to #666666.**

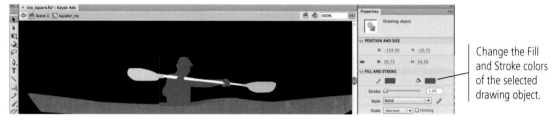

Change the Fill and Stroke colors of the selected drawing object.

4. **Click Scene 1 in the Edit bar to return to the main Stage.**

5. **Save the file and close it.**

6. **In the Project panel, double-click the `cko_med_rect.fla` file to open it.**

   Remember, earlier you checked the Link option for the kayaker_mc symbol, which means it is a shared asset in the entire project. The kayaker_mc symbol instance in the medium rectangle ad now shows the same change that you made when the square ad file was open.

The symbol instance is changed in the medium rectangle ad as well.

7. **Save the active file, than continue to the next exercise.**

   You do not have to open each project file for the changes to apply. When you publish the files in the next exercise, you will see the same change in the large rectangle ad.

PUBLISH THE AD FILES

Exporting a document refers to publishing it in a form that can be viewed in another application. The File>Export menu has three options: Export Image, Export Selection, and Export Movie.

- If you choose **Export Image**, you can save your file as a static graphic (with no animation) in formats such as GIF or JPEG.

- **Export Selection** saves the active selection as a Flash XML Graphics (FXG) file, which is an XML-based graphics file format for vector graphics.

- **Export Movie** allows you to create a file (or sequence of files) that includes animation, which can be placed into an HTML document created in another application. A number of formats are available in the Format/ Save As Type menu. Each format has distinct uses, advantages, and disadvantages.

  - **SWF Movie.** A Flash movie file can be placed into an HTML file, or it can be used in another Flash application.

  - **QuickTime.** Selecting this option opens the Export QuickTime MOV dialog box, where you can set options such as dimensions, Stage color, last frame reached, time elapsed, and others.

  - **Animated GIF.** Files exported in this format preserve all the animations in a single file. You can specify various options, such as dots per inch (dpi), image area, colors, and animation for the resulting GIF file. Setting the dpi is the same as setting the dimensions of the image; the number of times the animation needs to repeat can also be defined (0 creates an endless loop).

  - **JPEG Sequence.** Selecting this option allows the file to be exported in the JPEG format. The Match Movie option matches the size of the exported file with that of the original document. When you use any of the sequence options, each frame of the movie is exported as a separate image.

  - **GIF Sequence.** This format exports the files in GIF format, except the files are generated in a sequence for each frame animation. The animated GIF format exports a single file that contains all of the animations; this option generates a sequence of files.

  - **PNG Sequence.** This option saves the files in the Portable Network Graphics (PNG) format, which supports transparency for objects that might need to be placed on various backgrounds. You can specify options such as dimension, resolution, colors, and filters.

  - **Windows AVI (Windows only).** This option exports the movie in the Windows format and results in huge file sizes. Choosing this option opens the Export Windows AVI dialog box, where you can set options such as dimensions and quality of the movie.

  - **WAV Audio (Windows only).** This option exports only the sound file from the current document. You can specify the sound frequency and format of the sound file (stereo, mono, etc.), as well as ignore event sounds if you don't want them in the exported file.

You can also use the Publish option (File>Publish) to generate a Flash SWF file, as well as a number of other formats. When you publish a file, the resulting output is based on the active options in the Publish Settings dialog box.

1. **With `cko_med_rect.fla` open, choose File>Publish Settings.**

   Rather than simply choose File>Publish, you are using this dialog box to first review the settings that will apply when you publish the files.

2. **In the Publish options on the left side of the Publish Settings dialog box, make sure the Flash (.swf) option is selected.**

3. **Uncheck all other options in the left side of the dialog box.**

   If the HTML Wrapper option is checked, the publish process also generates an HTML file that includes the necessary code for opening the SWF file in a browser window. Because these ads will be distributed for insertion into other sites, the HTML option is not necessary.

Note:

The Win Projector and Mac Projector options generate platform-specific executable applications, which can function without the Flash Player.

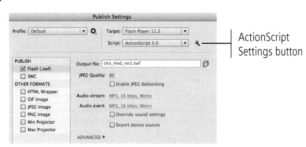

ActionScript Settings button

4. **At the top of the dialog box, click the ActionScript Settings button.**

5. **In the resulting Advanced ActionScript 3.0 Settings dialog box, make the Library Path options visible in the lower half of the dialog box.**

6. **Choose Merged into Code in the Default Linkage menu.**

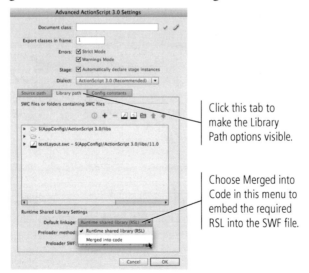

Click this tab to make the Library Path options visible.

Choose Merged into Code in this menu to embed the required RSL into the SWF file.

As we explained earlier, TLF text objects require an RSL file to function properly. To avoid problems with the file not being available, you are merging the RSL code directly into the published SWF file.

Keep in mind that this will significantly increase the size of each resulting SWF file; we are simply showing you how to embed the code for illustration purposes.

FLASH FOUNDATIONS

The Publish Settings dialog box (File>Publish Settings) contains all the necessary settings for publishing files from Flash.

- The **Profile** menu lists all available saved profiles, which save defined publish settings for easier access. You can use the attached Profile Options button to create, import, export, and manage saved profiles.

- The **Player** menu allows you to export the Flash movie to be compatible with an earlier version of the Flash Player, Adobe AIR 1.1, or various versions of Flash Lite.

- **Script** specifies which version of ActionScript (1, 2, or 3) is or will be used in the file. Clicking the ActionScript Settings button opens a dialog box where you can specify the path of an external ActionScript file.

The **Publish** list on the left side of the dialog box lists the available formats that can be exported; the Flash (.swf) and the HTML Wrapper formats are selected by default. When you click a specific format in the Publish list, options related to that format appear on the right side of the dialog box.

When **Flash (.swf)** is selected in the Publish list, options on the right side of the dialog box determine how the animation is exported.

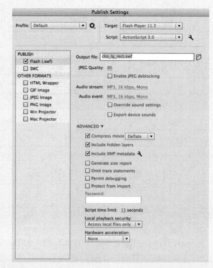

- The **Output File** field shows the default file name, based on the FLA file name, that will be used for the exported file. By default, exported files are created in the same location as the FLA file you are exporting; you can click the folder icon next to a file name to specify a different publishing location. (This is available for all publishing formats.)

- **JPEG Quality** specifies the quality of exported JPEG images. Lower values result in smaller file sizes, but also lower image quality; as the quality increases, so does the size of the file. The **Enable JPEG Deblocking** option helps smooth the appearance of JPEG files with very high levels of compression.

- The **Audio Stream** and **Audio Event** options show the current sound export settings for stream and event sounds (respectively). Clicking either link opens the Sound Settings dialog box, where you can change the compression format, quality, and bit rate options for each type of sound.

(A streaming sound plays as soon as enough data is downloaded; the sound stops playing as soon as the movie stops. An event sound does not play until it downloads completely, and it continues to play until explicitly stopped.)

- If **Override Sound Settings** is checked, the default options in the Audio Stream and Audio Event settings override any settings that are defined for individual sound files in the file's library.

- **Export Device Sounds** allows you to export sounds for mobile devices, encoded in the device's native format rather than a standard format.

- **Compress Movie** reduces the size of the exported file, which also reduces download time. Compressed files can only be played in Flash Player 6 or later.

- **Include Hidden Layers** allows you to export hidden layer information in the exported file.

- **Include XMP Metadata** exports any metadata that is entered in the File Info dialog box (File>File Info).

- **Generate Size Report** creates a text file with information about the amount of data in the final Flash content.

- **Omit Trace Statements** allows Flash to ignore trace options, which are ActionScript functions that display the results of certain code in the Output panel. (You'll use a Trace function in Project 6: Gopher Golf Game.)

- **Permit Debugging** allows you to debug your SWF file and allows other users to debug your SWF file remotely. You can also define a password, which other users will have to enter in order to debug the file.

- **Protect from Import** prevents others from importing your SWF file. This option also allows you to protect your SWF file with a password.

- **Password** is activated if you select the Protect from Import or the Permit Debugging option. You can specify a password that other users must enter in order to import the file or debug the movie.

- **Local Playback Security** options provide security for your application. If you select Access Local Files Only, your SWF file can interact only on the local machine. If you choose Access Network Only, your SWF file can communicate with resources only on the network and not on the local machine.

- **Hardware Acceleration** options can be used to speed up the graphics performance of the exported movie.

FLASH FOUNDATIONS

HTML Wrapper options relate to publishing a Flash document on the Web. In this case, you might need an HTML file that will embed your SWF file.

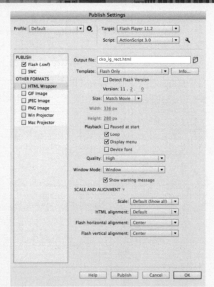

- The **Templates** menu contains various Flash templates in which an HTML file can be published. Selecting a template and then clicking the Info button shows a dialog box with information about the selected template.

- If you select **Detect Flash Version**, you can use the Version fields to define which is required. Detection code is embedded in the resulting file to determine if a user has the required version. If not, a link is provided to download the latest version of the Flash player. (Some templates do not support the Flash detection code.)

- The **Size** options define the dimensions of the resulting HTML file. You can choose Match Movie to use the size of the Flash Stage, or define specific width and height (in pixels or percent).

- The **Paused at Start** option keeps the SWF file from being played unless the user clicks to initiate the movie.

- The **Loop** option causes the Flash content to repeat after it reaches the final frame, so the movie plays in a continuous loop.

- The **Display Menu** option enables a shortcut when the user Control/right-clicks the SWF file in the browser. Deselecting this option shows only About Flash in the shortcut menu.

- Selecting the **Device Font** option displays users' system fonts instead of the fonts used in the SWF file, if those fonts are unavailable in the user's system.

- **Quality** specifies the quality of the SWF content embedded in the HTML file. Auto Low gives preference to document loading rather than quality, but also tries to improve the quality of the SWF file. Auto High treats loading and quality equally; when the loading speed is reduced, quality is compromised. The remaining three options are self-explanatory; lower quality settings mean higher compression, smaller file sizes, and faster download times.

- **Window Mode** sets the value of the wmode attribute in the object and embed HTML tags. Window does not embed window-related attributes in the HTML tags; the background of the Flash content is opaque. Opaque Windowless sets the background of the Flash content to opaque, which allows HTML content to appear on top of Flash content. Transparent Windowless sets the back-ground of the Flash content to transparent.

- The **Show Warning Messages** option displays all warning and error messages whenever a setting for publishing the content is incorrect.

- **Scale** controls the display of Flash content when you change the dimension of the Flash content in the HTML file. Default (Show All) displays the entire document in the specified area. No Border fits the document in the specified area and maintains the quality of the SWF file by avoiding distortion. Exact Fit fits the entire SWF file in the specified area, but compromises the quality of the SWF file. No Scale prevents the Flash content from being scaled in the HTML file.

- **HTML Alignment** aligns the Flash content in the browser window. The Default option displays the SWF file in the center of the browser; you can also choose Left, Right, Top, or Bottom.

- **Flash Vertical and Horizontal Alignment** set the alignment of Flash content within the HTML file.

7. **Click OK to close the Advanced ActionScript 3.0 Settings dialog box, then click OK again to close the Publish Settings dialog box.**

8. **Save the open file (cko_med_rect.fla) and close it.**

9. **Repeat Steps 1–8 for the other two ad files in the project.**

   Publish settings are file-specific, so you have to define these options for each file that will be published.

10. **If you don't see all three ad files in the Project panel, choose Refresh in the panel Options menu.**

11. **In the Project panel, check the Publish Items boxes for the three ad files.**

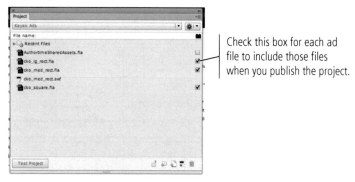

Check this box for each ad file to include those files when you publish the project.

12. **Open the panel Options menu and choose Publish Project.**

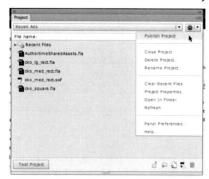

When the process is complete, all three SWF files now appear in the panel — one for each ad file you created.

13. **Close the Project panel.**

1. A(n) _____ timeline functions independently of other symbols in the same file.

2. _____ are required when an object needs to change in some way at a given point in time.

3. The _____ can be used to scale an object proportionally with respect to the object's defined transformation point.

4. A(n) _____ can be used if you need to change an object's color over time.

5. A(n) _____ is identified by blue frames and an arrow in the timeline.

6. _____ is simple text that does not change.

7. _____ is essentially a text area that can be populated with the contents of an external file.

8. _____ solves the potential problem of used fonts not being available on a user's computer.

9. The _____ option allows you to change objects' size based on the edited document settings.

10. _____ is the default extension for exported file movies.

1. Briefly explain how a movie clip symbol relates to the primary timeline of a file.

2. Briefly explain one method for creating multiple versions of the same file, with different file sizes.

3. Briefly explain the purpose of the AuthortimeSharedAssets.fla file in a Flash project.

Portfolio Builder Project

Use what you learned in this project to complete the following freeform exercise.
Carefully read the art director and client comments, then create your own design to meet the needs of the project.
Use the space below to sketch ideas; when finished, write a brief explanation of your reasoning behind your final design.

art director comments

Our agency has been hired to create a series of animated videos explaining scientific principles for children.

To complete this project, you should:

❏ Research the topics you are going to model and determine what (if any) data will be required to create a scientifically accurate illustration.

❏ Create or locate images or graphics that will result in a "kid-friendly" learning experience.

❏ Develop the animations so they can be placed into an existing Web site, using a 600 × 800 file size.

client comments

We would like you to create a series of videos over the next year (as the grant funds become available). The first one we want is an illustration of gravity. We're not entirely sure how it should look or function. We kind of like the legend of Isaac Newton sitting under an apple tree, when an apple fell on his head. If you could figure out how to make that work, great. If not, we're happy to consider other solutions.

As more funds become available, we're also going to want movies to illustrate other scientific principles. Our current list includes tectonic plate movement, tidal patterns, volcanic eruptions, and friction.

It will be easier to secure the secondary grants for later projects if we can include a specific plan for the different programs. Once you finish the gravity illustration (which we already have the money for), can you sketch out plans for at least two others?

project justification

_____

_____

_____

_____

_____

_____

_____

_____

_____

_____

_____

_____

_____

Project Summary

The ability to control object shape and movement is one of the most important functions in designing animations. In this project, you used a number of basic techniques for animating object properties, including size, color, and position. As you complete the next projects in this book, you will expand on the knowledge from this chapter, learning new ways to animate multiple properties at once. The skills and knowledge from this project, however, apply to any animation — understanding frame rate, keyframes, and timeline independence are essential to being a successful animator.

This project also introduced the concept of adding text to a movie. Many Flash projects will involve some text, even if that text is eventually converted into drawing objects; you now know how to create and format text to communicate a client's message directly within a Flash file.

Finally, you learned techniques for creating multiple variations of the same file, using built-in tools to automate as much of the work as possible. These techniques and skills can be helpful whenever you need multiple versions of a single file — whether you need to create multiple different-sized ads or different versions for various mobile device sizes.

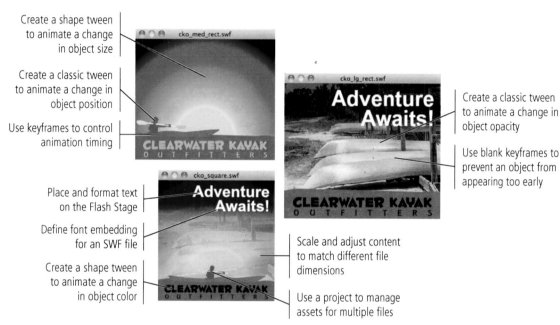

Create a shape tween to animate a change in object size

Create a classic tween to animate a change in object position

Use keyframes to control animation timing

Place and format text on the Flash Stage

Define font embedding for an SWF file

Create a shape tween to animate a change in object color

Create a classic tween to animate a change in object opacity

Use blank keyframes to prevent an object from appearing too early

Scale and adjust content to match different file dimensions

Use a project to manage assets for multiple files

Ocean Animation

Your client, Bay Ocean Preserve, wants to add an interactive animation to the kids' side of its Web site. As part of the Flash development team, your job is to build the required animations, and then add the necessary controls to make the buttons function as expected.

This project incorporates the following skills:

❑ Importing and managing artwork from Adobe Photoshop

❑ Importing symbols from external Flash file libraries

❑ Understanding the different types of Flash symbols

❑ Building frame-by-frame animations

❑ Creating motion tweens to animate various object properties

❑ Creating an animated armature

❑ Animating in three dimensions

❑ Preparing symbol instances for scripting

❑ Adding basic button controls to instances on the Stage

Project Meeting

client comments

Our organization focuses on natural resource conservation and habitat preservation on the central California coast. This area is home to a number of endangered species, and we work to educate people about observing those creatures without interacting and interfering with them.

We've been told that some kind of interactivity will be an important part of capturing a younger audience. Although we think cartoon fish dancing across the screen would minimize the seriousness of our message, we understand that we have to do something to make the site more interesting for children.

We were thinking about an "aquarium" screen saver we used to have, and we thought that kind of thing would be a good balance between user interactivity and pointless arcade games.

art director comments

Since the client clearly wants to avoid a cartoon look, I had the staff artist create some fish and other illustrations that are fairly realistic. I also found a good photo of a turtle that will work well with the other elements.

One of the animations — the swaying kelp forest — should play constantly, and will not be controlled by buttons.

Three animations will be controlled by the buttons. First, a fish hiding in a cave will blow bubbles. Second, a turtle will swim across the scene and get bigger, to create the effect of swimming closer. Finally, a school of different fish will swim in the other direction, in a "train" across the scene.

One other animation — the organization's logo — will play as soon as the file opens, and then not again until the entire file is reset.

The programming for this isn't very complicated, so you should be able to create it with Flash's built-in Code Snippets.

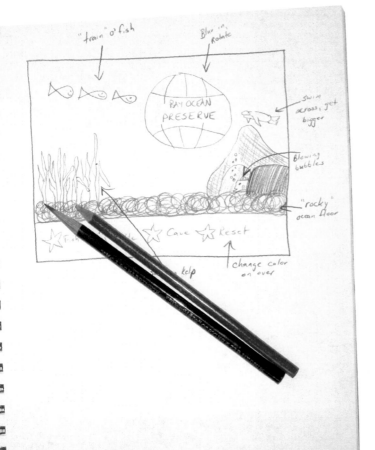

project objectives

To complete this project, you will:

- ❏ Create symbols from imported Photoshop artwork
- ❏ Import symbols from other Flash files
- ❏ Place and manage instances of symbols on the Stage, and control the visual properties of those symbols
- ❏ Control timing using keyframes
- ❏ Create animated movie clip symbols
- ❏ Generate motion tweens to animate changes in object properties
- ❏ Use the Motion Editor to numerically control properties at specific points in time
- ❏ Animate a bone armature
- ❏ Animate an object in three-dimensional space
- ❏ Add interactivity to button symbols

Stage 1 Importing Bitmaps and Symbols

In this project, much of the artwork was created in Adobe Photoshop — a common workflow. It's important to understand that artwork from a Photoshop file is imported into Flash as bitmap objects, which (as you learned in Project 1: Corvette Artwork) are raster images that can result in large file sizes. Fortunately, the Flash symbol infrastructure means that objects in a file's library are downloaded only once; you can use multiple instances of a symbol without increasing overall file size.

You should also keep in mind that the quality of a bitmap object is defined by its resolution. Bitmap objects can typically be reduced in size, but enlarging them much beyond 100% could significantly reduce the image quality.

Finally, it's important to realize that Flash is designed to create files that will be viewed on a digital screen. Flash recognizes the actual number of pixels in a bitmap image rather than the defined pixels per inch (ppi). If you import a 3″ × 3″ bitmap image that is saved at 300 ppi (typical of print-quality images), that image is 900 pixels × 900 pixels high. In Flash, the image is still 900 pixels × 900 pixels, but those same pixels occupy 12.5″ at a typical screen resolution of 72 ppi.

IMPORT ADOBE PHOTOSHOP ARTWORK

Importing a Photoshop file to the Flash Stage is very similar to importing an Illustrator file. Because of the different nature of the two applications, however, you have fewer options when you work with Photoshop files.

1. **Download FL6_RF_Project4.zip from the Student Files Web page.**

2. **Expand the ZIP archive in your WIP folder (Macintosh) or copy the archive contents into your WIP folder (Windows).**

   This results in a folder named **Aquarium**, which contains the files you need for this project. You should also use this folder to save the files you create in this project.

3. **In Flash, create a new Flash document for ActionScript 3.0.**

4. **Choose File>Save. Save the file in your WIP>Aquarium folder as a Flash file named ocean.fla.**

5. **Open the Flash Preferences dialog box and click PSD File Importer in the list of categories.**

6. **Choose Lossless in the Compression menu at the bottom of the dialog box, and click OK.**

   When you import a Photoshop file, you can define the compression settings for each resulting bitmap object. By default, Flash applies lossy compression, which can create smaller file sizes but can also degrade the quality of resulting images. By changing the setting in the Preferences dialog box, you are changing the default setting that will be applied to all bitmap objects that are created when you import a Photoshop file to the Flash Stage. This means you won't have to change the option for individual objects.

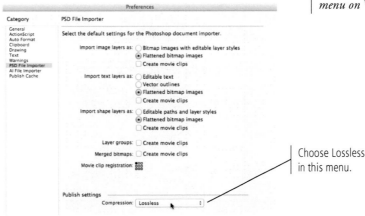

Choose Lossless in this menu.

Note:

*Learn more about Adobe Photoshop in the companion book of this series, **Adobe Photoshop CS6: The Professional Portfolio**.*

Note:

Preferences are accessed in the Flash menu on Macintosh or the Edit menu on Windows.

7. **Choose File>Import>Import to Stage. Navigate to the file ocean.psd in your WIP>Aquarium folder and click Open/Import.**

8. **In the resulting dialog box, click the first item in the list of Photoshop layers, and then review the options on the right.**

   Flash recognizes individual layers in the native Photoshop file; objects on each individual layer will be imported as separate bitmap objects. For each layer/bitmap object, you can choose to automatically create a movie clip instance from the imported artwork.

Each layer in the Photoshop file will be imported as a separate bitmap object.

Layer groups are recognized, and can be maintained in the imported artwork.

If you select multiple layers in the list, you can merge those layers into a single bitmap object.

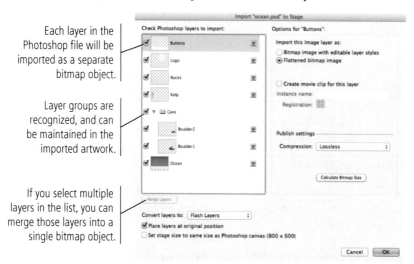

Note:

When artwork is created in Photoshop, each object that needs to be managed separately in Flash should be created on a separate Photoshop layer.

Photoshop File Import Preferences

FLASH FOUNDATIONS

You can define the default options for importing native Photoshop files in the PSD File Importer pane of the Preferences dialog box. (Remember, preferences are accessed in the Flash menu on Macintosh and in the Edit menu on Windows.) Most of these options will become clearer as you work through this project and learn more about the different types of symbols. In some cases, the options refer to specific features in the native application.

- The **Import Image Layers As** options determine whether applied layer styles are editable (if possible) in the Flash file. Photoshop layer styles include blending modes, drop shadows, beveling, and others.

- The **Import Text Layers As** options determine how Photoshop text layers are managed. These options are the same as the Import Text As options for importing Illustrator files. The default here, however, is to import Photoshop text layers as flattened bitmap images, which means you can't edit the text or the vector letter shapes.

- The **Import Shape Layers As** options determine how vector-based shape layers in the Photoshop file import into Flash. If you choose Editable Paths and Layer Styles, vector shapes in the Photoshop file are imported as vector paths; styles applied to the shape layer in Photoshop are imported as styles in the Flash file (if those styles are supported by Flash). If you choose Flattened Bitmap Images, Photoshop shape layers are flattened in Flash, which means you cannot access or edit the vector information within the Flash file.

- **Layer Groups** and **Merged Bitmaps** in the native Photoshop file can be automatically converted to a single movie clip if these options are checked.

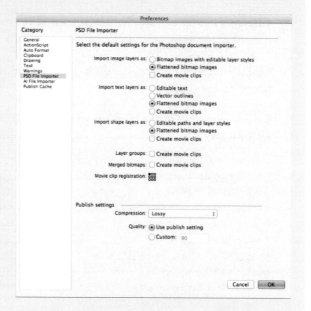

- **Movie Clip Registration** defines the default registration options for any objects you import as movie clip symbols.

- **Publish Settings** defines the type of compression applied and the quality of the resulting files.

9. **Click the Cave item in the list of layers and review the options.**

   If a Photoshop file includes a layer group, the group is recognized in the Import dialog box. You can choose to create a single movie clip from the entire layer group, and name the instance that is placed on the stage.

10. **Check the Create Movie Clip for this Layer option. In the Instance Name field, type ca**_**e_mc, and choose the bottom-right registration option.**

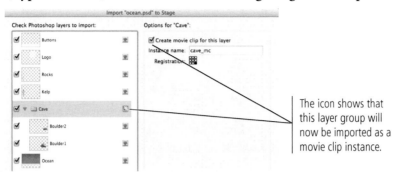

The icon shows that this layer group will now be imported as a movie clip instance.

Note:

If you don't choose the Create Movie Clip option, the layer group is maintained in the imported artwork as a layer group in the Flash timeline.

11. **Make sure Flash Layers is selected in the Convert Layers To menu. Check the options to place layers at their original position, and to set the stage size to match the Photoshop canvas.**

    These options have the same general purpose as the similar options when you import a native Illustrator file. (The Photoshop document area is called the canvas; the Illustrator document area is an artboard.)

12. **Click OK to complete the import process, and review the results.**

    If you had not chosen to make a movie clip instance from the layer group, those layers would have been imported as a layer group in the Flash timeline. The default Layer 1 from the original file is maintained at the bottom of the layer stack.

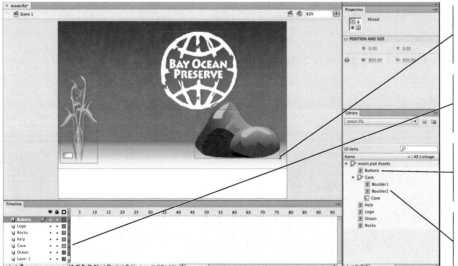

The symbol registration point indicates that this object is an instance of a movie clip symbol.

Because you created a movie clip instance from the layer group, only one layer is created to contain the instance.

Each resulting bitmap exists inside a folder that is named based on the imported Photoshop file.

Objects from the Cave layer group are nested in a folder that is named based on the layer group name.

13. **Select Layer 1 and click the Delete button at the bottom of the Timeline panel.**

14. **Save the file and continue to the next exercise.**

COPY ASSETS FROM EXTERNAL LIBRARIES

When objects already exist in a Flash file, it is a fairly simple process to copy them from one file to another. If both files are open, you can simply copy a symbol instance from the Stage of one file and paste it into the Stage of the other file; the necessary assets are automatically pasted into the library of the second file. You can also simply open the Library panel of an external file, which enables you to access the assets in that library without opening the second file's Stage.

1. **With ocean.fla open, choose File>Open. Navigate to creatures.fla (in the WIP>Aquarium folder) and click Open.**

   When more than one file is open, each file is represented by a tab at the top of the document window. You can click any document tab to make that file active.

Note:

Press Shift to select multiple consecutive items in a dialog box or panel. Press Command/ Control to select multiple non-contiguous items.

2. **In the Library panel, Shift-click to select the Fish1, Fish2, and Turtle items in the library. Control/right-click one of the selected items and choose Copy from the contextual menu.**

The document tabs show that there are two open files.

The Library panel shows the library for the currently active file.

3. **Click the Library menu at the top of the Library panel and choose ocean.fla to display that file's library.**

Use this menu to switch between the libraries of all open files.

4. **Control/right-click the empty area at the bottom of the Library panel (below the existing assets) and choose Paste from the contextual menu.**

   The Air bitmap item is also pasted because it is used in the Fish2 movie clip.

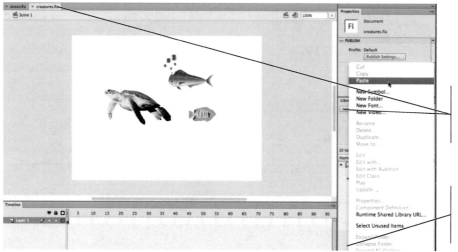

Although creatures.fla is active, the Library panel now shows the library of the ocean.fla file.

Control/right-click the empty area at the bottom of the Library panel to paste the copied items.

5. **Click the Close button on the creatures.fla document tab to close that file.**

None of the pasted symbols is added to the Stage; they are only placed in the ocean.fla file's library.

The three selected items from the creatures file are pasted into the ocean.fla library.

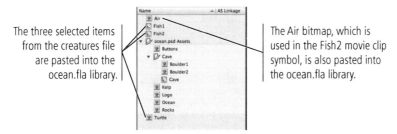

The Air bitmap, which is used in the Fish2 movie clip symbol, is also pasted into the ocean.fla library.

6. **With ocean.fla still open, choose File>Import>Open External Library.**

7. **Select buttons.fla (in the WIP>Aquarium folder) and click Open.**

You can use this option to open the library of another file without opening the external FLA file. The external library opens as a separate panel; the file name is included in the panel tab.

Note:

The keyboard command for opening an external library is Command/Control-Shift-O.

8. **Shift-click the Reset, Showcave, Showfish, and Showturtle button symbols.**

9. **Click any of the selected items and drag to the ocean.fla Library panel.**

The files don't disappear from the external library; they're simply duplicated in the ocean.fla Library panel. The Starfish bitmap object is also copied because it is used in the four button symbols.

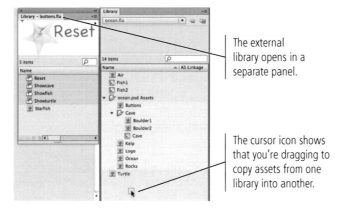

The external library opens in a separate panel.

The cursor icon shows that you're dragging to copy assets from one library into another.

10. **Click the Close button of the buttons.fla Library panel to close the external file's library.**

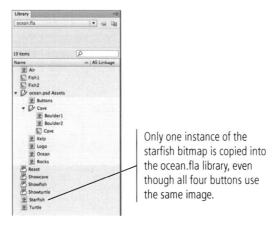

Only one instance of the starfish bitmap is copied into the ocean.fla library, even though all four buttons use the same image.

11. **Save the ocean.fla file and continue to the next exercise.**

ALIGN OBJECTS ON THE STAGE

The four buttons for this project need to be placed across the bottom of the Stage, aligned to appear equally distributed across the Stage area. The Align panel makes it very easy to position multiple selected objects relative to one another.

1. **With `ocean.fla` open, select the Buttons layer to make it active.**

2. **Drag instances of the four button symbols to the middle of the Stage, arranged so they do not overlap.**

3. **Select the placed Showfish button instance. Using the Properties panel, position the instance at X: 20, Y: 510.**

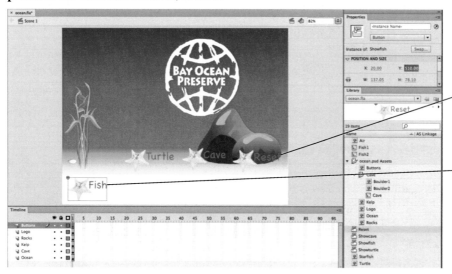

Place the symbols (from left to right) Showfish, Showturtle, Showcave, and Reset across the Stage.

Use the Properties panel to position the Showfish button instance.

4. **Using the Selection tool, Shift-click to select all four button instances on the Stage.**

5. **In the Align panel, turn off the Align To Stage option and then click the Align Bottom Edge button.**

The Align options position objects based on the selected edge. Because you used the Align Bottom Edge button, the selected objects are all moved to the bottom edge of the bottommost object in the selection.

Align Bottom Edge button

Align To Stage should not be checked.

The selected objects move to align with the bottom edge of the bottommost selected object.

Note:

Many of the steps required in the first stage of this project reinforce the techniques you already learned if you completed the first three projects in this book. Creating symbols and positioning objects on the Stage are fundamental skills that will be important in virtually any Flash project.

6. **Select only the Reset button and set the X position to 605.**

7. **Select the four placed button instances.**

8. **In the Align panel, click the Space Evenly Horizontally button.**

This option calculates the overall space across the selection, then shifts the objects so the same amount of space appears between each object in the selection. Because the buttons are different widths, this option creates a better result than the distribution options.

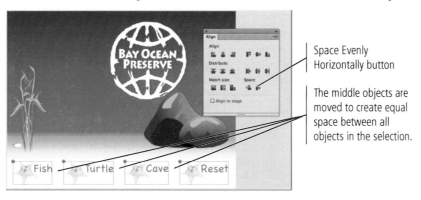

Space Evenly Horizontally button

The middle objects are moved to create equal space between all objects in the selection.

9. **Save the file and continue to the next exercise.**

TRANSFORM SYMBOLS AND INSTANCES

Placed instances of a symbol are unique objects, which means they can be manipulated separately without affecting other instances of the symbol. Each instance remains linked to the primary symbol, however, so transforming the actual symbol affects all placed instances of that symbol.

1. **With ocean.fla open, Control/right-click the Kelp object on the Stage and choose Convert to Symbol from the contextual menu.**

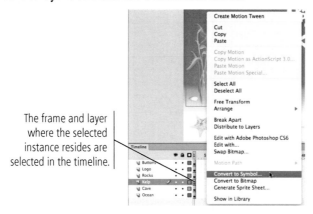

The frame and layer where the selected instance resides are selected in the timeline.

2. **In the resulting dialog box, type Seaweed in the Name field and choose Movie Clip in the Type menu. Choose the bottom-center registration point and click OK.**

Remember, the name of the actual symbol is only used internally while you develop the file. The specific names of instances on the Stage are more important, as you will see later when you add ActionScript to control the various pieces of this movie.

The Properties panel now shows that the selected object is an instance of the Seaweed symbol, which has been added to the Library panel.

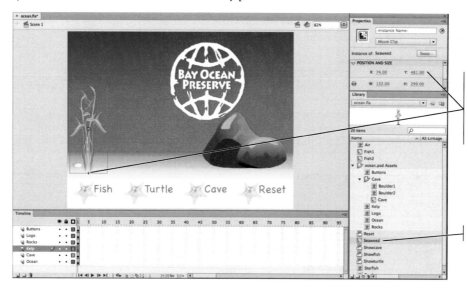

You defined bottom-center registration for this symbol, so measurements are based on the bottom-center point of the instance.

There's the new symbol.

3. **Using the Selection tool, Option/Alt-click the existing instance and drag right to clone a second instance of the Seaweed symbol.**

4. **Repeat Step 3 to clone one more instance.**

5. **Deselect everything on the Stage. Choose the Free Transform tool in the Tools panel and click the middle Seaweed instance to select it.**

The Free Transform tool is used to change the size or shape of an object. Remember, all transformations are applied around the transformation point.

Note:

If the Free Transform tool is active when you click and drag to move an object, make sure you don't click the object's transformation point before you drag the object.

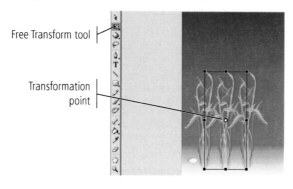

Free Transform tool

Transformation point

6. **Drag the transformation point until it snaps to the bottom-center bounding box handle.**

7. **Click the top-center bounding box handle and drag down to make the selected instance shorter.**

Transformations applied to individual instances have no effect on the original symbol or on other placed instances.

Note:

You can also press Option/Alt and drag a center handle to transform the object around the opposite bounding-box handle, without moving the transformation point.

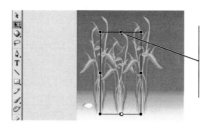

Because the bottom-center is the transformation point, dragging the top-center handle makes the instance shorter without moving the bottom of the instance.

8. **In the Library panel, double-click the Seaweed symbol icon to enter into the symbol.**

The crosshairs in the bottom center identify the **symbol registration point**, or the location of X:0, Y:0 for placed instances; all measurements for placed instances begin at this location.

Even though you changed the transformation point of one placed instance on the main Stage, the transformation point remains in the center of the object on the symbol Stage. The transformation point is specific to each placed object or instance.

The Edit bar shows you are now working on the Seaweed symbol Stage.

The transformation point is particular to each instance of an object or symbol.

The symbol registration point appears on the symbol Stage.

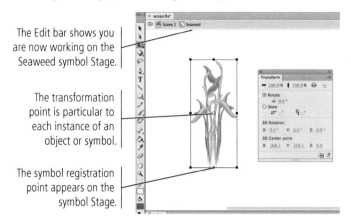

9. **With the Free Transform tool still selected, drag the transformation point until it snaps to the bottom-center bounding box handle of the object on the symbol's Stage.**

10. **Show the Transform panel (Window>Transform).**

11. **In the Transform panel, make sure the Link icon is not active and then change the Scale Height value to 80%.**

If the Link icon shows two solid chain links, changing the height of the object would proportionally affect the width of the object (and vice versa). Because you only want to change the height, you need to break the Link icon.

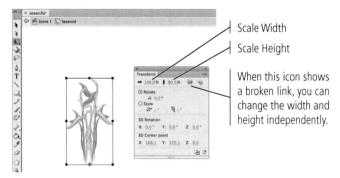

Scale Width

Scale Height

When this icon shows a broken link, you can change the width and height independently.

12. **In the Edit bar, click Scene 1 to return to the main Stage.**

When you modify the original symbol, the changes ripple through all placed instances.

Editing the actual symbol affects all three placed instances on the Stage.

The instance you resized on the main Stage is still proportionally sized.

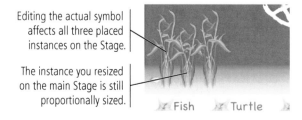

13. **Save the file and continue to the next exercise.**

WORK WITH LAYERS IN AN IMPORTED MOVIE CLIP SYMBOL

When you imported the project artwork, the two pieces of the cave were created on separate layers — in a layer group — because you need to place a fish inside the cave, stacked between the two boulders. If the two boulders had existed on a single layer, the result would have been a single bitmap object, which could not be broken apart in Flash to accomplish the desired effect.

1. **With ocean.fla open, use the Selection tool to click the Cave movie clip instance on the Stage.**

2. **In the Properties panel, change the X position of the instance to 850.**

   If the cave objects had been placed in the correct position in the Photoshop file, the imported bitmap objects would have been clipped at the Stage edge. If you create artwork in Photoshop, make sure pieces are entirely inside the Canvas edge if you don't want them clipped when they are imported into Flash.

   Note:

   Remember, the symbol registration point marks the location of the defined X and Y values on the main Stage.

   Because you defined the bottom-right corner as the symbol's registration point, that point is positioned at X: 850.

3. **Double-click the Cave movie clip instance on the Stage to edit the symbol in place on the Stage.**

   Because this movie clip (and instance) was created based on a layer group in the imported artwork, the layers that made up the group are maintained in the symbol timeline.

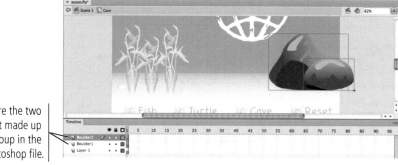

   These are the two layers that made up the layer group in the original Photoshop file.

4. **Change the name of Layer 1 to Fish.**

5. **Drag an instance of the Fish2 movie clip symbol onto the Stage, near the existing cave pieces.**

   Because the Fish layer is below the two Boulder layers, the Fish2 instance should be at least partially hidden by the two pieces of the cave.

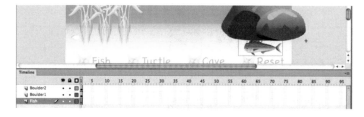

6. **In the Timeline panel, drag the Fish layer above the Boulder 1 layer.**

   The Fish2 instance is now between the two pieces of the cave.

7. **Drag the Fish2 instance so only the head appears to the left of the front rock.**

   In the second stage of this project, you will learn a number of techniques for creating animations — including making this fish blow bubbles that rise up and off the Stage.

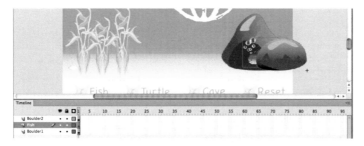

8. **Click Scene 1 in the Edit bar to return to the main Stage.**

9. **Save the file and continue to the next exercise.**

 USE THE SPRAY BRUSH

Although you could simply place multiple instances of the Rock bitmap without converting it to a formal symbol, you can use the Spray Brush tool to quickly and easily add numerous instances of the object.

1. **With ocean.fla open, Control/right-click the Rock bitmap on the Stage. Choose Convert to Symbol in the contextual menu.**

2. **In the resulting dialog box, type Gravel in the Name field. Choose Graphic in the Type menu, select the center registration point, and click OK.**

   You are going to create many instances of this symbol as decoration on the Stage. Because you don't need to address the instances with script, you don't need to name the instances. You also don't need to animate anything inside the Gravel symbol, so the Graphic type is appropriate.

3. **Choose the Spray Brush tool (nested under the Brush tool) in the Tools panel.**

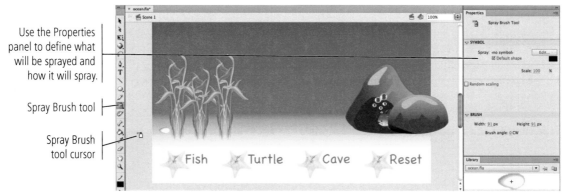

Use the Properties panel to define what will be sprayed and how it will spray.

Spray Brush tool

Spray Brush tool cursor

4. **In the Properties panel, click the Edit button in the Symbol area. Choose Gravel in the resulting dialog box, and then click OK.**

5. **Check the Random Scaling, Rotate Symbol, and Random Rotation options. Change the Brush Width and Height values to 75 pixels.**

6. **In the timeline, make sure the Rocks layer is active. Zoom out so you can see the entire Stage, and some of the surrounding area on the left and right.**

7. **Click the left edge of the Stage over the brown area of the gradient. Hold down the mouse button and drag slowly to the right across the entire Stage.**

The dozens (or even hundreds, depending on how you dragged) of instances do not significantly increase the file size because the symbol needs to download only once.

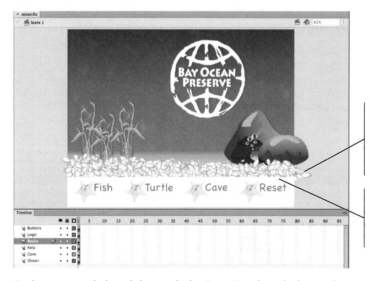

Using the Random Scaling, Rotate Symbol, and Random Rotation options, you used a single symbol to create a field of ocean rocks.

Because the Buttons layer is higher than the Rocks layer, the bottom edges of the sprayed instances are hidden.

Each time you click and drag with the Spray Brush tool, the resulting instances are grouped. If you click again to add more rocks, you will have more than one group of rock instances.

8. **Save the file and continue to the next exercise.**

Note:

As you learned in Project 1: Corvette Artwork, you have to enter into the group (double-click) to access the individual instances.

ORGANIZE YOUR LIBRARY WITH FOLDERS

Folders make it easy to organize the assets that make up a movie, and easy to find the assets you need.

1. **With ocean.fla open, click the Library panel to activate it.**

2. **In the Library panel, change the name of the ocean.psd Assets folder to Bitmaps.**

    Changing a library folder name has no effect on placed instances of the symbols inside that folder.

3. **Select the two bitmap objects in the nested Cave folder and move them into the first-level Bitmaps folder.**

    Moving symbols to a new location in the Library panel has no effect on placed instances of the moved symbols.

4. **Select the three bitmap objects in the first level of the library and drag them into the Bitmaps folder.**

5. **Drag the nested Cave folder from the Bitmaps folder to the main level of the Library panel.**

6. **Change the Cave folder name to Movie Clips, then drag the three movie clip symbols from the main level of the Library into the Movie Clips folder.**

    Drag the nested folder to the empty area at the bottom of the panel to move it to the first level of the library.

    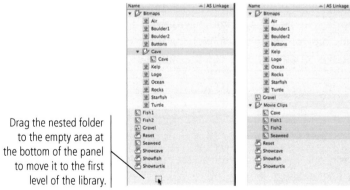

7. **Create a new folder named Buttons at the first level of the library. Drag the four button symbols into the new folder.**

   New folders are created at the same nesting level as the current selection. To create a new folder at the first level of the library, click the empty area at the bottom of the panel to deselect any files before clicking the New Folder button.

8. **Create another new folder named Graphics at the main level of the Library folder. Drag the one remaining graphic symbol into the folder.**

Note:

In this case, we're using simple names for the symbol folders. The names you assign to symbol (or layer) folders aren't functional; they just need to make sense to you (or other people working with your files).

9. **Save the file and continue to the next stage of the project.**

Stage 2 Animating Symbols

In Project 2: Talking Kiosk Interface, you created simple animation by swapping symbols at specific points in time; although nothing technically changed position, replacing one object with another is still considered "animation." In Project 3: Animated Internet Ads, you used classic tweening to move objects across the Stage; this technique creates smooth motion by defining the start point keyframe, end point keyframe, and path shape. In this project, you learn several different techniques for creating various types of animation, including animating specific properties of an object.

 CREATE A BASIC FRAME ANIMATION IN A MOVIE CLIP SYMBOL

Movie clips are animated symbols that reside on a single frame of the timeline. In its simplest form, a movie clip can include a solitary fish swimming across the ocean; at its most complex, a movie clip can include fully interactive elements in a video game. In this exercise, you create the most basic type of animation — a frame animation.

1. **With ocean.fla open, choose the Selection tool and make sure nothing is selected on the Stage.**

2. **In the Properties panel, change the FPS hot text to 15 frames per second.**

   Using 15 fps for a movie — especially one that's going to run on the Web — provides decent quality. The default 24 fps is not necessary for standard computer viewing, and could create files that require too much processing power for some users.

   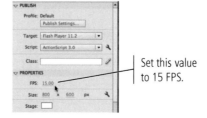

   Set this value to 15 FPS.

3. **On the Stage, double-click the Cave symbol instance to enter into the symbol Stage (edit in place).**

4. **Double-click the placed Fish2 instance to enter into the nested symbol. Click away from the selected objects, and then select only the air bubbles.**

   All of the bubbles together are a single bitmap object; you don't need to select each individual bubble shape.

   When you first enter into a symbol, all objects in the symbol are automatically selected. You are going to move only the bubbles, which are an instance of the Air bitmap object that is placed inside the Fish2 symbol. You are editing in place on the main Stage so you can see when the bubbles are entirely outside the Stage area.

5. **In the timeline, click Frame 7 to select it, then press F6 to insert a new keyframe.**

   As you already know, **keyframes** are special frames where something happens to an object: it appears or disappears, changes size, moves to another position, changes color, and so on. When you add a new keyframe, regular frames are automatically added immediately before the previous keyframe; objects on the preceding keyframe will remain in place until the playhead reaches the new keyframe.

Note:

You can also Control/ right-click a frame in the timeline and choose Insert Keyframe from the contextual menu, or choose Insert>Timeline> Keyframe.

You are working on the Fish2 symbol, which is nested inside the Cave symbol.

When you add a keyframe, regular frames are automatically added directly before it.

6. **With Frame 7 selected in the timeline, deselect the fish graphic, and then reposition the air bubbles graphic directly above the original position.**

   When you select a keyframe, all objects on that keyframe are automatically selected (just as when you first enter into the symbol). You can Shift-click the fish instance to deselect that object, leaving only the air bubbles object selected.

With the Frame 7 keyframe active, move the bubbles up from their previous position.

7. Repeat Steps 5–6 six more times, adding a new keyframe every seven frames and moving the air bubbles up until they are outside the top edge of the Stage.

Because you're editing the symbol in place, you can see how far you need to move the bubbles until they are outside the Stage area. If you edited this symbol on its own Stage, you would have to guess about positioning, return to the main Stage to test your guess, return to the symbol to add more frames, return to the main Stage to test again, and so on.

At Frame 49, the bubbles are entirely off the Stage.

Note:

If you change an object on a regular frame in the middle of a movie, all frames between the two surrounding keyframes (or the nearest preceding keyframe and the final frame) reflect that change.

8. Select Frame 56 on the timeline and press F5 to add a regular frame.

This regular frame at Frame 56 extends the timeline by half a second, which prevents the bubbles from reappearing in the fish's mouth (Frame 1) immediately after they move past the top of the Stage (Frame 49).

9. Click the playhead above the timeline and drag left and right.

Scrubbing the playhead allows you to look at specific sections of a movie over time so you can see if they work the way you expect.

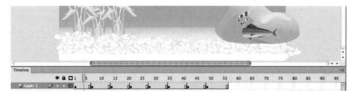

10. Click Scene 1 in the Edit bar to return to the main Stage.

11. Press Command-Return/Control-Enter to test the movie.

You can see the air bubbles moving up in a continuous loop. Even though the primary Stage has only one frame, the movie clip's timeline continues to play as long as the movie remains open.

The animation plays in the Flash Player window.

Note:

Remember from Project 3: Animated Internet Ads, you can't preview the animation inside a movie clip on the main Stage. You have to test the file in the Flash Player window to see the animation.

12. Close the Flash Player window and return to Flash.

13. Save the file and continue to the next exercise.

CREATE A MOTION TWEEN

Creating the appearance of continuous, fluid movement requires a slightly different position or shape (depending on what you are animating) on every frame in an animation. Rather than defining each individual frame manually — which could take days, depending on the length of your animation — you can let Flash define the frames that are in between two keyframes (the tween frames).

Flash incorporates technology that makes it very easy to define smooth animations by simply moving a symbol object around on the Stage.

1. **With `ocean.fla` open, create a new layer named `Turtle` immediately above the Logo layer. Select the Turtle layer as the active layer.**

2. **Drag an instance of the turtle bitmap image from the Library panel onto the Stage. Use the Transform panel to scale the instance uniformly to 50%, and then position it beyond the right edge of the Stage, higher than the Cave instance.**

3. **Control/right-click the turtle instance and choose Convert to Symbol from the contextual menu.**

4. **In the resulting dialog box, type `Swimmer` in the Name field, choose Movie Clip in the Type menu, and choose the center registration point.**

5. **Click the Folder link. Select the Existing Folder radio button and choose Movie Clips in the list. Click Select to return to the Convert to Symbol dialog box, and then click OK to create the new symbol.**

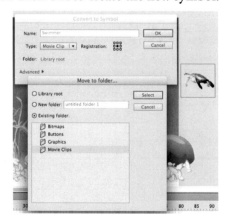

6. **Click Frame 90 on the Turtle layer and press F5 to add a new regular frame.**

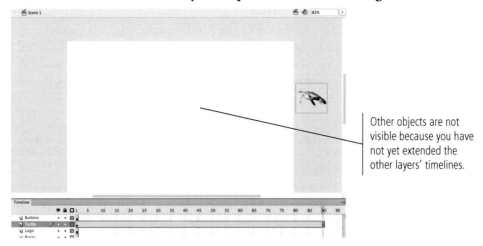

Other objects are not visible because you have not yet extended the other layers' timelines.

7. **Control/right-click any frame on the Turtle layer and choose Create Motion Tween.**

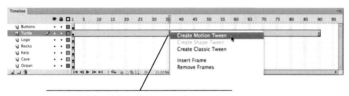

Control/right-click anywhere on the timeline between keyframes to add a motion tween between keyframes (or between a keyframe and the final frame on that layer).

Note:

We enlarged our workspace by collapsing docked panels to icons so we could see the entire Stage and all 90 frames in the timeline.

8. **Click Frame 90 to select that frame.**

Flash creates a motion tween in the frames between keyframes. Because Frame 1 is the only keyframe on this layer, the motion tween is created between Frame 1 and the last frame on the layer.

9. **Select the Turtle image located at the right of the Stage, and drag it off the left edge of the Stage.**

A new keyframe is automatically added on Frame 90 to mark the new position of the Swimmer symbol. A line — the motion path — shows the path of movement from the symbol's position on Frame 1 to its position on Frame 90. The small dots along the motion path correspond to the frames within the tween.

When you edit symbols on a tween layer, the position of the playhead is crucial. When you change any property of an object on a tween layer, a property keyframe is automatically inserted at the current frame. Flash generates the tween frames based on the change in the property value between the active keyframe and the previous one.

Note:

The motion path line corresponds to the color of the layer containing the path.

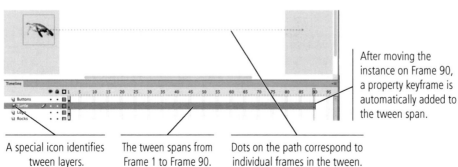

After moving the instance on Frame 90, a property keyframe is automatically added to the tween span.

A special icon identifies tween layers.

The tween spans from Frame 1 to Frame 90.

Dots on the path correspond to individual frames in the tween.

10. Click Frame 1 to move the playhead back to the beginning of the timeline, and then press Return/Enter to play the timeline on the Stage.

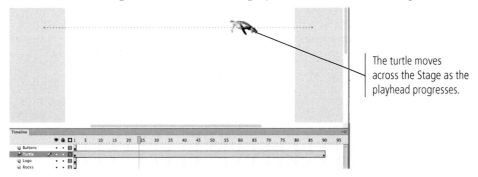

The turtle moves across the Stage as the playhead progresses.

11. Move the playhead to Frame 90, then click the turtle instance to select it. Scale the selected instance to 200% of its current size. If necessary, use the Selection tool to reposition the resized turtle so it is entirely outside the edge of the Stage.

The term "motion path" is deceptive because you can animate much more than just motion when you apply a motion tween. By scaling the object on Frame 90, you told Flash to change both the symbol size and position as the timeline progresses.

Note:

Because you scaled the object to 50% before you created the Swimmer symbol, resizing this instance to 200% restores the bitmap to its original size.

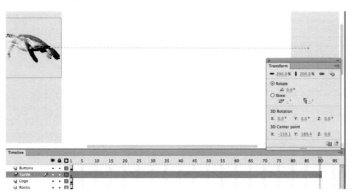

12. Return the playhead to Frame 1, and then press Return/Enter to play the timeline on the Stage.

Now the turtle gets larger as it moves across the stage, creating the effect of the turtle swimming closer. Flash automatically calculates the appropriate position and size of the symbol for all frames between the Frame 1 and Frame 90 keyframes.

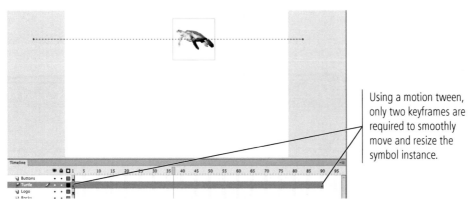

Using a motion tween, only two keyframes are required to smoothly move and resize the symbol instance.

13. Save the file and continue to the next exercise.

FLASH FOUNDATIONS

Onion skinning, accessed through a set of buttons at the bottom of the timeline, is a technique that allows you to view more than one frame of an animation at a time.

- Clicking the **Onion Skin** button toggles the feature on or off.

- Clicking the **Onion Skin Outlines** button turns all visible skins to outlines (or wire frames). Combining outlines and onion skins allows you to clearly see the components of your animations without fills or (true-weight) strokes.

- Clicking the **Edit Multiple Frames** button allows you to edit multiple frames at the same time: moving an entire animation, for example, or simply changing single frames within a tween. Without this feature, you would have to move objects one frame at a time. With the feature, you can see previous or subsequent frames, which often helps when you're fine-tuning an animation and you need to move an object in one frame relative to its position in other frames.

- Clicking the **Modify Markers** button allows you to select from a range of predefined skins, or turn onion skinning off. You can choose to have onion skins span two frames, five frames, or all frames. You can also manually adjust the onion skin markers and bypass these presets.

The following illustrations show a simple motion tween that moves the oval symbol across the Stage, from left to right.

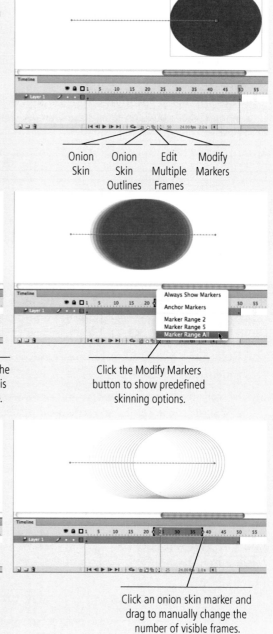

Onion Skin   Onion Skin Outlines   Edit Multiple Frames   Modify Markers

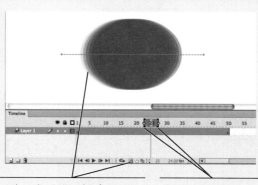

When the Onion Skin feature is active, you can see multiple frames at once (around the playhead).

These markers show the range of frames that is visible on the Stage.

Click the Modify Markers button to show predefined skinning options.

When Onion Skin Outlines is active, frames within the visible onion skin display as wireframes.

Click an onion skin marker and drag to manually change the number of visible frames.

EDIT THE SHAPE OF THE MOTION PATH

As you learned in the previous exercise, moving an object and changing its size (or other properties) can be as simple as creating a motion tween and adjusting the symbol at specific frames on the timeline. You don't need to manually create keyframes because Flash adds them for you whenever you change the symbol at a particular point in the timeline. In this exercise, you work with the motion path line, which can be edited like any other line in Flash — giving you precise control over the course of a tween.

1. **With `ocean.fla` open, click the Swimmer symbol on the Stage to select it and reveal the related motion path.**

2. **Move the Selection tool cursor near the center of the motion path until you see a curved line in the cursor icon. Click near the path and drag up to bend the motion path.**

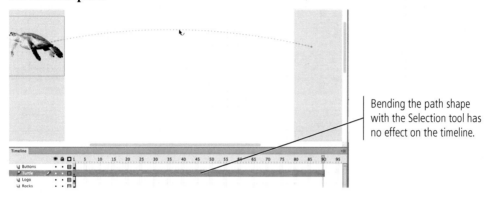

Bending the path shape with the Selection tool has no effect on the timeline.

3. **Return the playhead to Frame 1 on the timeline, and then press Return/Enter to play the timeline.**

   The turtle now follows the new shape of the motion path.

4. **Click Frame 45 of the Turtle layer to select that frame. Using the Selection tool, click the Swimmer symbol instance and drag down.**

   Flash automatically adds another keyframe to the motion tween to mark the instance's position at that point; the motion path bends again to reflect the defined position for the turtle at the selected frame. Flash adds an anchor point to the path at the new keyframe.

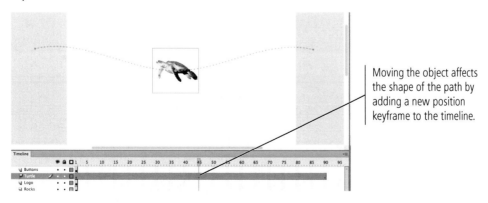

Moving the object affects the shape of the path by adding a new position keyframe to the timeline.

5. **Choose the Subselection tool in the Tools panel, and then click the new anchor point in the middle of the path.**

   Remember from Project 1: Corvette Artwork, that the Selection tool selects entire paths. The Subselection tool selects the anchor points and handles that make up a shape.

Controlling Animation Speed with Easing

Physical objects are subject to physical laws; in the real world, friction, momentum, and mass (among other things) affect how an object moves. A bouncing ball is a good example of these laws. If you throw a ball at the ground, how hard you throw the ball determines its beginning speed. When the ball hits the ground, it transfers energy to the ground and then rebounds, causing the ball to move away from the ground (its first bounce), at which point it is moving slightly faster than when you threw it. As the ball arcs through the bounce it slows down, then starts to drop and hits the ground again, repeating the process in ever-decreasing arcs until it finally gives up its energy and then stops. The speed of the ball changes when the energy behind the ball changes.

In animation terms, these changes in speed are called **easing**. In Flash, you can control easing in the Properties panel when a tween is selected in the timeline.

- Positive Ease values decrease the distance of movement on subsequent frames, causing the object to slow down as it moves through the tween.

- Negative Ease values increase the distance of movement on subsequent frames, causing a moving object to speed up through the tween.

- Ease values closer to 100 or −100 result in greater apparent changes in speed.

The accompanying illustrations show a simple 50-frame motion tween that moves a circle symbol across the Stage.

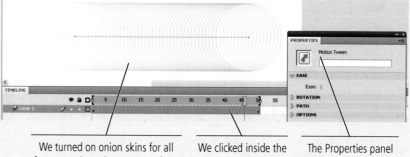

We turned on onion skins for all frames to show the position of the symbol at each frame in the tween.

We clicked inside the tween span to select the motion tween.

The Properties panel shows options related to the selected tween.

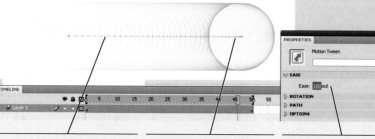

By increasing the Ease value, the object moves farther on earlier frames than on later frames.

This creates the effect of the object slowing down over the course of the animation.

The Properties panel shows the positive Ease value as "out" — it slows down.

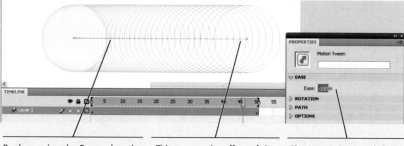

By decreasing the Ease value, the object moves farther on later frames than on earlier frames.

This creates the effect of the object speeding up over the course of the animation.

The Properties panel shows the negative Ease value as "in" — it speeds up.

6. **Click the selected anchor point and drag left.**

Moving the anchor point changes the shape of the motion path, just as it does when you edit a regular Bézier curve. You can also adjust the handles of the point to change the shape of the motion path between the two connecting anchor points (the selected point and the point at the left end of the path).

When you change the position of the anchor point, notice that the number of dots (representing frames in the tween) on either side of the point remains unchanged. Because you effectively shortened the left half of the path, the same number of frames display over a shorter distance than the same number of frames to the right of the selected point. In effect, you made the turtle swim faster in the first half of the animation (moving a longer distance) and slower in the second half (moving a shorter distance).

Note:

You can use the Convert Anchor Point tool to convert a smooth anchor point on a motion path to a corner anchor point, allowing you to change directions in the tween.

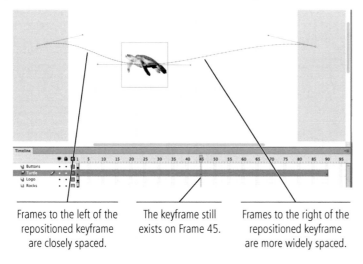

Frames to the left of the repositioned keyframe are closely spaced.

The keyframe still exists on Frame 45.

Frames to the right of the repositioned keyframe are more widely spaced.

Note:

By default, motion paths are created with a non-roving keyframe property, which means the anchor points along the path are attached to specific keyframes in the timeline.

7. **Control/right-click anywhere within the motion tween (in the timeline), and choose Motion Path>Switch Keyframes to Roving in the contextual menu.**

When you choose this option, the dots along the path redistribute to equal spacing across the entire length of the tween, and the keyframe from Frame 45 is removed from the layer timeline. The shape of the path is not affected.

A roving-property keyframe is not attached to any particular frame in the tween. This type of keyframe allows you to create a custom-shaped motion path with consistent speed throughout the tween.

Note:

If you convert keyframes to roving, frames are redistributed along the entire span of the tween. If you then convert the frames back to non-roving, the location of keyframes added to the tween is determined by the location of anchor points on the path.

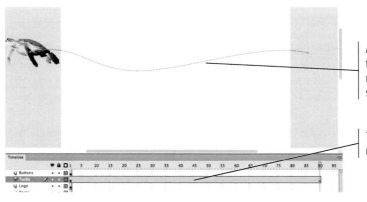

After choosing the Roving option, the frames along the path are redistributed to be equally spaced across the entire path.

The property keyframe is removed from the timeline.

8. **Save the file and continue to the next exercise.**

Motion Tween Properties and Presets

In Flash CS6, the motion tween includes all information for the animation, including the length of the animation and specific object properties at various points along the path. A motion path is actually a specific type of object rather than simply a guide; the Properties panel shows a number of options that relate to the selected motion tween.

Because a motion tween is an actual object, you can attach any symbol to the path by simply dragging a new symbol onto the Stage when the motion tween layer is selected.

Use this field to define a name for the motion path instance.

Use Ease values to speed up or slow down an animation over time.

Use this option to rotate a symbol X number of times as it moves along the motion path.

Check this option to rotate the object so its bottom edge follows the contour of the path.

Use these options to change the position and size of the overall path.

Check this option to synchronize the number of frames in a tween within a graphic symbol to match the number of frames on the timeline where the graphic symbol is placed.

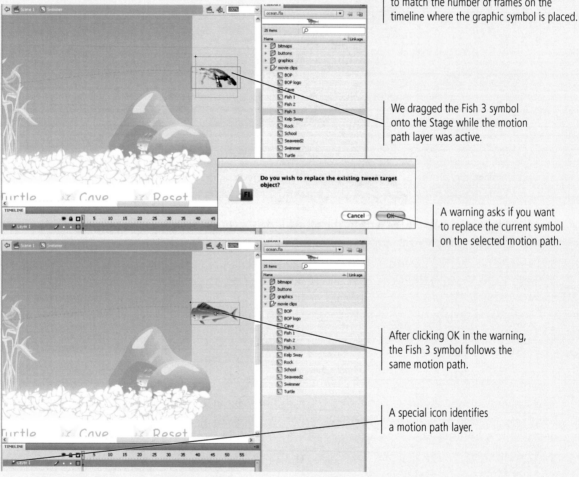

We dragged the Fish 3 symbol onto the Stage while the motion path layer was active.

A warning asks if you want to replace the current symbol on the selected motion path.

After clicking OK in the warning, the Fish 3 symbol follows the same motion path.

A special icon identifies a motion path layer.

Flash includes a number of predefined motion presets (Window>Motion Presets), which you can use to add common animations to your files. You can also save your own motion presets by Control/right-clicking an existing motion path and choosing Save as Motion Preset from the contextual menu. (User-defined presets are stored and accessed in a Custom folder in the Motion Presets panel. Custom presets do not include previews.)

COPY AND PASTE FRAMES

Your turtle currently swims from right to left across the Stage. When the animation loops, however, it would seem to miraculously jump back to the right and swim across again. For a more realistic effect, you are going to make a copy of the motion path animation and reverse it so the turtle swims back across the Stage before the animation loops.

1. **With ocean.fla open, Control/right-click anywhere in the Turtle layer motion tween and choose Copy Frames from the contextual menu.**

    You could also use the options in the Edit>Timeline submenu, but the standard Edit menu commands (and the related keyboard shortcuts) do not work when you want to copy or paste frames in the timeline.

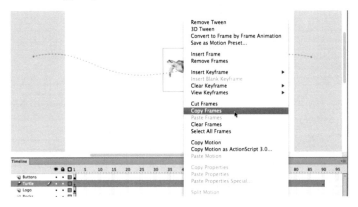

2. **Control/right-click Frame 95 of the Turtle layer and choose Paste Frames.**

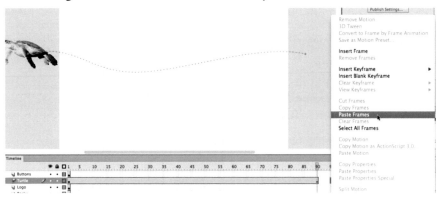

> **Note:**
>
> *You are allowing five extra frames between the time the turtle leaves and then re-enters the Stage area (ostensibly enough time for it to turn around before it swims back).*

    You pasted an exact copy of the selected frames (the motion tween) — including the position of the symbol at various keyframes. In other words, the turtle is on the right at Frame 95 and on the left at Frame 184 (the end of the pasted animation).

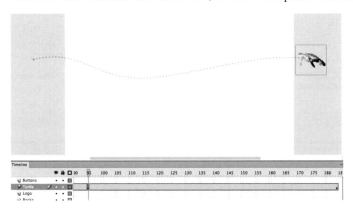

3. **Control/right-click anywhere between Frame 95 and Frame 184 and choose Reverse Keyframes.**

Reversing the keyframes moves the turtle to the left at Frame 95 and the right at Frame 184.

4. **Select the Frame 95 keyframe in the timeline. Click the turtle instance with the Selection tool, then choose Modify>Transform>Flip Horizontal.**

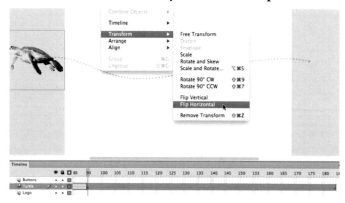

For the turtle to realistically swim back across the Stage, you have to flip the symbol instance to face in the correct direction.

5. **Click Frame 184 to make that the active frame. With the turtle on the Frame 184 keyframe selected, choose Modify>Transform>Flip Horizontal again.**

The turtle now faces to the right throughout the entire second half of the animation.

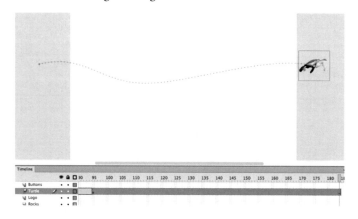

6. **Add new regular frames to Frame 184 of the remaining layers on the timeline.**

Remember, you have to manually extend each layer so they will all exist throughout the length of the entire animation.

7. **Click Frame 1 to reposition the playhead, then press Return/Enter to play the movie on the Stage.**

The air bubbles in the Cave movie clip instance do not move because that animation exists only inside the movie clip timeline.

8. **Save the file and continue to the next exercise.**

USE THE MOTION EDITOR

In addition to making changes on the Stage, you can also use the Motion Editor panel to define specific changes to specific properties at specific points in time. In this exercise, you use the Motion Editor to animate the seaweed with a tween that creates the effect of a smooth, swaying motion.

1. **With `ocean.fla` open, double-click the Seaweed symbol icon in the Library panel to enter into the symbol Stage.**

2. **In the timeline, select Frame 40 and press F5 to insert a new regular frame.**

   You can also Control/right-click the frame and choose Insert Frame from the contextual menu.

3. **Control/right-click anywhere between Frame 1 and Frame 40 and choose Create Motion Tween.**

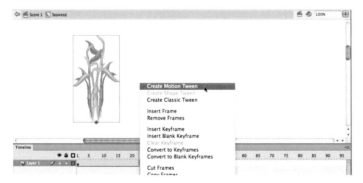

4. **Read the resulting message.**

   Motion tweens only work with symbol instances. As you see in the warning dialog box, Flash can automatically convert the placed bitmap instance to a movie clip symbol.

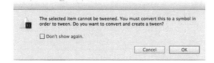

5. **Click OK in the message to create a symbol from the selected object.**

   You are not creating this tween on the main Stage because you want the animation to loop continuously regardless of the position of the playhead on the main timeline. Even though you are already inside of a symbol, you need to create a nested symbol structure for the motion tween to work properly.

6. **In the Library panel, change the name of Symbol 1 to Seaweed Sway, and then move the symbol into the Movie Clips folder.**

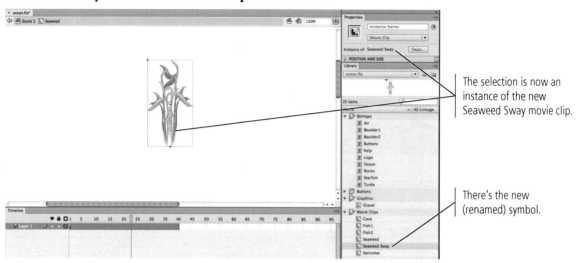

The selection is now an instance of the new Seaweed Sway movie clip.

There's the new (renamed) symbol.

7. **Choose the Free Transform tool in the Tools panel. Select the object on the Stage, and drag the transformation point to the bottom-center handle.**

8. **Open the Motion Editor panel (Window>Motion Editor). If it is grouped with the Timeline panel, click the Motion Editor tab and drag it away from the group so you can see both panels at once.**

9. **Move the playhead to Frame 10. In the Motion Editor panel, scroll until you find the Skew X option. Click the hot text for that option and change the value to 5°.**

   The concept here is the same as in the previous exercise: select the frame, and then change the object properties to what you want at that particular point in time. The difference is that you're using the Motion Editor panel to easily define numeric values for specific properties at specific points in time.

Note:

We docked the Motion Editor panel above the Timeline panel so we could see both panels and the symbol Stage.

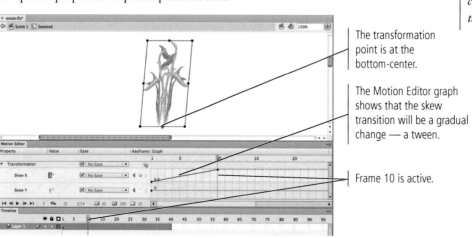

The transformation point is at the bottom-center.

The Motion Editor graph shows that the skew transition will be a gradual change — a tween.

Frame 10 is active.

FLASH FOUNDATIONS

The Motion Editor panel can control any property that can be animated in a motion tween. All available properties are listed in the left side of the panel. The right side shows a graph with the number of frames defined in the bottom-left corner; the active frame, and the value of each property at that frame, appears at the left edge of the graph. (You can change the number of viewable frames in the bottom-left corner of the dialog box to see an overview of the entire path animation.)

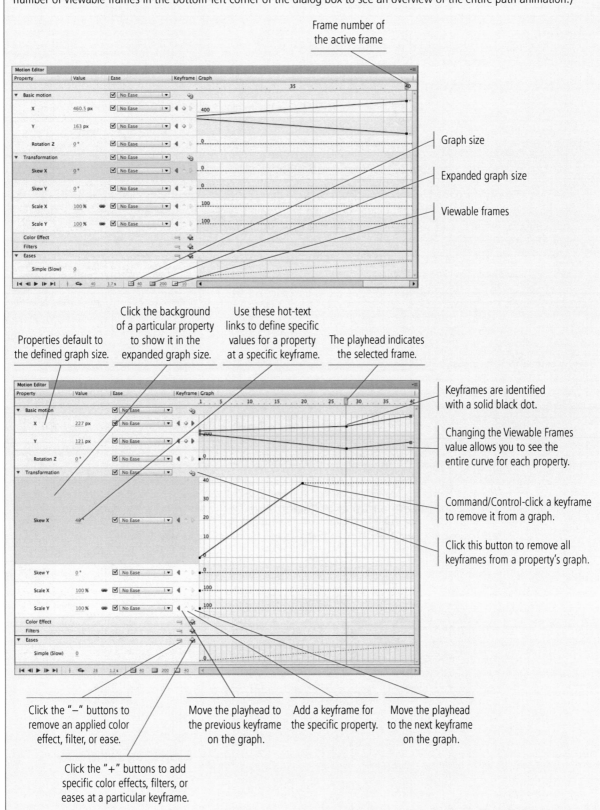

Frame number of the active frame

Graph size

Expanded graph size

Viewable frames

Properties default to the defined graph size.

Click the background of a particular property to show it in the expanded graph size.

Use these hot-text links to define specific values for a property at a specific keyframe.

The playhead indicates the selected frame.

Keyframes are identified with a solid black dot.

Changing the Viewable Frames value allows you to see the entire curve for each property.

Command/Control-click a keyframe to remove it from a graph.

Click this button to remove all keyframes from a property's graph.

Click the "−" buttons to remove an applied color effect, filter, or ease.

Move the playhead to the previous keyframe on the graph.

Add a keyframe for the specific property.

Move the playhead to the next keyframe on the graph.

Click the "+" buttons to add specific color effects, filters, or eases at a particular keyframe.

10. **Move the playhead to Frame 30 in the timeline. In the Motion Editor panel, change the Skew X value to –5°.**

You don't need to manually move the skew back to 0°; the tween frames do that for you.

Note:

Selecting a different frame in either panel changes the selected frame in the other panel.

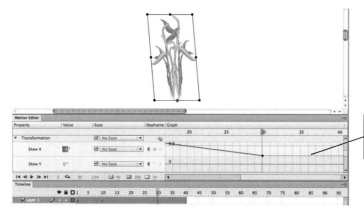

The dotted lines show the projected value of the property at specific frames.

As you can see by the dotted line after the Frame 30 keyframe, the Skew X transition would stay at –5° until the animation looped, when it would jump back to 0°. To avoid this jump, you need to manually set the skew back to 0° at the last keyframe.

11. **Move the playhead to Frame 40 in the timeline. In the Motion Editor panel, change the Skew X value to 0°.**

Note:

The object must be selected to change its properties. Selecting the frame in the timeline also selects the object on that frame.

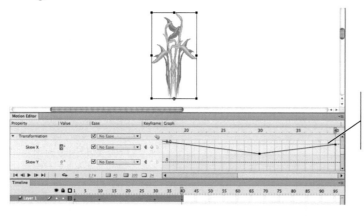

The animation will now flow smoothly from –5° to 0° before it loops back to Frame 1.

12. **Click Scene 1 in the Edit bar to return to the main Stage.**

13. **Press Command-Return/Control-Enter to test the movie in a Player window.**

Because you created the animation inside of the Seaweed movie clip symbol, all three instances of the symbol sway continuously as long as the animation is open.

14. **Close the Player window and return to Flash. Save the file and continue to the next exercise.**

Graphics vs. Movie Clips

Both graphics and movie clips can include animation. However, there are two fundamental differences in the capabilities of the two symbol types.

First, movie clip symbol instances can be named, which means they can be addressed by code. You can write scripts to control the timeline within a movie clip symbol independently of other objects in the file. Graphic symbol instances can't be named, which means you can't affect them with code.

Second, if you create animation inside of a graphic symbol, the timeline where you place the instance determines how much of the graphic symbol's animation plays. In other words, frames in the graphic symbol must correspond to frames on the parent timeline (they are "timeline dependent").

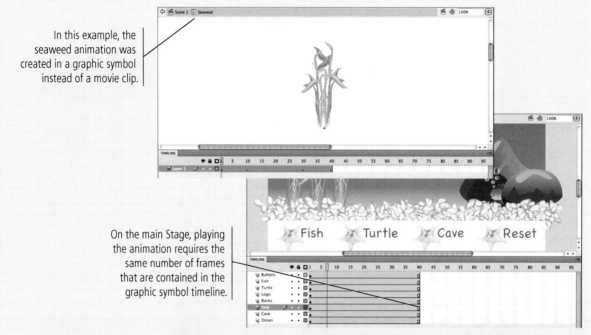

In this example, the seaweed animation was created in a graphic symbol instead of a movie clip.

On the main Stage, playing the animation requires the same number of frames that are contained in the graphic symbol timeline.

In the example here, the 40-frame seaweed animation was created in a graphic symbol (as you can see in the Edit bar above the symbol Stage). For the instances on the main Stage to play properly, you need to extend all of the layers to 40 frames (above right).

If the parent timeline includes more frames than the graphic symbol (as in the example on the right, where the main timeline has 94 frames) the graphic symbol's timeline will play slightly less than 2.5 times before looping back to the beginning — causing a visible jump in the animation.

If you have a number of animations of different length, you should use movie clip symbols, which function independently of the timeline where they are placed and can loop continuously regardless of the length of other animations on the same parent timeline.

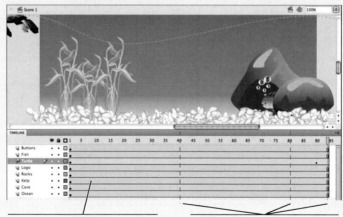

If you extend the timeline to create other animations, the graphic symbol animation repeats as long as the timeline allows.

The 40-frame seaweed animation would play 2 full times plus 14 frames before the main timeline looped back to Frame 1.

ANIMATE EFFECTS AND FILTERS

In addition to changing the common symbol properties — position, size, etc. — a motion tween can be used to animate a number of other options. Effects and filters, which can add visual interest to most objects on the Stage, can also be animated to change over time. In this exercise, you are going to cause the client's logo to fade into view over time, changing from blurry to clear and fully visible.

1. **With `ocean.fla` open, choose the Selection tool. Control/right-click the logo instance on the Stage and choose Convert to Symbol.**

2. **Name the new symbol `BOP`, choose Movie Clip as the type, and choose the center registration point. Use the Folder link to place the new symbol inside the Movie Clips folder. Click OK to create the symbol.**

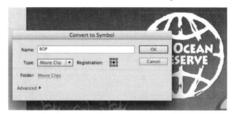

3. **Using the Selection tool, double-click the Logo instance on the Stage to edit the symbol in place. Select Frame 60 on the Layer 1 timeline and press F5 to add a new regular frame.**

   Because movie clips are self-contained animations, every movie clip in the file can last a different amount of time. You also need to be able to control this animation separately from other animations, which is why you are creating the tween inside of the symbol.

4. **Control/right-click between Frame 1 and Frame 60 and choose Create Motion Tween from the contextual menu.**

 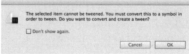

5. **Click OK in the resulting warning. Rename the new Symbol 1 as `BOP Animated`, and then move the symbol into the Movie Clips folder.**

6. **Move the playhead to Frame 1 in the timeline, and then click the symbol instance on the Stage to select it.**

7. **In the Properties panel, open the Style menu in the Color Effect section and choose Alpha. Drag the resulting slider all the way to the left to change the Alpha value to 0.**

    If you don't see the Style menu, click the arrow to the left of the Color Effect heading.

    The Alpha value controls an object's opacity; a value of 0 means the object is not visible.

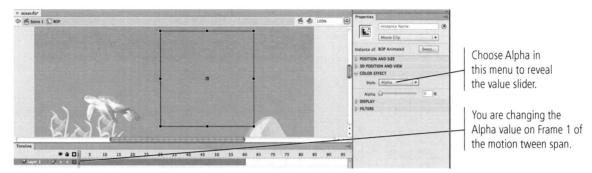

Choose Alpha in this menu to reveal the value slider.

You are changing the Alpha value on Frame 1 of the motion tween span.

8. **Move the playhead to Frame 30 on the Layer 1 timeline, and then click the symbol registration point on the Stage to select the symbol instance.**

    Because the current Alpha value is 0, you can't see the actual object to select it; you have to rely on the registration point to select the instance on the Stage.

9. **In the Properties panel, change the Alpha value back to 100.**

10. **With the instance selected on the Stage, open the Motion Editor panel (if necessary) and move the playhead to Frame 1.**

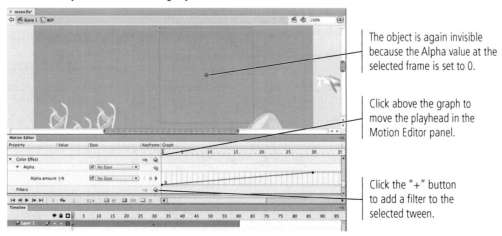

The object is again invisible because the Alpha value at the selected frame is set to 0.

Click above the graph to move the playhead in the Motion Editor panel.

Click the "+" button to add a filter to the selected tween.

11. Scroll through the Motion Editor properties until you find the Filters option. Click the "+" button for Filters and choose Blur from the resulting menu.

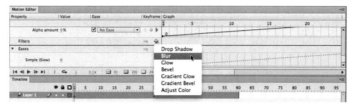

12. Make sure the Blur X and Blur Y values are linked, and then change the Blur X value to 30 px.

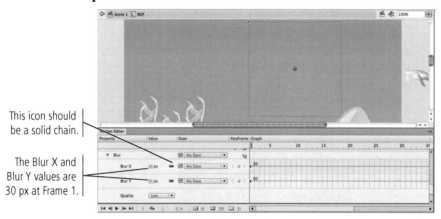

This icon should be a solid chain.

The Blur X and Blur Y values are 30 px at Frame 1.

13. Move the playhead to Frame 30 in the timeline. In the Motion Editor panel, click the Add Keyframe button for the Blur X property, and then change the Blur X value to 0 px.

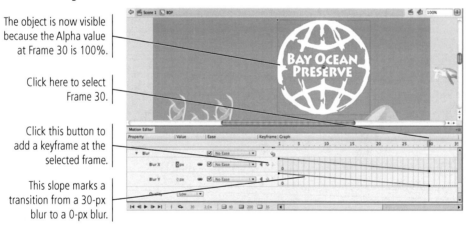

The object is now visible because the Alpha value at Frame 30 is 100%.

Click here to select Frame 30.

Click this button to add a keyframe at the selected frame.

This slope marks a transition from a 30-px blur to a 0-px blur.

14. Return the playhead to Frame 1, and then press Return/Enter to preview the animation.

The BOP Animated movie clip gradually becomes clearer and more visible between Frame 1 and Frame 30.

15. Save the file and continue to the next exercise.

ANIMATE IN 3D

As you know, you can change the X (left/right) and Y (up/down) properties to move an object around the Stage. Flash CS6 adds the third dimension — the Z dimension (near/far or front/back) — so you can animate movie clips in three dimensions.

1. **With ocean.fla open, make sure you are editing the BOP movie clip in the Edit in Place mode.**

2. **Move the playhead to Frame 45 in the timeline, and then choose the 3D Rotation tool in the Tools panel.**

   When a movie clip is selected with the 3D Rotation tool, a visual indicator overlays the selected instance. Each axis (X, Y, or Z) is controlled by a different aspect of the overlay.

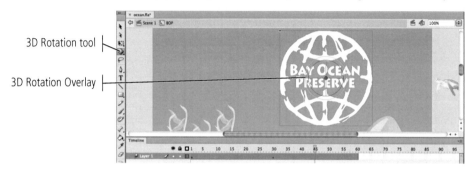

3D Rotation tool

3D Rotation Overlay

   If you look carefully, you can see the related axis in the cursor icon when you move over a specific element in the overlay graphic.

The green line represents the Y axis.

The red line represents the X axis.

The blue circle represents the Z axis.

3. **Click the Y Axis overlay and drag right to rotate the movie clip around the Y axis.**

The shaded area of the overlay shows how far you have rotated the object.

4. **In the Motion Editor panel, make sure the graph view shows at least 60 frames.**

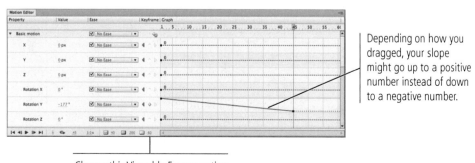

Depending on how you dragged, your slope might go up to a positive number instead of down to a negative number.

Change this Viewable Frames option to 60 to show the entire graph.

5. **Review your Rotation X, Rotation Y, and Rotation Z properties.**

Simply dragging the overlay allows you to rotate the object, but it is not guaranteed to produce the results you want. Before you finalize the animation, you should use the Motion Editor graph to verify that what you get is what you want (a single, full revolution).

6. **In the Motion Editor panel, make sure the Rotation Y property is –180 (one half of an entire revolution), and the other two rotation properties are 0.**

The negative value causes the rotation to occur from left to right. Positive values cause the rotation to occur from right to left.

Note:

If you use a mouse with a scroll wheel, be very careful scrolling through the Motion Editor panel. The menus and hot-text values in the panel are also scrollable. If your cursor moves over one of these values or menus as you try to scroll, you will inadvertently change the values or menu where you scroll.

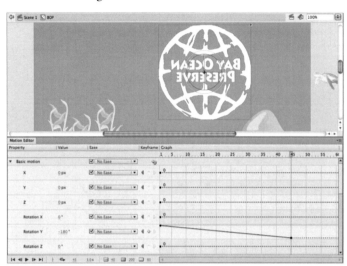

7. **Move the playhead to Frame 60 in the timeline. In the Motion Editor panel, click the Add Keyframe button for the Rotation Y property and change the value for that keyframe to –360.**

Using the negative number, you are causing the object to make one full revolution; the constant downslope in the Motion Editor panel confirms this. If you had simply used 0, the object would revolve halfway, then go back in the opposite direction for the second half of the revolution. Because the slope is steeper from Frames 45–60, the second half of the rotation will occur more rapidly than the first half of the rotation.

8. **Select Frame 30 in the Motion Editor, and then click the Rotation Y Add Keyframe button. Change the value of the new keyframe to 0.**

The two halves of the revolution will now occur at the same speed; none of the revolution will take place until the logo is entirely visible and unblurred.

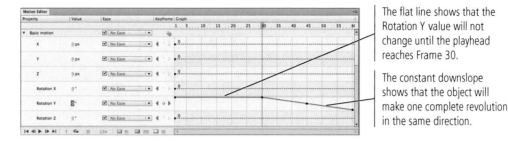

The flat line shows that the Rotation Y value will not change until the playhead reaches Frame 30.

The constant downslope shows that the object will make one complete revolution in the same direction.

9. **Click Scene 1 in the Edit bar to return to the main Stage.**

10. **Press Command-Return/Control-Enter to test the animation.**

Because the Rotation Y value is a continuous slope, the logo completes one entire revolution in the same direction between Frame 30 and Frame 60.

11. **Save the file and continue to the next exercise.**

 ## WORK WITH THE BONE TOOL

The Bone tool can be used to link symbols to one another in a "chain" effect, or to define complex motion for simple objects. As the name implies, you use the Bone tool to create "bones" and "joints" where the bones connect. In this exercise, you use the Bone tool to create a school of fish that swims in a line across the Stage.

1. **With `ocean.fla` open, create a new layer named `Fish` above the Turtle layer in the main timeline.**

2. **Choose Insert>New Symbol. Name the new symbol `School` and choose Movie Clip as the symbol type. Use the Folder link to place the new symbol inside the Movie Clips folder. Click OK to create the new symbol.**

When you use the New Symbol dialog box to create a new symbol from scratch, there is no registration point option because there is no existing art that must be placed in reference to the symbol registration point. When you add objects to the symbol, you can position them relative to the symbol registration point on the Stage.

3. **Using the Selection tool, drag a copy of the Fish1 movie clip onto the symbol Stage. Align the top-right corner of the instance to the symbol registration point.**

When you use the Insert>New Symbol command, the Stage automatically switches to a new blank symbol Stage.

4. **Press Option/Alt, click the Fish1 instance, and then drag left to clone a second instance of the first Fish1. Repeat this process to add two more Fish1 instances.**

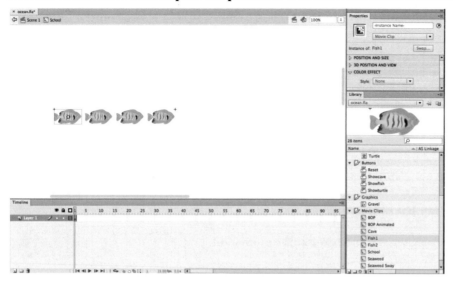

5. **Click Scene 1 in the Edit bar to return to the main Stage.**

6. **With Frame 1 of the Fish layer selected, drag an instance of the School movie clip symbol onto the Stage. Position the School instance in the top half of the Stage, but entirely outside the left edge of the Stage.**

7. **Double-click the placed School instance to edit the symbol in place on the Stage.**

   You're going to make this school of fish swim across the top of the movie. Editing the symbol in place allows you to see how far you need to move the fish to ensure they go all the way across the ocean.

8. **Choose the Bone tool in the Tools panel.**

9. **Click the first Fish symbol instance, hold down the mouse button, and drag to the middle of the second Fish instance.**

   When you use the Bone tool, a new Armature layer is added to the timeline. The objects connected to the armature joints — called IK nodes — are moved from their original layer to the Armature layer.

Note:

*The process underlying the Bone tool is called **inverse kinematics**.*

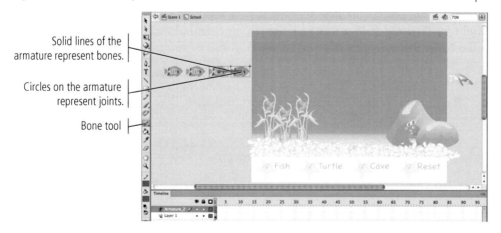

Solid lines of the armature represent bones.

Circles on the armature represent joints.

Bone tool

10. **Click the second Fish instance, hold down the mouse button, and drag to the middle of the third Fish instance.**

11. Click the third Fish instance, hold down the mouse button, and drag to the middle of the fourth Fish instance.

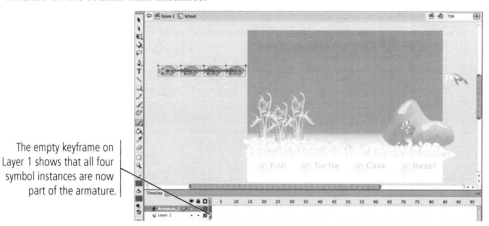

The empty keyframe on Layer 1 shows that all four symbol instances are now part of the armature.

12. Save the file and continue to the next exercise.

 ANIMATE THE BONE ARMATURE

Now that you have the armature structure in place, you can use it to animate the instances' movement in both time and space.

1. With **ocean.fla** open, make sure you are editing the School symbol instance in place on the Stage.

2. In the timeline, click Frame 45 of the Armature layer to select it, and then press F5 to insert a new regular frame.

3. Choose the Selection tool in the Tools panel, and then choose Edit>Select All to select all four instances on the armature.

4. Hold down the Option/Alt key, click any of the selected instances, and then drag the selection until the front fish is about halfway across the Stage and approximately one inch down.

You can Option/Alt-drag instances to move them to a new position without rotating the object or bending the armature.

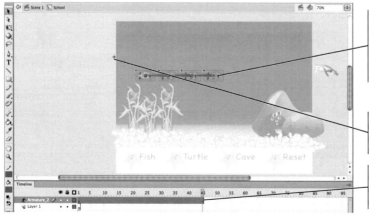

Option/Alt-drag to move the selected objects (and the armature) without bending the armature or rotating the objects.

The registration point shows the symbol's original starting point.

Changing an object on the armature automatically adds a keyframe.

5. **Deselect all instances, click only the back Fish instance on the armature, and then drag up to the top of the Stage.**

As you drag the Fish instance without pressing Option/Alt, notice how other instances on the armature are affected, and how the different objects (including the one you drag) are rotated around the joints.

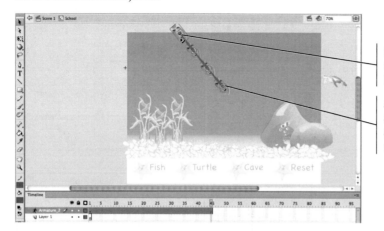

Click and drag to move the object and bend the armature.

All instances move according to the new armature shape, and they rotate around the joints.

6. **Press Option/Alt, click the second Fish instance, and drag down and right until this instance is partially on top of the first Fish instance.**

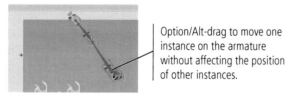

Option/Alt-drag to move one instance on the armature without affecting the position of other instances.

7. **Repeat Step 6 for the other two instances on the armature.**

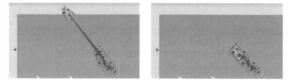

8. **In the timeline, insert a new regular frame at Frame 90 to extend the timeline of the armature tween. Select all four Fish instances on the armature, press Option/Alt, and move all four objects past the right edge of the Stage and up near the top of the Stage.**

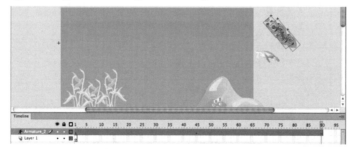

9. Deselect the four objects, select only the fourth Fish instance, and then Option/Alt-drag down to change the direction of the armature.

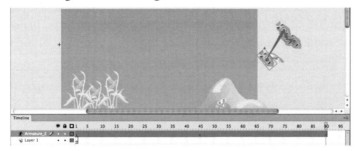

10. Using the Transform panel, change the rotation of the selected fish as necessary so it appears to point up and away from the Stage.

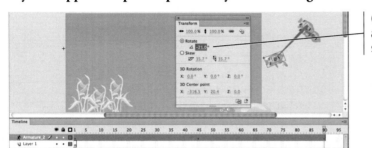

Change the Rotate value to adjust the orientation of the selected symbol instance.

11. Repeat the process from Steps 9–10 to change the relative position of the fish in the school.

We spread out the fish so they no longer overlap, and rotated each to point in the same general direction.

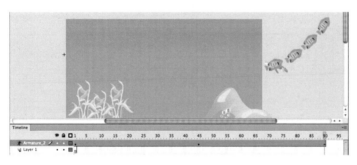

Note:

You can use the Transform panel to precisely control the position and rotation of a specific instance on an armature.

12. Click Scene 1 on the Edit bar to return to the main Stage.

13. Press Command-Return/Control-Enter to test the movie, including the armature animation.

14. Close the Player window and return to Flash. Save the file and continue to the next stage of the project.

Stage 3 **Programming Basic Timeline Control**

You now have all of the pieces in place for the ocean scene, including a number of animations that play automatically when the movie opens. According to the project specs, however, most of these animations should not play until a user clicks the appropriate button at the bottom of the screen. For everything to work properly, you need to complete several additional steps to accomplish the following goals:

- Play the logo animation only once when the movie first loads.

- Play the school of swimming fish when the Fish button is clicked.

- Play the swimming turtle when the Turtle button is clicked.

- Show the cave with the bubbly fish when the Cave button is clicked.

- Hide the cave, stop the turtle and school of fish, and replay the logo animation when the Reset button is clicked.

CONVERT A MOTION TWEEN TO A MOVIE CLIP

At this point, all but the swimming turtle animations are contained inside of various movie clip symbols. In order to add code that controls the turtle animation independently of the main timeline, you need to move the symbol instance and motion tween into a symbol, and then place an instance of that symbol on the Stage.

1. **With ocean.fla open, double-click anywhere in the Frame 1–94 tween on the Turtle layer.**

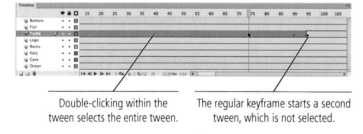

Double-clicking within the tween selects the entire tween.

The regular keyframe starts a second tween, which is not selected.

2. **Press Shift, then click in the Frame 95–184 tween to add it to the selection.**

Because Frame 95 is a regular keyframe and not a property keyframe within the tween, the two tweens are technically treated as separate. You have to manually select both tweens to copy them into a symbol.

3. **Control/right-click inside the selected frames and choose Copy Frames in the contextual menu.**

You are copying both selected tweens.

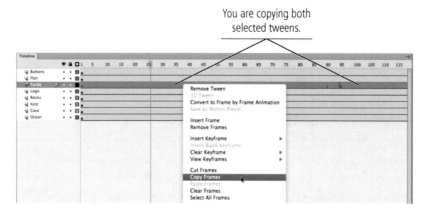

4. **Choose Insert>New Symbol. In the resulting dialog box, type Swimming Turtle in the Name field and choose Movie Clip in the Type menu. Use the Folder link to place the new symbol in the Movie Clips folder, then click OK.**

5. **Control/right-click Frame 1 of the new symbol's timeline and choose Paste Frames from the contextual menu.**

This pastes the full set of contents of the selected frames — including the turtle instance and the motion tweens — inside the symbol.

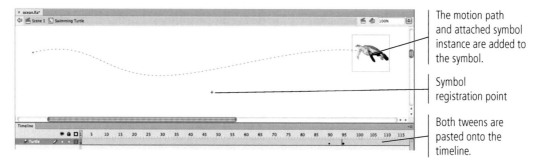

The motion path and attached symbol instance are added to the symbol.

Symbol registration point

Both tweens are pasted onto the timeline.

6. **Click anywhere within the Frame 1–94 tween, then choose Edit>Select All to select the motion path and the turtle symbol instance.**

This command selects both the motion path and the attached instance, so you can drag the entire piece as a single group.

7. **Use the Selection tool to drag the selection until the right end of the motion path aligns to the symbol registration point.**

The existing Swimmer symbol uses the center registration point. You're going to swap symbols, and the registration point in this symbol will align to the position of the previous one. For the tween to work as it does on the main timeline, you need to place the right end of the motion path at the registration point. (This will make more sense shortly).

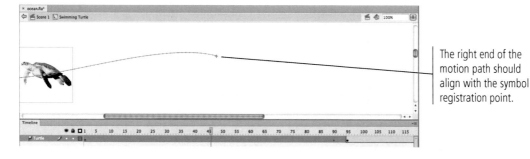

The right end of the motion path should align with the symbol registration point.

8. **Repeat Steps 6–7 for the Frame 95–184 tween.**

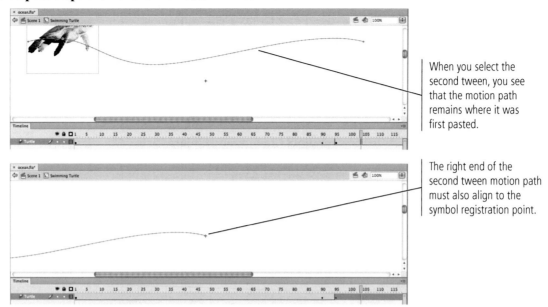

When you select the second tween, you see that the motion path remains where it was first pasted.

The right end of the second tween motion path must also align to the symbol registration point.

9. **Click Scene 1 in the Edit bar to return to the main Stage.**

10. **Control/right-click the Frame 1–94 tween on the Turtle layer and choose Remove Tween from the contextual menu.**

   The tweens now exist in the new Swimming Turtle movie clip symbol, so they are no longer needed on the main timeline.

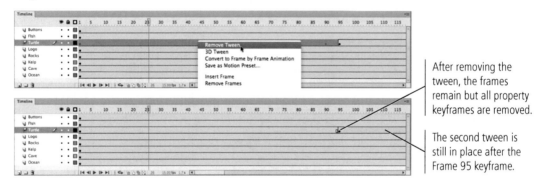

After removing the tween, the frames remain but all property keyframes are removed.

The second tween is still in place after the Frame 95 keyframe.

11. **Repeat Step 10 for the Frame 95–184 tween on the same layer.**

12. **Click Frame 2 of the topmost layer to select it.**

13. **Scroll the Timeline panel as necessary until you see Frame 184. Press Shift, then click the last frame on the bottommost layer to select all frames from 2–184 on all layers.**

Note:

After removing both motion tweens, the Turtle layer is converted back to a regular layer.

Click Frame 2 of the top layer...

...then Shift-click Frame 184 of the bottom layer to select all contiguous frames between the two you click.

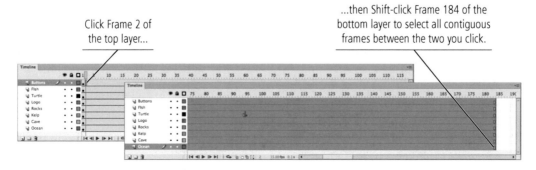

14. **Control/right-click anywhere within the selected frames and choose Remove Frames from the contextual menu.**

    Because all animation in this movie occurs within the timelines of various movie clip symbols, you don't need 184 frames on each layer of the main timeline.

    You can't simply press the Delete key to remove frames. You must use the contextual menu (or the related commands in the Edit>Timeline submenu).

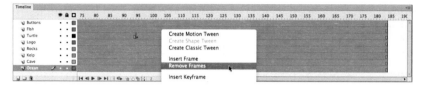

    Because all of the animations are now contained within movie clip symbols, the main timeline now has only a single frame for each layer.

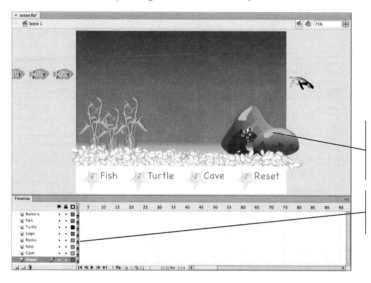

    All objects are still in place because they exist on the Frame 1 keyframes of their respective layers.

    Each layer on the main timeline now has only a single frame.

15. **Click the existing turtle instance on the Stage to select it.**

    As you can see in the Properties panel, this is currently an instance of the Swimmer movie clip. You need to replace it with the Swimming Turtle movie clip.

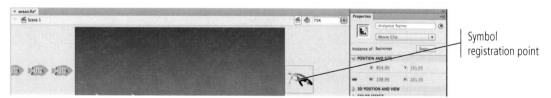

    Symbol registration point

16. **In the Properties panel, click the Swap button. Choose the Swimming Turtle movie clip in the resulting dialog box and click OK.**

When you swap symbols, the registration point of the new symbol is put in exactly the same spot as the registration point of the replacement symbol — which is why you moved the right end of the motion path to align with the symbol's registration point.

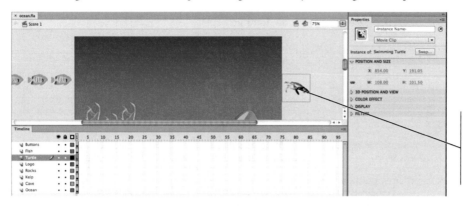

Although you can't see the motion path, you know the center of the turtle was aligned to the right end of the path at Frame 1.

17. **Save the file and continue to the next exercise.**

Prepare Symbol Instances for ActionScript

As you know, when you drag a symbol from the Library panel to the Stage, you create an instance of the symbol. **Named instances** are instances that have been assigned a unique identifier or name, which allows them to be targeted with ActionScript code.

1. **With ocean.fla open, make sure you are working on the main Stage.**

2. **Using the Selection tool, click the School instance to the left of the Stage. In the top field of the Properties panel, type school_mc.**

Use the Properties panel to assign instance names.

Note:

You don't need to name the seaweed instances because those will not be targeted with scripts.

3. **Define names for the rest of the placed instances as follows:**

Remember, the Cave instance was named cave_mc when you created it during the import process.

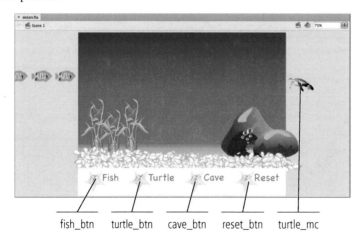

fish_btn   turtle_btn   cave_btn   reset_btn   turtle_mc

Note:

The "_mc" and "_btn" naming convention is common in the world of Flash development. This convention allows programmers to easily recognize the type of a particular instance when they add scripts to the file.

4. **In the Timeline panel, click the Logo layer name to select the layer.**

   Remember, selecting a layer reveals the bounding boxes for all objects on the layer.

5. **Click the symbol registration point to select the logo.**

   Because the logo object has an Alpha value of 0, this is the easiest way to select the instance so you can name it.

6. **In the Properties panel, type `bop_mc` as the instance name.**

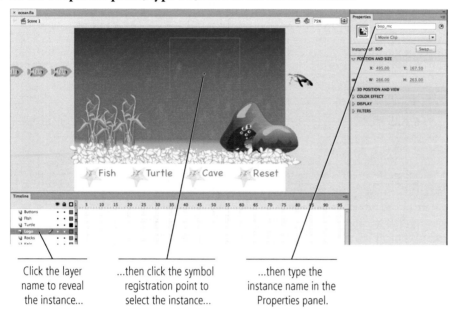

Click the layer name to reveal the instance...

...then click the symbol registration point to select the instance...

...then type the instance name in the Properties panel.

7. **Save the file and continue to the next exercise.**

 ## ADD MOVIE CLIP CONTROLS

If you completed Project 3: Animated Internet Ads, you saw that the Code Snippets panel makes it relatively easy for non-programmers to add basic code to a Flash movie. Items in the panel, written in plain English, automatically add whatever code is necessary to perform the listed function. In this exercise, you will use code snippets to determine what is visible when you first open the movie.

1. **With `ocean.fla` open, open the Code Snippets panel from the Window menu.**

   Different types of common commands are available, grouped into logical sets or folders.

2. **Expand the Actions folder in the Code Snippets panel. Click Stop a Movie Clip to select it, then click the i icon on the right side of the panel to get more information about that snippet.**

Note:

Beginning in Project 5: Gator Race Game, you will work in depth with ActionScript code. Although the Code Snippets panel is supposed to make coding more user friendly for non-programmers, the various options in the panel will make more sense after you have at least a basic understanding of how ActionScript works. You should explore the different Code Snippets on your own, especially after completing the next project.

3. **Select the Swimming Turtle instance on the stage and then double-click the Stop a Movie Clip item.**

Adding the snippet automatically opens the Actions panel, showing the new code.

Code snippets include instructions in the form of comments, which are enclosed by /* and */.

The actual command uses dot syntax, defining what instance is being affected and what will happen to that instance.

The code is added to the selected frame on a new layer named Actions.

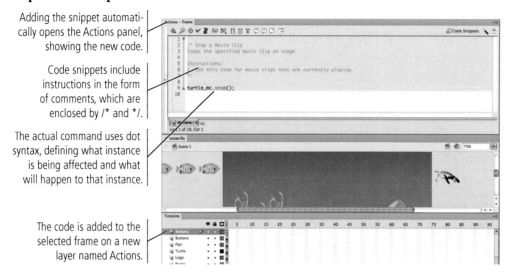

It might seem that by first selecting the object, you are attaching script to that object. Instead, you are telling Flash which object you want the command to address. In ActionScript 3.0, all scripts are placed on the timeline frames rather than attached to specific objects on the Stage. In the Timeline panel, a new layer named Actions is added to the top of the layer stack. (Although not required, this separate layer for the code is a common convention among developers.)

In the Actions panel, which opens automatically when you add the snippet, you can see that the stop command has been added to Frame 1 of the Actions layer. The command references turtle_mc, which is the instance name you defined. In other words, this command stops the turtle_mc instance from playing. The instance is stopped as soon as the main timeline reaches the command; because the command is on Frame 1 of the main timeline, the instance is stopped as soon as the movie opens.

Note:

The format or syntax of the added code is called **dot syntax**: *it first defines the object you are addressing, then adds a dot, then defines what you want to do to that object.*

4. **Select the School instance on the Stage, and then double-click the Stop a Movie Clip item in the Code Snippets panel.**

When you first select the school instance, the Actions panel warns that code can't be attached directly to objects on the Stage.

The code (including comments) is added to Frame 1 of the Actions layer, after the existing code.

This tab shows where the code is added (Layer:Frame).

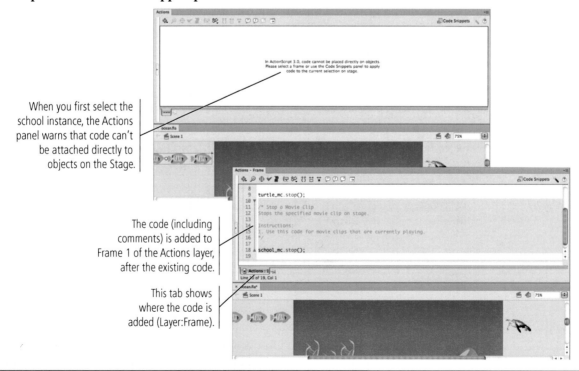

5. **Select the Cave instance and then double-click the Stop a Movie Clip item.**

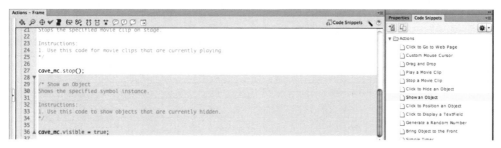

6. **Select the Cave instance again and double-click the Show an Object item.**

   Unfortunately, there is no snippet to simply hide an item without requiring the user to click something. (The Click to Hide an Object item is not appropriate because you want to hide the instance as soon as the movie opens, and not as a reaction to the user's click.)

   However, the added Show an Object statement shows that the value "true" is attached to the visible property of the instance. To make the instance *not* visible, you simply have to change the property's value in the code.

7. **In the Actions panel, change the word true to `false` on Line 36.**

8. **Press Command-Return/Control-Enter to test the movie.**

   The three movie clips are stopped, and the cave is hidden. However, the logo animation still plays continuously and you want it to play only once when the movie opens.

9. **Close the Player window and return to Flash.**

10. **In the Library panel, double-click the BOP movie clip symbol icon to enter into the symbol's Stage.**

You can't stop the BOP instance on the main Stage because you want the animation to play one time when the movie first opens. To accomplish this goal, you need to add a stop command to the end of the movie clip timeline.

11. **Move the playhead to the last frame in the timeline, then click the logo instance on the Stage to select it.**

12. **Expand the Timeline Navigation folder in the Code Snippets panel, then double-click the Stop at this Frame item to add the necessary code.**

Timeline Navigation snippets can be used to control the timeline (and thus, the playback) of specific symbols. The Stop at this Frame command affects the active timeline at the selected frame, so a specific instance is not referenced in the resulting code.

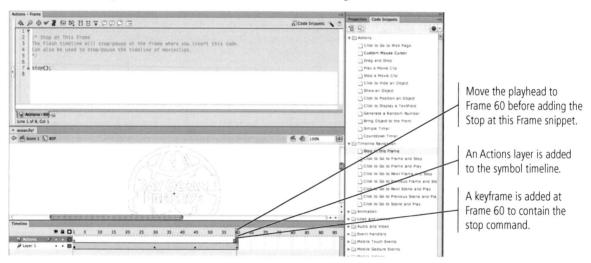

Move the playhead to Frame 60 before adding the Stop at this Frame snippet.

An Actions layer is added to the symbol timeline.

A keyframe is added at Frame 60 to contain the stop command.

13. **Click Scene 1 in the Edit bar to return to the main Stage, then press Command-Return/Control-Enter to test the movie.**

The logo animation now plays only once and then stops.

14. **Close the Player window and return to Flash, then save the file and continue to the next exercise.**

ADD EVENT HANDLERS TO BUTTONS

As you saw in the previous exercise, ActionScript 3.0 requires code to be attached to a frame on the timeline. To affect a specific object on the Stage, you have to use the defined instance names as reference in the code. Programming a button requires more complex code called an **event handler**, with (at least) two referenced objects — the event that triggers the action, and the name of the function that is affected by the event.

The Code Snippets panel includes options for creating event handlers with the proper syntax, although defining what occurs as a result of the event might require a few workaround steps. Even using code snippets, it is helpful if you are familiar with the basics of ActionScript code.

1. **With ocean.fla open, expand the Timeline Navigation folder in the Code Snippets panel.**

2. **Select the Fish button instance on the Stage, then double-click the Click to Go to Frame and Play item in the Code Snippets panel.**

The added code is not attached to the selected instance; it is added to Frame 1 of the existing Actions layer, after all code that you already added.

The selected instance becomes the object that can trigger the function.

The specific trigger (CLICK) is defined inside the event listener.

This is the event listener statement.

This is the function that is called when the defined event occurs.

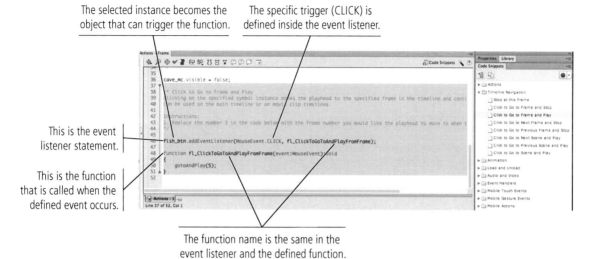

The function name is the same in the event listener and the defined function.

Although you do not need to know every detail of ActionScript code to use Code Snippets, there are a few important points that you should understand:

- The first line of added code defines what will happen to trigger the function (the **event listener**). Inside the parentheses, the MouseEvent.CLICK statement says that the following function will be called when the fish_btn instance is *clicked*.

- The first line of code includes a **function name** immediately before the closing parenthesis. That same name is defined at the beginning of the following function so the file knows which function to play when the defined button is clicked.

- The **function body** — between the two braces — defines what occurs when the event is triggered.

3. **Place the insertion point before the gotoAndPlay command inside the function body and click the Insert Target Path button at the top of the Actions panel.**

The statement inside this function currently says "go to Frame 5 and play the timeline". Because the statement does not address a specific instance, the code will be interpreted to mean the timeline on which the code is placed (in this case, the main Stage timeline). You want the function to play the School movie clip instance, so you have to add the appropriate reference.

Insert Target Path button

Click here to place the insertion point.

4. Choose school_mc in the Insert Target Path dialog box and click OK.

This dialog box lists every nameable instance on the Stage, so you can choose from the list instead of trying to remember the exact name you defined for a specific object.

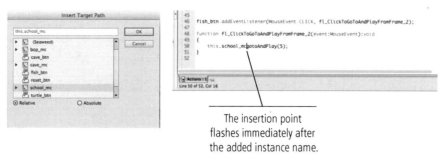

The insertion point flashes immediately after the added instance name.

Note:

If an item in the Insert Target Path dialog box appears in parentheses, the instance is not yet named; selecting it will prompt you to define an instance name.

The word "this" in the instance name refers to the timeline where the code is written. The overall statement is essentially saying, "On *this* timeline, you will find something called school_mc. Tell school_mc to execute its gotoAndPlay() method."

Note:

The word "this" is automatically included when you use the Insert Target Path dialog box. It is not strictly necessary in this case because the instances are all on the main timeline of the file you are building, but you do not need to remove it from the code.

5. Type a period (dot) immediately after the instance name to separate it from the gotoAndPlay command.

Remember, dot syntax requires a period separating the different parts of code — in this case, the instance that will be affected by the gotoAndPlay command.

6. Change the number inside the parentheses to 1.

This number defines the frame number of the instance that will be called when a user clicks the button. You want the instance to start at the beginning, so you are changing the frame reference to 1.

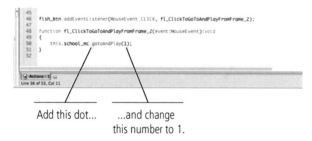

Add this dot...   ...and change this number to 1.

7. Repeat Steps 2–6 to create an event handler for the Turtle button that plays the turtle_mc movie clip instance from Frame 1.

Notice that a sequential number is added to the function name, both in the event listener statement and in the first line of the function. Every function in a file must have a unique name so it can be called when necessary.

8. Repeat Steps 2–6 to create an event handler for the Cave button that plays the cave_mc movie clip instance from Frame 1.

```
60
61    turtle_btn.addEventListener(MouseEvent.CLICK, fl_ClickToGoToAndPlayFromFrame_3);
62
63    function fl_ClickToGoToAndPlayFromFrame_3(event:MouseEvent):void
64    {
65        this.turtle_mc.gotoAndPlay(1);
66    }
67
68    /* Click to Go to Frame and Play
69    Clicking on the specified symbol instance moves the playhead to the specified frame in the timeline and conti
70    Can be used on the main timeline or on movie clip timelines
71
72    Instructions:
73    1. Replace the number 5 in the code below with the frame number you would like the playhead to move to when t
74    */
75
76    cave_btn.addEventListener(MouseEvent.CLICK, fl_ClickToGoToAndPlayFromFrame_4);
77
78    function fl_ClickToGoToAndPlayFromFrame_4(event:MouseEvent):void
79    {
80        this.cave_mcgotoAndPlay(1);
81    }
82
```

9. **Select the cave_mc instance on the Stage (not the Cave button) and double-click the Show an Object item in the Actions folder of the Code Snippets panel.**

The Cave button needs to show the instance before it plays, so this button function needs two lines of code. However, the Code Snippets panel was not designed to add code inside of an existing function. The Show an Object snippet is added at the end of the existing code, *after* the function that is called when a user clicks the cave_btn instance. As a work-around, you have to add the necessary command and then paste it into the function body.

The "show" command is added outside of the existing function.

10. **Select the line of code that makes the cave_mc instance visible (Line 90 in the above example) and press Command/Control-X to cut the selected code.**

You have to use the keyboard shortcuts to copy (Command/Control-C), cut (Command/Control-X), or paste (Command/Control-V) code in the Actions panel. The menu commands do not work while you are active in the Actions panel.

11. **Place the insertion point immediately after the opening brace in the function (Line 79 in our example) and press Return/Enter to add a new line in the function body. Press Command/Control-V to paste the code that you cut in Step 10 to the function body.**

12. **Delete the extra lines of comments at the end of the code.**

Note:

The comments are the gray lines that are surrounded by /* and */. After you moved the actual code into the function body (Steps 10–11), the comments from the original code are unnecessary. Deleting them helps keep the code pane as clean as possible.

13. **Press Command-Return/Control-Enter to test the movie.**

Test the buttons that you just programmed. Each should play the relevant movie clip.

14. **Close the Player window and return to Flash, then save the file and continue to the next exercise.**

COMBINE MULTIPLE EVENT HANDLERS IN A BUTTON

The final element of this project is the Reset button, which needs to accomplish a number of things. As the name suggests, clicking this button should restore the movie to exactly what happens when it first opens. Because the symbols in this movie are controlled with code, you need to add more code that defines what happens when this button is clicked.

1. **With `ocean.fla` open, select the Reset button on the Stage.**

2. **Double-click the Click to Go To Frame and Stop item in the Timeline Navigation folder of the Code Snippets panel.**

3. **Inside the function body, add a reference to the school_mc instance before the gotoAndStop command, and change the referenced frame inside the parentheses to `1`.**

4. **Select the line inside the function body (Line 96 in our example) and copy it.**

5. **Place the insertion point at the beginning of the existing function body (Line 96) and paste the copied code three times.**

6. **Change the second and third lines to reference the turtle_mc and cave_mc instances, respectively.**

7. **Change the fourth line to reference the bop_mc instance, and change the command to `gotoAndPlay`.**

   When a user clicks the Reset button, the logo animation should replay from the first frame. It will replay only once because you already added the stop command inside the movie clip's timeline.

These lines will stop the first three animations and effectively hide the first two on the Stage.

This command will cause the logo animation to play once.

8. **Select the Reset button on the Stage again, and double-click the Click to Hide an Object item in the Actions folder of the Code Snippets panel.**

9. **In the resulting function, change the referenced instance to `cave_mc`.**

   By default, this snippet hides the object that triggers the function. Because you want to hide the cave and not the Reset button, you need to change the instance name inside of the function body.

10. Cut the function body (Line 113 in our example) from the code and then paste it inside the body of the previous function.

In this case, it is not necessary to have two separate event handlers for the same button.

Note:

You don't need to hide the turtle and school instances because Frame 1 of those movie clips exists entirely out of the Stage area so they won't be visible when their timelines are reset.

Change the reference to cave_mc...

...then cut this line from this function...

...and paste it into this function.

11. Delete all code related to the second reset_btn event handler (Lines 103–115 in our example above).

Because you combined this function body with the other event handler for the same button, this code is no longer necessary.

12. Press Command-Return/Control-Enter to test the movie.

Test the buttons that you just programmed. The Reset button should stop and hide all animations except for the swaying seaweed.

13. Close the Player window and return to Flash, then save the file and close it.

fill in the blank

1. Objects from a Photoshop file should be created on _____ if they need to be managed separately when imported into Flash.

2. The _____ tool places multiple instances of a selected symbol; you can use the Properties panel to randomize the instances that are placed.

3. The _____ defines the point around which object transformations are made.

4. The X and Y position of a symbol instance is based on the _____.

5. You can use the _____ panel to define numeric scale, skew, and rotation values for the selected object.

6. Animation in a _____ requires the same number of frames on the timeline where the instance is placed.

7. Animation in a _____ plays regardless of the number of frames in the timeline where instances are placed.

8. _____ is the format required by ActionScript 3 code.

9. Using ActionScript 3, code is attached to a specific _____, and uses instance names to address specific objects.

10. In ActionScript, a(n) _____ includes a statement defining the instance that triggers an event and the function that is called when the defined event occurs.

short answer

1. Briefly explain the concept of "tweening."

2. Briefly explain the difference between a graphic symbol and a movie clip symbol.

3. Briefly define an event handler.

Use what you learned in this project to complete the following freeform exercise.
Carefully read the art director and client comments, then create your own design to meet the needs of the project.
Use the space below to sketch ideas; when finished, write a brief explanation of your reasoning behind your final design.

art director comments

The media director for the Chicago Wild Animal Park is re-branding the facility from the "City Zoo" image it has had for the past twenty years. He has hired you to create a series of animated icons for the park's new interactive Web site.

To complete this project, you should:

❏ Use the drawing tools to develop icons for the main areas of the facility. Create each of the icons in the same shape and size, and use the same general style for each.

❏ Add some kind of animation to each icon. Use any combination of frame animations, shape tweens, and/or motion tweens.

client comments

We've gotten rid of the cages and created realistic natural habitats for the animals. Our main goals now are rehabilitation, preservation, and education. We're going to have educational programs and exhibits throughout the facility, but we don't want people to be scared off by the idea of learning!

We have many international visitors, so most of our collateral — including our new Web site — is based on images that can be understood in any language. Although there will be text as well, the icons should very clearly indicate what users will find when they click on any specific one (even if they can't read the words).

We need a series of six animated icons that will label the different areas of the facility. The six main sections are: the tropics, the desert, the Arctic, the forest, the ocean, and the sky. There will also be a special children's section that needs its own icon.

project justification

_____

_____

_____

_____

_____

_____

_____

_____

_____

_____

_____

_____

_____

Project Summary

This project incorporated artwork that was created in Adobe Photoshop, which is a common development workflow. You also worked with symbols that were created in another Flash file, which is also a common collaborative process.

The second stage of this project focused on different methods of creating animation — frame-by-frame to move something in jumps, motion tweening to move objects smoothly, tweening to change only certain properties over time, and even tweening to rotate an object in three-dimensional space. To create these animations, you have also learned a number of techniques for transforming objects on the Stage; the Transform panel, the Free Transform tool, the Properties panel, and the Motion Editor panel all play valuable roles in Flash development.

Finally, you were introduced to the object-oriented model of ActionScript 3 when you added button controls using the Code Snippets panel. With very little (if any) knowledge of coding or programming, you were able to use the built-in functionality to meet the project's interactive requirements.

Import artwork from an Adobe Photoshop file

Import symbols from an external Flash library

Create frame animations to move objects over time

Animate Alpha properties, graphic filters, and 3D attributes

Animate multiple symbols on an armature

Define movie clip symbols to create tween animations

Define a graphic symbol and spray multiple instances

Add code to control the playback of various movie clip instances.

Gator Race Game

Your client publishes children's books. The marketing manager has found that the more time kids spend on the company's Web site, the more sales the company makes. To entice children to visit the site more often and for longer periods, the client hired your agency to build an interactive children's game featuring Baggy Gator, one of the publisher's most well-known characters.

This project incorporates the following skills:

❑ Preparing objects on the Stage for ActionScript

❑ Getting help in the Actions panel

❑ Using functions to control movie clip timelines and object properties

❑ Creating event handlers and custom functions

❑ Working with variables and arrays

❑ Scripting loops and conditional statements

❑ Adding sound with ActionScript

Project Meeting

client comments

We're adding a number of kid-friendly tools to our Web site, including a talking introduction screen and interactive games. The more time kids spend on our site, the more they see (and learn to recognize) different characters from our books — and the more they want to read the books where those characters are involved.

We want the first game to feature Baggy Gator and his friends. We thought a foot race would be familiar enough for young children to understand and non-violent enough for parents to allow.

There are a few technical requirements that we think are important:

- The game should be no longer than 10 seconds to accommodate children's short attention spans.

- There should be constant action to keep the kids interested and engaged.

- There should be sound effects while the race is running.

art director comments

The client mentioned three specific requirements, so make sure you keep those in mind while you develop this piece.

- The entire movie needs to be 10 seconds long. Translate that to frames when you combine all the pieces.

- The movie needs to have constant action. Different characters need to "take the lead" at different points in the race. If you move one character every second, that should meet this requirement.

- Add sound effects. Races have audiences, and audiences cheer. I think a cheering crowd sound will help kids to get into the spirit of the race. Let's also add some "running feet" sounds to make the running animation seem more realistic.

I asked our illustrator to build the necessary graphics for the project. As our resident coding expert, I need you to put all the pieces together and make the movie work.

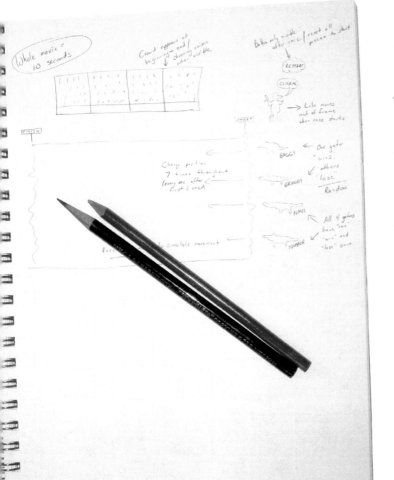

project objectives

To complete this project, you will:

- ❏ Define instance names
- ❏ Add motion to the background
- ❏ Use functions to control movie clip timelines
- ❏ Create scripts to control object properties
- ❏ Write event handlers and functions
- ❏ Create an event listener
- ❏ Define a custom function
- ❏ Script with variables and random numbers
- ❏ Create an array to store racer positions
- ❏ Define a loop to evaluate array elements
- ❏ Use an if condition to call movie clip frames
- ❏ Program the Restart button functionality
- ❏ Test the movie and change cursor styles
- ❏ Create access for sound files in the library
- ❏ Develop scripts to call sounds

Stage 1 Preparing for ActionScript

The best way to start a complicated project such as this one is to develop a clear plan of what needs to take place. You need to complete a number of different tasks to create the final game. Some of these tasks, such as the various actions for each character, can be accomplished with frame animations and tweens. Other tasks require more complicated interaction between the elements of the movie. Fortunately, ActionScript provides the mechanism for creating and controlling this type of interaction.

This project was designed to give you an idea of what you can do with ActionScript. To enhance your marketability as a professional Flash programmer, we encourage you to continue your ActionScript education beyond this book.

REVIEW THE FILE STRUCTURE

When you work as part of a development team, it's always best to review the various elements in a file before you begin any programming activities.

1. Download **FL6_RF_Project5.zip** from the **Student Files Web page.**

2. **Expand the ZIP archive in your WIP folder (Macintosh) or copy the archive contents into your WIP folder (Windows).**

    This results in a folder named **Gators**, which contains the files you need for this project. You should also use this folder to save the files you create in this project.

3. **In Flash, choose File>Open and navigate to the WIP>Gators folder. Select gator_race.fla and click Open.**

    The basic file for this movie has already been created, with most of the pieces in place. Each element (characters, buttons, and the racetrack background) is a different movie clip on a separate layer. Your task is to turn the various symbols into an interactive game.

4. **Choose View>Magnification>Show All to see all contents of the movie.**

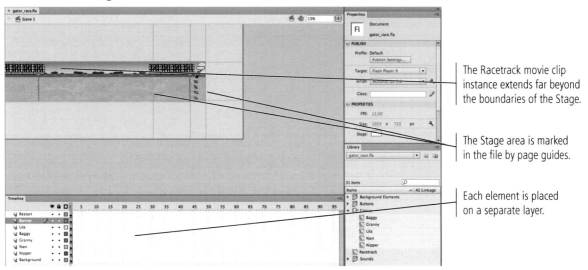

The Racetrack movie clip instance extends far beyond the boundaries of the Stage.

The Stage area is marked in the file by page guides.

Each element is placed on a separate layer.

5. **Choose View>Magnification>Fit in Window to enlarge the movie so the Stage area (but not the entire movie) is visible.**

6. **In the timeline, hide the Background layer.**

   The Racetrack movie clip (which occupies the Background layer) is larger than the actual Stage, so the actual Stage area is visible when the Background layer is hidden. The original designer placed guides to mark the edges of the Stage, which will be helpful as you create the interactive elements of the game.

   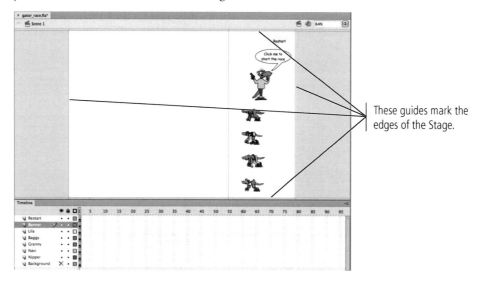

   These guides mark the edges of the Stage.

7. **Show the Background layer again.**

8. **Open the Library panel and expand the Gators folder.**

9. **Select the Baggy movie clip in the Library panel and click the Play button in the preview area.**

   Each racer movie clip was created with frame animations to simulate the effect of getting set for the race (Frame 5), running (Frames 10–17), crying (Frames 20–23), and cheering (Frames 26–29).

   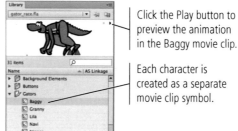

   Click the Play button to preview the animation in the Baggy movie clip.

   Each character is created as a separate movie clip symbol.

   The crying frame animation will be called if the character loses the race. The cheering animation will be called if the character wins the race.

10. **Select the Lila movie clip in the library and click the Play button.**

    Lila is the race starter. This movie clip includes a frame animation with Lila tapping her foot while waiting for a user to start the race (Frames 1–20). Beginning at Frame 21, Lila looks up from her watch, and then raises the starting pistol and fires (Frame 24).

11. **Save the file and continue to the next exercise.**

DEFINE INSTANCE NAMES

Instance names are required if you want to target individual objects using ActionScript. You need to provide an instance name for a symbol instance before you can refer to that object in your code.

1. **With gator_race.fla open, use the Selection tool to click the Lila movie clip instance on the Stage.**

2. **In the Properties panel, type lila_mc in the Instance Name field.**

Use this field to name the selected instance.

Instance of the Lila movie clip on the Stage

3. **Using the same technique, assign unique instance names to the remaining elements, using the instance names listed in the following table.**

Object to Select	Instance Name
Baggy movie clip instance	char1_mc
Granny movie clip instance	char2_mc
Navi movie clip instance	char3_mc
Nipper movie clip instance	char4_mc
Racetrack movie clip instance	racetrack_mc
Banner movie clip instance	banner_mc
Restart button instance	restart_btn

Note:

Although it might seem as though you should use the character names as the instance names, using the numerals 1–4 to name the instances will be important later when you write the ActionScript to make the game work properly.

The characters "_mc" and "_btn" in the instance names help differentiate movie clip instances from button instances. Although not strictly necessary, many Flash programmers use this naming convention.

restart_btn

banner_mc

lila_mc

char1_mc

char2_mc

char3_mc

char4_mc

racetrack_mc

4. **Save the file and continue to the next exercise.**

ADD MOTION TO THE BACKGROUND

The Background movie clip is currently a movie clip symbol, with additional nested movie clip animations. The individual racer movie clips and the race starter movie clip already contain frame animations. After you add the necessary looping, the frame animations will make it appear as though the racers are running along the track. The overall effect is not complete, however, because the background does not yet move in time with what appears to be running gators. You correct that problem in this exercise.

1. **With gator_race.fla open, double-click the racetrack_mc instance on the Stage to edit the movie clip symbol in place.**

2. **Select Frame 6 of the Track layer and insert a new keyframe.**

    To add a new keyframe, choose Insert>Timeline>Keyframe, press F6, or Control/right-click the frame and choose Insert Keyframe from the contextual menu.

    When the race starts, the running gators take five frames to "get set" before they begin running. Adding this keyframe keeps the track in place until the gator animations actually start "running."

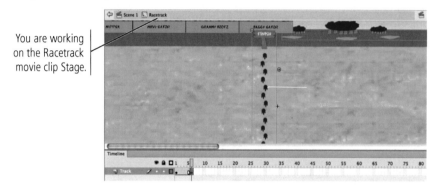

You are working on the Racetrack movie clip Stage.

3. **Select Frame 100 of the Track layer and add a new regular frame.**

    To add a new regular frame, choose Insert>Timeline>Frame, press F5, or Control/right-click the frame and choose Insert Frame from the contextual menu.

    Later in this project, you will add ActionScript to change the racers' positions eight times (i.e., every second); the final movement occurs at Frame 96. Shortly after the final change of position, the race "ends" at Frame 100.

Note:

We will not continue to repeat the complete instructions for adding a new frame or keyframe to the timeline.

4. **Control/right-click anywhere between the keyframe at Frame 6 and the regular frame at Frame 100 and choose Create Motion Tween from the contextual menu.**

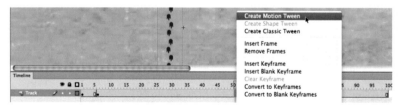

5. **Read the resulting warning and click OK to convert the objects on the Stage to a single movie clip symbol.**

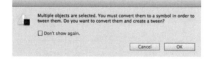

6. In the Library panel, rename the new Symbol 1 `Racetrack_Group`.

7. **Select Frame 100 on the Track layer. Using the Selection tool, click the Racetrack_Group movie clip on the Stage to select it. Press Shift and drag right until the finish line graphic is on top of the guide where the starting line was located (as shown in the following image).**

   If you switch between the Frame 1 and 100, the background jumps from its original location to the ending position.

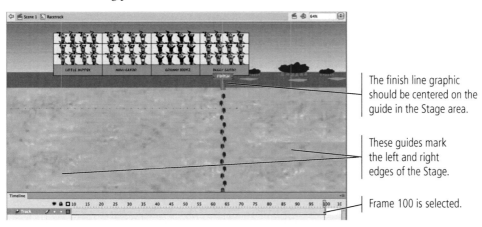

The finish line graphic should be centered on the guide in the Stage area.

These guides mark the left and right edges of the Stage.

Frame 100 is selected.

8. **Select Frame 120 on the Track layer and add a new regular frame.**

   If the entire movie lasts 10 seconds, at 12 fps, 120 frames are required. You want the race to end before the actual movie file, which is why you placed the finish line at Frame 100 instead of Frame 120.

   Adding the new frame extends the tween after the last frame. The symbol on the tween remains in the same position on Frame 120 as it was on Frame 100; a keyframe is added at Frame 100 to mark that position.

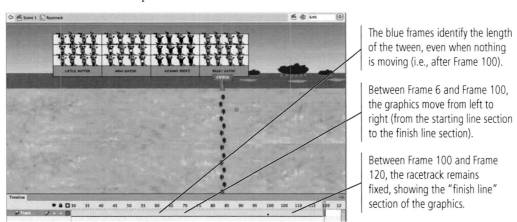

The blue frames identify the length of the tween, even when nothing is moving (i.e., after Frame 100).

Between Frame 6 and Frame 100, the graphics move from left to right (from the starting line section to the finish line section).

Between Frame 100 and Frame 120, the racetrack remains fixed, showing the "finish line" section of the graphics.

9. **Click Scene 1 on the Edit bar to return to the main Stage.**

10. Press Command-Return/Control-Enter to test the movie.

As the movie plays, the background moves smoothly to the right. When the playhead reaches the final frame of the background timeline, the timeline returns to Frame 1 and automatically plays again.

If you watch the individual characters, you see the different animations included in each movie clip. At this point, each character simply plays from start to finish, and then starts over again. Later in the project, you will add ActionScript to control which frames play at specific points throughout the movie.

The racetrack moves...

...but the characters remain fixed in place.

11. Close the Player window and return to Flash.

12. Save the file and continue to the next exercise.

DEFINE FRAME LABELS

As you have already learned in previous projects, you can use frame numbers to direct the playhead to a specific point in the timeline. Numbers are vague, however, so it is often helpful to use a more meaningful method to describe a specific frame or series of frames. In this exercise, you are going to use frame labels to identify various parts of the individual characters' movie clip timelines. Later you will use those labels to move the timeline to the correct location on the symbols' timelines.

1. With gator_race.fla open, double-click the Baggy symbol icon in the Library panel to enter into that symbol.

Remember, each racer's movie clip includes frame animations for getting set (Frame 5), running (Frames 10–17), crying when the character loses (Frames 20–23), and cheering when the character wins (Frames 26–29).

2. Add a new layer named Labels to the top of the symbol timeline.

You are creating the frame labels on a separate layer in this movie clip so that you can copy the Labels layer and paste it into the other racers' timelines, rather than manually defining the same labels in each movie clip.

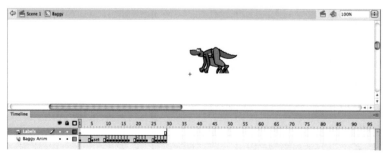

3. **Select the Frame 1 keyframe on the Labels layer, then type waiting in the Label Name field of the Properties panel.**

   A red flag in the Timeline panel indicates a frame that has a defined label. The actual label appears to the right of the keyframe where the label is applied.

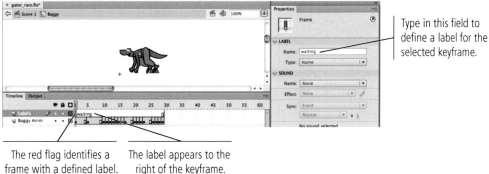

   Type in this field to define a label for the selected keyframe.

   The red flag identifies a frame with a defined label.

   The label appears to the right of the keyframe.

4. **Insert a new keyframe at Frame 5 of the Labels layer.**

   Because the Frame 1 keyframe is blank, the new keyframe on Frame 5 is also blank. The frame label from the previous keyframe is not copied to the new keyframe.

5. **With the Frame 5 keyframe selected, type set in the Label Name field of the Properties panel. Press Return/Enter to apply the label.**

   If another keyframe appears too soon after a keyframe with a defined label, some or all of the defined label might not be visible in the Timeline panel.

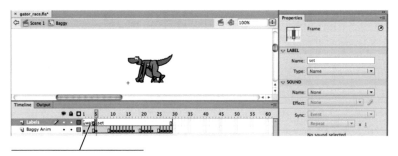

   Labels are only visible up to the next keyframe on the layer.

6. **Repeat Steps 4–5 to add keyframes and define the following labels on the Labels layer:**

Keyframe	Label
Frame 10	running
Frame 20	crying
Frame 26	cheering

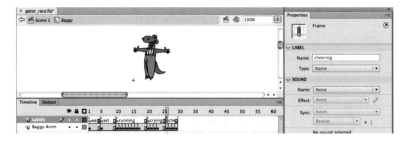

7. **Control/right-click the Labels layer name in the Timeline panel. Choose Copy Layers from the contextual menu.**

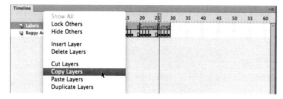

8. **In the Library panel, double-click the Granny symbol icon to enter into that symbol.**

   You do not need to return to the main movie timeline (Scene 1) before entering into a different symbol.

9. **Control/right-click the Granny Anim layer in the Timeline panel, then choose Paste Layers in the contextual menu.**

   You are pasting the layer into the Granny symbol timeline.

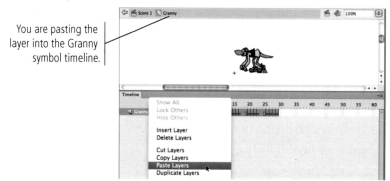

   The Labels layer that you defined in the Baggy symbol is now added to the Granny symbol. Because each of the racers' timelines has the same structure, the keyframes and labels you already defined are appropriate for each racer's symbol.

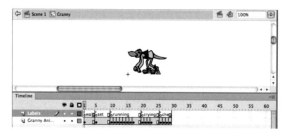

10. **Repeat Steps 8–9 to paste the Labels layer into the Navi and Nipper symbols.**

   Unless you copy something else in the interim, you do not need to re-copy the Labels layer before pasting it into each symbol.

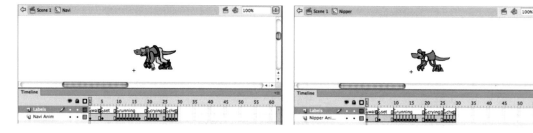

11. **Double-click the Lila symbol icon in the Library panel to enter into that symbol.**

This symbol has a different set of motions than the racers, so it requires different labels.

12. **Using the same process as you used in the Baggy symbol, add a new Labels layer to the Lila movie clip timeline.**

- **Define waiting as the label for the Frame 1 keyframe.**
- **Add a keyframe at Frame 21 and define fire as the label for that keyframe.**

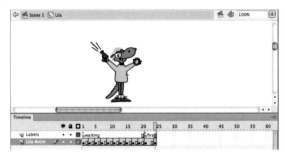

13. **Click Scene 1 in the Edit bar to return to the main Stage.**

14. **Save the file and continue to the next exercise.**

 MOVE THE STARTER OFF THE STAGE

In the Lila movie clip, the race starter looks up from her watch at Frame 21, and then fires the gun at Frame 24. After Lila starts the race, she needs to move off the Stage to complete the illusion of the racers running away from the starting line. To create the desired effect, you need to add a motion tween within the Lila movie clip.

1. **With gator_race.fla open, double-click the lila_mc instance on the Stage to edit the movie clip in place.**

2. **Insert a new regular frame at Frame 35 of the Lila Anim layer, and then create a motion tween between Frames 24 and 35.**

You are working in the Lila symbol.

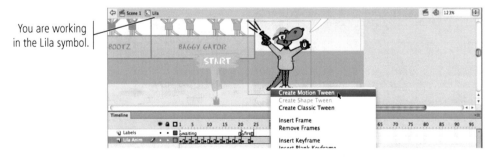

3. **Read the resulting warning, and then click OK to convert the objects on the Stage to a single movie clip symbol.**

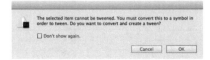

4. **In the Library panel, rename the new Symbol 1 Lila_Move and drag it into the Gators library folder.**

5. **Select Frame 35. Click the lila_mc instance on the Stage, and then Shift-drag the movie clip right until the instance is outside the Stage boundary.**

You might need to scroll or zoom out to show the gray area to the right of the Stage (where you want to place the Lila instance).

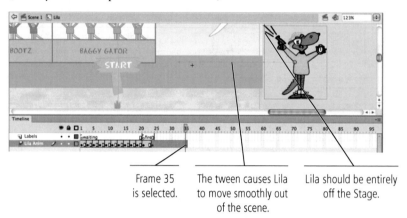

Frame 35 is selected.

The tween causes Lila to move smoothly out of the scene.

Lila should be entirely off the Stage.

Note:

You are not moving the banner because it should simply disappear when the race starts (instead of moving to the right). Later, you will add ActionScript to control the visibility of this element.

6. **Click Scene 1 on the Edit bar to return to the main Stage.**

7. **Press Command-Return/Control-Enter to test the movie.**

Lila now moves out of the scene after she fires the starting pistol. The timelines still loop, however, so Lila reappears in the scene after Frame 35.

Lila moves out of the scene after firing, but her timeline automatically loops after Frame 35, so she reappears.

8. **Close the Player window and return to Flash.**

9. **Save the file and continue to the next stage of the project.**

Stage 2 Working with ActionScript 3

ActionScript moves basic animation to a higher level, allowing the end user to interact with a movie. As you add more complex scripts, users can move objects around, change their colors, toggle soundtracks on and off, and even play games. Experienced developers use ActionScript to enable Flash to communicate with other programming languages (including XML) to create complex, data-driven Flash sites for electronic commerce, news feeds, and various forms of entertainment.

ActionScript 3, the current version, is a complex language that adds incredible programming power to Flash. In Project 4: Ocean Animation, you got a brief introduction to ActionScript 3 when you used built-in code snippets to add user controls in the ocean animation. Although it can be intimidating at first — especially for graphic designers with no programming experience — it is well worth the effort to understand the basics of ActionScript 3, if for no other reason than to enhance your understanding of what you can create with this powerful tool.

To complete the gator race game, you need to create a number of functions that control the various elements of the game (the race characters and the Background movie clip, as well as the button that allows the user to start the game). You will use ActionScript 3 to attach the necessary scripts to specific frames in the timeline — both in the main Stage and within the individual movie clips.

Planning Your Script

It is better to prepare your ActionScript in a logical way before you add a single line of code into the Actions panel. You might use an entire notebook to plan your movie before you begin coding, but doing so will save hours of troubleshooting time later. There are five questions you can use as a guide for applying ActionScript:

- **What are you trying to do with the script?** You should know what you are trying to achieve before you begin. Something as basic as the difference between tinting a solid color and tinting a gradient could change what script you use; be specific about what you want the script to do.

- **What are you trying to affect?** ActionScript can modify or reference the entire movie, external movies, external scripts, movie clips, buttons, frames, keyframes, and even shapes. Before you start scripting, knowing what you want to reference changes what script is available.

- **What script is most appropriate to create the desired result?** Once you know what you are trying to do and what will be affected, you can begin to determine what actions might achieve your desired result. Programming languages often use terms that you won't find in conversational language, but ActionScript is easier than most in this regard. At times you might have to think in synonymous terms to find the script you need. For example, you could research how to *start* a movie clip for quite some time to no avail, but if you researched how to *play* a movie clip, you would quickly find it.

- **What makes this script occur in the movie?** After you know what script you want to use, the next step is to determine what will trigger the action. Does the action happen when a button is clicked? Does it occur when the playhead crosses the fourth frame of a movie clip? Many actions are designed to execute when the movie first starts.

- **Where should the script be attached?** Even though you know what script to use, what it should do, what it references, and what triggers the action, if you put the script in the wrong place, it could either fail to work, or at the very least, require different syntax.

Understanding ActionScript Terminology

The terminology of ActionScript — and the vocabulary used to refer to scripting elements — can be unfamiliar and even intimidating to people whose backgrounds are more design oriented, rather than grounded in computer programming. Keep in mind the following definitions as you work through the rest of this project:

- When working with ActionScript, everything in Flash is treated as an **object**; this includes not only actual objects, but also frames, timelines, scenes, etc.

- Objects in Flash are divided into **classes** (they are "classified"); put another way, an object is defined as an instance of a class. Different classes of objects (Bitmap, Movie Clip, etc.) have different available properties, methods, and events because, for example, a movie clip can do different things than a bitmap image.

- In the Flash Actions panel, classes are grouped into **packages**, which simply provide a convenient way to manage large groups of related elements.

- **Properties** are the characteristics of an object, such as position, color, and visibility.

- **Methods** are the things an object can do, such as sending the playhead to a specific frame.

- **Events** are the triggers that an object can react to, such as a button being clicked or the playhead crossing a certain frame on the timeline.

- An **event listener** (also called an **event handler**) tells Flash when to react, and what to do when Flash "hears" the triggering event. Essentially, the code says, "When A happens, do B." In this plain-English example, the entire statement is the event listener; A is the event and B is the function that will occur.

- **Function** is simply a generic term for a block of code, which doesn't do anything ("execute") until it is called. A single function can be called as many times as necessary to accomplish the intended goal.

> **Note:**
>
> *Technically speaking, ActionScript includes a generic "object" class, which is simply the broadest possible class.*

> **Note:**
>
> *Professional developers often use "method" to refer to functions that are pre-built into a class definition, and "function" for custom functions that they write on a timeline.*

Understanding Dot Syntax

Syntax refers to the structure of the language, including placing all characters in the proper order. Leaving out a single character (such as a semicolon) can render the code unreadable (known as a **syntax error**). Scripting in ActionScript 3 uses **dot syntax** to apply script commands to different objects. For example:

char4.y = 350;

Because you aren't attaching a script to a specific object, you first have to tell the script which object you want to affect with a specific action. The first part of the script, then, identifies the object you are targeting. In this example, the script affects an object with the instance name **char4**.

The second part of the script (**y = 350**) is the property you want to affect with the action. In this case, the line of code moves the **char4** object vertically (the y axis) on the Stage. The dot that separates the object from the action is where the term "dot syntax" originates.

USE FUNCTIONS TO CONTROL MOVIE CLIP TIMELINES

Before you can begin programming the functionality of the game, you need to add some basic controls to the movie clips in the file. Specifically, you need to:

- Stop the Racetrack movie clip from moving until the race starts.

- Prevent the Racetrack movie clip from looping when it reaches the final frame.

- Loop the Lila animation so she continues to tap her foot until a user clicks to start the race (in other words, prevent Lila from firing the starting pistol before a user starts the race).

- Prevent the individual racers from running before a user starts the race.

- Loop the three animations in each racer's movie clip so each character can run continuously, cry continuously, or cheer continuously, depending on the current context of the game.

By adding simple functions to control the timelines of the movie clips, you can easily address all of these issues. Functions, as explained earlier, are blocks of code that do not run until they are called. The structure (syntax) of a function is shown here:

function name (parameter1, parameter2, …) : returntype {}

Functions start with the name of the function, followed by the parameters that the function accepts. The function name can be predefined in ActionScript (such as "stop") or user-defined. When you define custom functions, keep in mind the ActionScript is case-sensitive; "countGators" is not the same thing as "CountGators" or "countgators."

Parameters (also known as **arguments**) are the values the function needs as input — for example, the specific frame to go to for the gotoAndPlay function. The number of parameters varies, depending on the function you're using; some functions, such as the stop function, do not accept or require any parameters.

The function parameters are placed inside parentheses. The parentheses are required for all functions, even if a specific function does not accept any arguments — for example, stop().

The third section is the **return type** of a particular function (for example, a certain function might return a bit of text from a specific text field). The **void** keyword means the function does not return any value — in essence, you are voiding the return type of the function. (If this section is left blank, it is assumed the function returns no value.)

The braces (curly brackets) represent the opening and closing of the **function body**. All code that needs to execute when a custom function is called is written between this pair of braces. This will be important later when you define your own function to perform a number of different actions at once.

Note:

You will create your own functions later in this project.

1. **With gator_race.fla open, double-click the racetrack_mc instance to edit the movie clip in place on the Stage.**

2. **Create a new layer named Actions at the top of the layer stack in the Timeline panel for the movie clip symbol.**

   It is common practice among developers to use a dedicated Actions layer in a complex project, making the scripts easier to find and manage. You cannot add actions to frames on a motion tween layer, so in this case you *must* add a separate layer to attach the actions.

Note:

Press Option-F9/F9 to open the Actions panel.

The Actions Panel in Depth

The Actions panel is used to create scripts in your Flash files.

The **Actions toolbox** contains all of the functions of the ActionScript language, sorted into **packages**, which contain **classes** that share the same basic characteristics.

Actions toolbox

Click this button to collapse or expand the Actions toolbox.

The **Script navigator** allows you to navigate through the various elements (movie clips, frames, etc.) in the file.

Script navigator

The **Script pane** is where you write the code.

This tab shows the selected layer and frame.

Script pane

Above the Script pane, a series of useful tools can help make it easier to write your own scripts.

A **Add a New Item to the Script** opens a menu with the same elements that are available in the Actions toolbox.

B Clicking the **Find** button opens the Find and Replace dialog box, which you can use to find specific text in the document, and then replace that text with other text.

C **Insert a Target Path** lets you choose which object you want to affect with a function or event, with either an absolute or relative path.

D **Check Syntax** helps you check the syntax of your code so errors will not display when the movie runs. If the script has syntax errors, a dialog box shows you the error location and description. (It is important to note that this option does not verify that your code is *functionally* correct; it only verifies the code syntax.)

E **Auto Format** allows you to format and indent the code as you enter it, which helps make the code more readable in the window. Flash tries to do this by default; however, if you move blocks of code around, you might change the default indents or spacing; clicking this button corrects those problems. (This option is available only if the code does not contain any syntax errors.)

F When you type a period in the Script pane, Code Hints appear to show relevant options for the code you are typing. If Code Hints do not appear, you can click the **Show Code Hint** button to trigger them.

G **Debug Options** helps you debug code, allowing you to set breakpoints into the code for finding errors. (This is generally used by more experienced programmers.)

H **Collapse Between Braces** shrinks the code between braces (for example, curly braces or brackets), which helps you navigate through the code.

I For easier navigation, you can select a particular block of code and then click the **Collapse Selection** option. The whole block of code collapses.

J **Expand All** expands all pieces of collapsed code.

K Clicking the **Apply Block Comment** button places /**/ into the code. Anything you type between the two asterisks is considered a comment. If you select an existing piece of code in the Script pane before clicking this button, the selected text will be surrounded by the comment characters (e.g., /*selected text*/). Comments are used to add explanations or other non-script elements, as well as to disable specific code during testing; they are ignored when the script is compiled.

L Clicking **Apply Line Comment** adds two forward slashes (//) to the script. Text on the line following this notation will be ignored when the script is compiled.

M Clicking **Remove Comment** un-comments the selected code by removing the comment notation (/* */ or //). The previously commented code is reactivated.

N Clicking **Show/Hide Toolbox** expands and collapses the Actions toolbox on the left side of the panel.

O Clicking **Code Snippets** opens the Code Snippets panel (refer to Project 4: Ocean Animation, for more on working with code snippets).

P **Build Scripts mode** prompts you for the various elements needed to create a script, using various fields and other notes above the Script pane.

Q The **Help** feature is especially useful in learning the use of specific functions, events, etc. This is also a good way to learn about what methods, properties, and events are available to any of the ActionScript classes, and to get documentation on those class elements.

FLASH FOUNDATIONS

When you work with ActionScript, you should be aware of the various ways Flash provides assistance in the Actions panel.

Clicking a package in the toolbox expands that package to show the classes contained within the package. Clicking a specific class expands that class and shows folders of the events, methods, and properties (if applicable) for that class.

You can then click the Events, Methods, or Properties folder to show the relevant options for the selected class.

Double-clicking a specific item in the Actions toolbox adds that item to the script in the Script pane; this is designed to help you construct scripts with correct syntax.

Click a package icon to access the various classes in the package.

Click a class icon to access the events, methods, and properties for that class.

Click the Events, Methods, or Properties icon to access the related commands.

Double-click an item in the list to add the item to the script.

If you know which item you want to add to your script, but aren't sure which package and class to open, you can use the Index at the bottom of the Actions toolbox. The Index lists all possible items alphabetically. The host package and class are shown in parentheses next to each item.

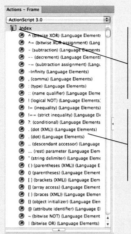

Open the Index for a complete alphabetical listing of all commands, keywords, and other elements of the ActionScript 3 language.

In the Index list, the parenthetical shows the package (and class, where relevant) where the item can be found.

If you move your mouse cursor over an item in the Actions toolbox, a tool tip provides helpful information about the potential use of the class, method, or property.

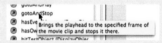

If you add a command that requires a target object, the "not_set_yet" warning is automatically added in red.

After double-clicking certain items in the Actions toolbox, a pop-up message might appear, showing the syntax rules for the added item.

If you type a command that has more than one possible use, the available options appear in a pop-up menu.

If you don't remember the exact name of the symbol instances in a file, you can use the **Insert Target Path** button to select the object you want the code to address.

The resulting dialog box lists all instances in the file. You can select a specific instance then click OK to reference that instance in your code.

- Relative references refer to an instance's position in relation to the active timeline.

- Absolute references refer to an instance's position in relation to the main movie timeline (Scene 1), which is the "root".

(Instances without names appear in parentheses; if you try to target an unnamed instance, Flash warns you that you must first define an instance name for that object.)

3. **Select the keyframe at Frame 1 of the Actions layer, and then open the Actions panel (Window>Actions).**

4. **In the Actions toolbox, expand the flash.display package, then expand the MovieClip class, and then expand the Methods folder.**

   The flash.display package contains many of the classes of objects you see on the Stage (including movie clips), so the events, methods, and properties of movie clips are accessed in the flash.display.MovieClip class.

5. **Double-click "stop" in the Actions toolbox to insert the command for the respective movie clip.**

   Typically, functions need to begin with a statement that determines which object should be affected by the function.

   When you use the Actions toolbox to add a method without first defining a target object, the inserted code includes the text "not_set_yet" in red to remind you to define the target of the method.

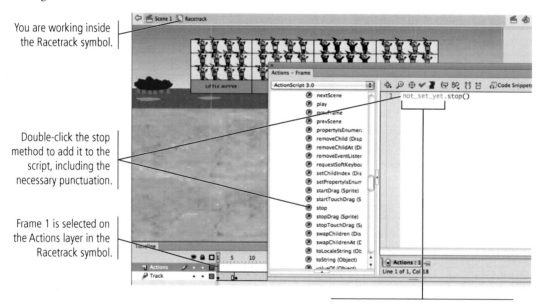

You are working inside the Racetrack symbol.

Double-click the stop method to add it to the script, including the necessary punctuation.

Frame 1 is selected on the Actions layer in the Racetrack symbol.

This text shows that a target object for the stop method has not yet been defined.

6. **In the Script pane, highlight the text "not_set_yet" and then type this.**

   The reference "**this**" refers to the current timeline — in this case, the Racetrack movie clip timeline. The stop function stops the timeline at a particular frame. In plain English, the code says:

At Frame 1	the racetrack_mc timeline will	stop playing
[where the action was created]	**this**	**stop()**

Note:

A small "a" appears in the frame where an action is attached, allowing you to more easily locate the actions in non-dedicated layers.

7. **Place the insertion point at the end of Line 1 in the Script pane, and type a semicolon.**

   A semicolon is required at the end of each line of code. If you forget to add this character at the end of a line, your code might result in an error.

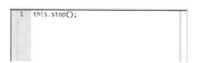

```
1   this.stop();
```

8. **Add a new keyframe at Frame 120 of the Actions layer, and then click inside the Actions panel Script pane to place the insertion point on Line 1.**

   When the Actions panel is open, selecting a new keyframe in the timeline shows the script associated with the selected keyframe. Simply click in the Script pane to edit the script for the currently selected keyframe.

9. **Type `this.stop();` on Line 1 of the Script pane.**

   Although you can always use the Actions toolbox to add elements of ActionScript code, in the case of basic functions such as this one, it is usually easier to simply type the command directly in the Script pane.

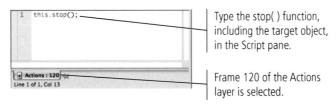

Type the stop( ) function, including the target object, in the Script pane.

Frame 120 of the Actions layer is selected.

Note:

*If you don't define a specific object, the code is interpreted to apply to "this", or the active timeline where the script is placed. The **this.** in this code is not strictly necessary, but we include it here to reinforce the idea of defining the target object for a specific method.*

   If you don't add the stop function at the final keyframe, the Racetrack movie clip would still play only once, but the timeline would automatically return to Frame 1 — which shows the starting line — before stopping. To prevent this, you have to add the stop function at both the beginning and ending frames.

10. **Click Scene 1 in the Edit bar to return to the main timeline.**

11. **Press Command-Return/Control-Enter to test the movie.**

    Now when you test the movie clip, the racetrack remains stationary. Later in the project, you will add code that causes the racetrack to start moving when Lila fires the starting pistol (which happens only after a user clicks the lila_mc instance).

12. **Close the Player window and return to Flash.**

13. **Save the file and continue to the next exercise.**

USE FUNCTIONS TO LOOP FRAME ANIMATIONS

When you tested the movie at the end of the previous exercise, the Background movie clip did not move, but the different characters' timelines did move. You have to use the stop function to prevent the characters from moving until you want them to move.

   You also need to use basic functions to create the loops for the different animations contained in each character's movie clip — running, crying, cheering, and (in Lila's case) foot tapping.

1. **With `gator_race.fla` open, double-click the Baggy movie clip icon in the Library panel (Gators folder) to enter into the movie clip Stage.**

2. **Add a new layer named Actions to the top of the symbol's timeline.**

   As with the Labels layer, you are going to define code in this symbol that must also be applied to the other racer symbols' timelines. By placing the required code on a separate layer, you will then be able to copy and paste the entire Actions layer — including the code — from one symbol to another.

3. **In the timeline, click the Frame 1 keyframe on the Actions layer to select it.**

4. **In the Actions panel (Window>Actions), click in the Script pane to place the insertion point.**

   From this point forward, we will assume the panel is open. Remember, you can access all panels in the Window menu.

5. **Type `this.stop();` in the Script pane.**

   As in the previous exercise, you are controlling the timeline of the current movie clip, so the "**this**" reference is sufficient.

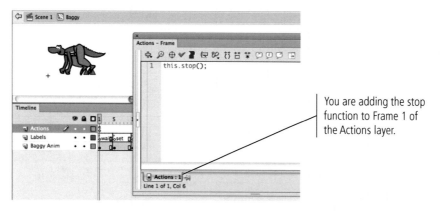

You are adding the stop function to Frame 1 of the Actions layer.

6. **Add a new keyframe to Frame 17 of the Actions layer.**

   The "running" animation exists from Frame 10 to Frame 17. For the character to run as long as necessary without creating extra frames, you have to create a loop that sends the playhead from Frame 17 back to Frame 10. The gotoAndPlay function is used for this purpose — moving the playhead from one frame to another in the timeline.

7. **Type `this` on Line 1 in the Script pane.**

8. **In the Actions toolbox, expand flash.display>MovieClip>Methods.**

9. **Double-click gotoAndPlay in the Actions toolbox.**

   Double-clicking inserts the gotoAndPlay function into the Script pane. Because you used the Actions toolbox, Flash adds the required dot between the object and the method.

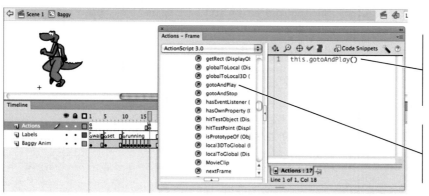

Double-clicking an item in the Actions toolbox adds the necessary code, including punctuation (except the terminal semicolon).

The gotoAndPlay function is a method of the MovieClip class, which is part of the flash.display package.

Note:

Color coding in the Code pane gives you an idea if your code syntax is correct. Different code elements are different colors, depending on the settings in the ActionScript pane of the Preferences dialog box. Default colors are:

Identifiers (events, methods, and properties):
   Blue

Keywords:
   Dark red

Comments:
   Gray

Variable names, parameters, and values ("foreground" elements):
   Black

Strings:
   Green

10. **Type** `"running"` **inside the parentheses of the gotoAndPlay function, and type a semicolon at the end of the line of code.**

Because you defined frame labels in the first stage of this project, you can use more-meaningful labels as the method parameter, rather than simply using the frame number. Keep in mind that when you use frame labels as method parameters, the labels must be enclosed in quotation marks.

This code creates a loop so, when the Baggy movie clip reaches Frame 17, the playhead moves back to the frame labeled "running" (Frame 10, the first frame of the "running" animation) — resulting in a continuous running effect using only eight frames.

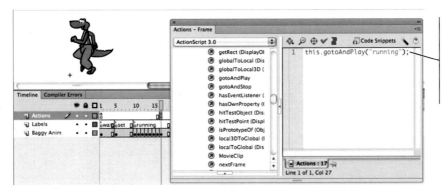

The entire line of code includes the object path, the method of the defined object, the function parameters (the frame where you are sending the playhead), and the semicolon.

11. **Add a new keyframe at Frame 23 of the Actions layer, and select the new keyframe in the timeline. On Line 1 in the Script pane, type** `this.go`

When you type in the Script pane, a Code Hint menu shows the properties and methods that are defined for this MovieClip object. As you type, the menu automatically scrolls to the first command that matches what you type.

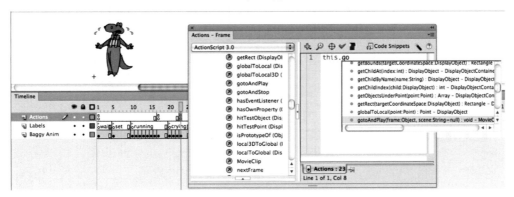

12. **Press Return/Enter to apply the gotoAndPlay function from the pop-up menu.**

The opening parenthesis of the gotoAndPlay statement is automatically added to the code; a tool tip hint shows you the syntax rules for the added item. The bold element in the hint identifies the parameter that you are currently defining.

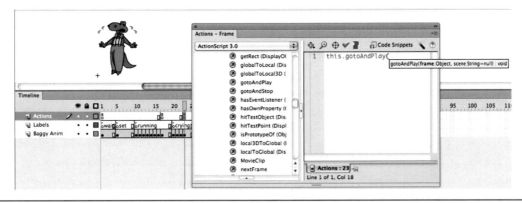

13. **With the insertion point after the opening parenthesis, type** `"crying");`.

Like the stop command, you can simply type the gotoAndPlay command rather than using the Actions toolbox. As you type, the Actions panel provides assistance in the form of pop-up Code Hint menus and tool tips.

When you write ActionScript, remember that capitalization counts. Make sure you capitalize words in the functions exactly as we show them here. "GoToAndPlay" is not the same as "gotoAndPlay."

Note:

As you become more comfortable working with ActionScript, you'll also become more comfortable typing your own code.

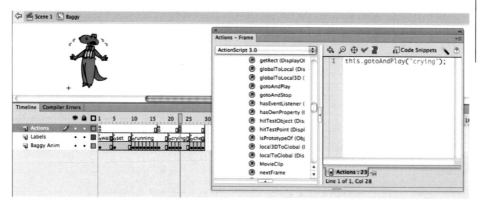

14. **Add a new keyframe at Frame 29 of the Actions layer, and select the new keyframe in the timeline. On Line 1 in the Script pane, type**

```
this.gotoAndPlay("cheering");
```

You can use the Code Hint menu to add the gotoAndPlay command, or simply type the entire statement.

Note:

Because the last frame includes a function to loop the playhead, you don't need to add the stop function to the last frame of the racers' movie clips.

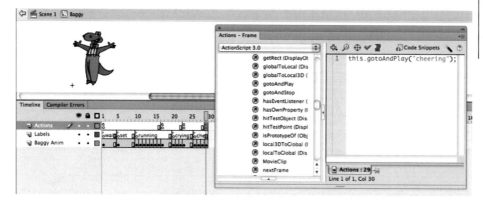

15. **Control/right-click the Actions layer name and choose Copy Layers in the contextual menu.**

16. **In the Library panel, double-click the symbol icon for the Granny symbol.**

17. Control/right-click the Labels layer name and choose Paste Layers in the contextual menu.

Remember, all the racer symbols have the same timeline structure. This means the code you defined for one racer can also be used for the other racer symbols.

The pasted layer is added above the layer where you opened the contextual menu.

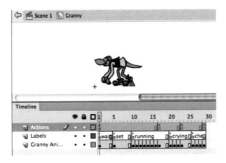

18. Repeat Steps 16–17 to paste the Actions layer into the Navi and Nipper symbol timelines.

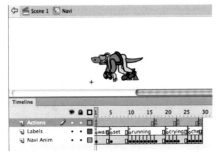 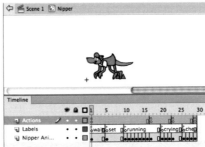

19. **Double-click the Lila symbol icon in the Library panel to enter into it.**

  - **Add an Actions layer at the top of the timeline, and create new keyframes at Frame 20 and Frame 35.**

  - **Add the gotoAndPlay() function to Frame 20 of the Lila movie clip, sending the playhead back to the frame labeled "waiting".**

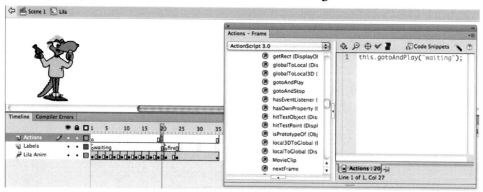

  - **Add the stop() function to the final frame of the Lila movie clip.**

    Because the final frames of the Lila movie clip are not looped, you need to stop the Lila movie clip when it reaches the final frame.

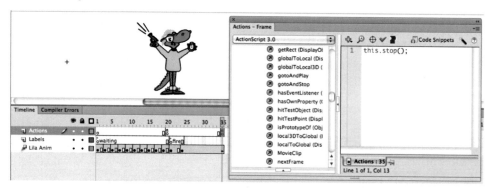

20. **Click Scene 1 in the Edit bar to return to the main Stage.**

21. **Press Command-Return/Control-Enter to test the movie.**

    When the movie opens, the background and racers remain fixed in position, and Lila taps her foot while waiting to be clicked. You can see the Restart button, which should not appear until the race is finished. Before programming the actual race functionality, you should add the necessary code to hide the Restart button until the race is over.

Note:

The Lila movie clip does not need a stop function in Frame 1 because she should start tapping her foot as soon as the movie opens.

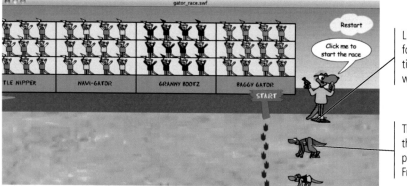

Lila continuously taps her foot because the movie clip timeline returns to Frame 1 when it reaches Frame 20.

The background and all the characters are fixed in place by the stop functions on Frame 1 in each movie clip.

22. **Close the Player window and return to Flash.**

23. **Save the file and continue to the next exercise.**

USE SCRIPTING TO CONTROL OBJECT PROPERTIES

In addition to controlling the playhead, you can use ActionScript to change the properties of specific objects — including visibility — at different times in a movie. In this case, the Restart button should be hidden when the race starts, and then become visible when the race is over. To accomplish this goal, you use the main movie timeline to add scripts at different points in the race.

1. **With `gator_race.fla` open, add a new layer named Actions at the top of the layer stack on the main timeline.**

2. **Select Frame 1 of the Actions layer, and then place the insertion point in the first line of the Actions panel Script pane.**

3. **Type `restart_btn.` in the Script pane.**

   As long as you know the exact name of the instance you want to address, you can simply type it in the Script pane and then add the method you want to use.

4. **In the Actions toolbox, expand the DisplayObject class in the flash.display package, then expand the Properties folder for that class.**

   The DisplayObject class includes all objects that can be placed in the display list (i.e., be visible in the movie). The button is visible on the Stage, so it falls into the DisplayObject class.

   Visibility is an object property, so you can access that option in the Properties folder.

5. **Double-click visible in the list.**

Visible is a property of the DisplayObject class.

6. **Type `=false;` at the end of the function line.**

   An option such as visibility has only two possible options — true (visible) or false (not visible). You could also use the Language Elements>Statements, Keywords & Directives folder in the Actions toolbox to add the property value to the code, but it is easier to simply type "true" or "false" to change the visibility property.

The command in the Script pane does not affect the object's appearance on the development Stage.

Note:

*The true-or-false value of the visible property is called a **Boolean value**.*

7. **Select Frame 120 of all layers on the main timeline. Press the F5 key to insert new frames to all layers at once.**

   As stated earlier in the project, the entire movie needs to last 10 seconds — or 120 frames. Before you can use the main timeline to program different actions at various points in the movie, you first have to create the necessary frames in the main timeline.

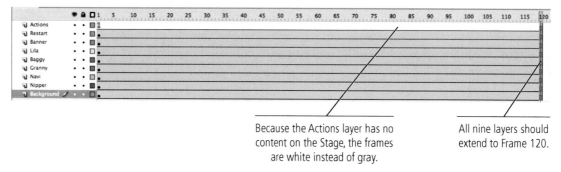

   Because the Actions layer has no content on the Stage, the frames are white instead of gray.

   All nine layers should extend to Frame 120.

8. **Add a keyframe to Frame 105 of the Actions layer.**

9. **With Frame 105 of the Actions layer selected in the timeline, place the insertion point in the first line of the Script pane.**

10. **Repeat the same basic process from Steps 3–6 to add a function that sets the visibility property of the Restart button to true.**

   The Restart button will only become visible when the Background timeline reaches Frame 105 (i.e., after the race is over).

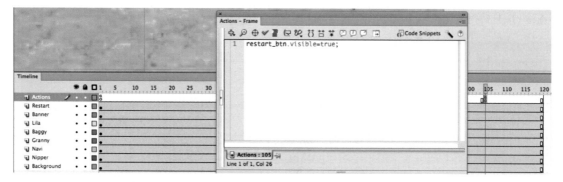

11. **Press Command-Return/Control-Enter to test the movie.**

   If you watch for ten seconds, you will see the Restart button reappear shortly before it disappears again. This is because the main timeline loops in the same way any other timeline loops — playing from start to finish and then returning to the beginning.

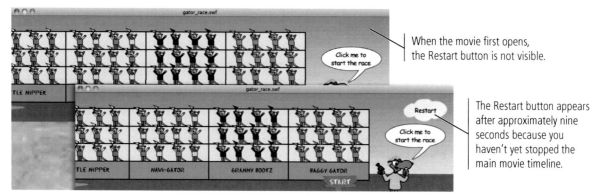

   When the movie first opens, the Restart button is not visible.

   The Restart button appears after approximately nine seconds because you haven't yet stopped the main movie timeline.

12. **Close the Player window and return to Flash.**

13. **Select Frame 1 of the Actions layer. Click in the Actions panel Script pane to place the insertion point at the end of the existing script (on Line 1).**

14. **Press Return/Enter to add a new line, and then press Return/Enter again to move to Line 3.**

    This extra line is called **white space**, which visually separates different pieces of code within the Script pane.

15. **Type `stop();` on Line 3 of the Script pane.**

    As we explained previously, if you don't define a specific target, the active timeline ("this") is implied. This statement stops the active timeline from automatically playing when the file opens.

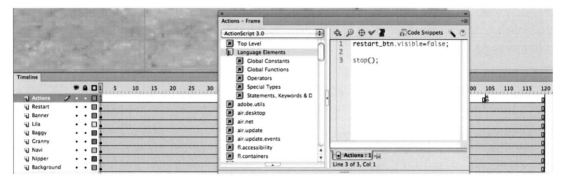

16. **Add a keyframe to Frame 120 of the Actions layer.**

    New scripts require keyframes, which means you have to add a keyframe at Frame 120 before you can add the stop function.

17. **With Frame 120 of the Actions layer selected, click in the Script pane and add the stop function.**

    This prevents the timeline from looping back to Frame 1, which means the Restart button remains visible.

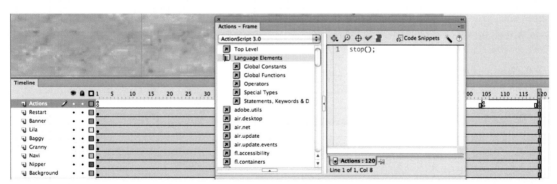

18. **Save the file and continue to the next stage of the project.**

Stage 3 Creating Custom Functions

If you play the movie now, nothing will happen — which is exactly what you want. The action in this movie is entirely dependent on a user clicking the Lila movie clip instance. When a user clicks the Lila movie clip instance, several things need to happen:

- The main timeline needs to start moving. You stopped the main timeline at Frame 1 in an earlier exercise; when a user clicks the Lila instance, the main timeline needs to move past the stop function and start moving (in other words, go to Frame 2).

- The Lila movie clip needs to stop looking at her watch and fire the pistol, and then move to the right and off the Stage area. This animation is already prepared within the movie clip; you already created the tween to move the character off the Stage area after firing the pistol, and defined a frame label to identify the beginning of that animation.

- Each racer needs to get "set" and start "running". You already created the loops and defined frame labels to identify the beginning of each loop, so the characters can run continuously until something stops them.

- The background needs to start moving to simulate the effect of the runners moving across the track. You already created the tween that moves the background to the right, but you also added a stop function that prevents the background from moving when the file opens.

- The "Click me" banner needs to disappear.

Because all these things need to happen when a user clicks Lila, the most logical solution is to define a custom function that triggers when the user clicks the character.

CREATE AN EVENT LISTENER

Creating an event is essentially a two-stage process:

- Add an event listener to a specific object (for example, a button users can click to make something happen), and define the event handler that names the function that will be called and run when the event listener is triggered.

- Define the actual function that is called by the event handler (i.e., exactly what to do when the triggering event occurs).

1. **With gator_race.fla open, add a new layer named Event Handlers to the top of the layer stack on the main timeline.**

2. **Select Frame 1 in the Event Handlers layer. Click in the Actions panel Script pane to place the insertion point.**

    As mentioned earlier, most programmers insert a specific layer for coding events. You can even use separate layers for different types of scripts — as we are doing here by using a separate layer to contain the event handlers in this movie.

3. **In the Script pane, add a reference to the lila_mc instance on the active timeline.**

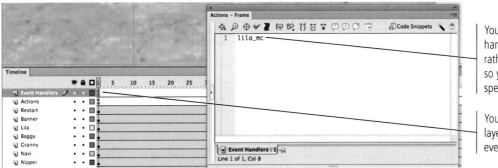

You are attaching an event handler to a specific instance rather than the main timeline, so you need to define the specific instance name.

You are using a separate layer to manage the event handler code.

4. **In the Actions toolbox, expand flash.events>EventDispatcher>Methods.**

The flash.events package consists of all the classes that are necessary for performing specific actions when an event occurs.

5. **Double-click addEventListener in the list to add it to the script.**

To perform a set of actions when a particular object is clicked, you must first add an event listener to that object. The defined object listens for a specific event, which will trigger a specific function when that event occurs.

The addEventListener method requires two arguments between the parentheses.

- The first argument specifies the event that will trigger the action (in other words, the event to listen for). If you want to perform an action when the user clicks an object, for example, define MouseEvent.CLICK as the first argument.

- The second argument is the (user-defined) name of the function that will be called when the defined event occurs.

Note:

The addEventListener method is available to most members of the DisplayObject class (buttons, movie clips, etc.).

Note:

When you use the Actions toolbox to place the addEventListener code, the Code Hint menu automatically appears. You can simply ignore it and type inside the parentheses.

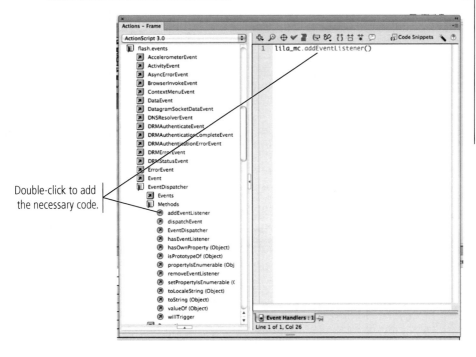

Double-click to add the necessary code.

6. **Place the insertion point between the function parentheses.**

7. **Expand flash.events>MouseEvent>Properties, and then double-click CLICK to insert the required code.**

Because you want to initiate an event on the *click* of the mouse, you are using the CLICK event of the MouseEvent class. (The constant "CLICK" defines which type of MouseEvent should trigger the function.)

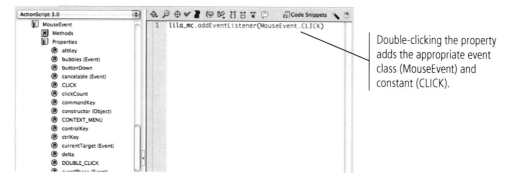

Double-clicking the property adds the appropriate event class (MouseEvent) and constant (CLICK).

8. **After CLICK and before the closing parenthesis, type `, raceStart` and add a semicolon after the closing parenthesis.**

The word **raceStart** is the name of the function that you will define in the next exercise.

A user-defined function can have any name. Keep in mind, however, that capitalization counts in ActionScript. RaceStart is not the same function as raceStart or racestart.

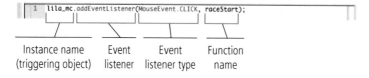

| Instance name (triggering object) | Event listener | Event listener type | Function name |

9. **Save the file and continue to the next exercise.**

 ## DEFINE A CUSTOM FUNCTION

Because stop, gotoAndPlay, and similar functions are so common, they are built into the ActionScript 3 dictionary. Remember, however, that a function is simply a block of code that you want to execute at a certain time — whether at a certain frame or as the result of a button being clicked. Functions can be as basic or complex as necessary — they can execute a single action (such as stopping a timeline) or multiple actions at once.

To complete this project, you need to control nine different elements when a user clicks the Lila instance. Rather than defining nine separate functions for a single event listener, you can combine all nine actions into the single custom function that you named raceStart in the previous exercise.

1. **With `gator_race.fla` open, select Frame 1 of the Event Handlers layer. Click to place the insertion point at the end of Line 1 in the Script pane.**

2. **Press Return/Enter twice to move the insertion point to Line 3.**

3. **Expand Language Elements>Statements, Keywords & Directives>Definition Keyword in the Actions toolbox.**

The Language Elements package contains all syntax and code of the ActionScript language; it contains all keywords and other expressions used to write ActionScript code.

4. Double-click function in the list to add it to the script.

Clicking this option in the Actions toolbox adds the function keyword, as well as the required punctuation (parentheses and braces). This is sometimes referred to as the **function skeleton**.

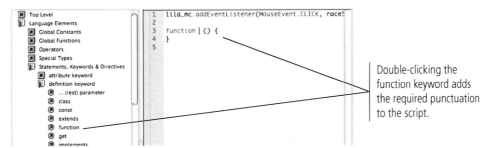

Double-clicking the function keyword adds the required punctuation to the script.

5. Type `raceStart` before the opening parenthesis of the function.

This code links the function to the event handler you defined in the previous exercise. In other words, you're writing the function that will execute when the event handler triggers.

```
1  lila_mc.addEventListener(MouseEvent.CLICK, raceStart);
2
3  function raceStart () {
4  }
5
```

6. Type `event:` between the opening and closing parentheses.

After typing a colon, Flash presents a list of available options in the Code Hint menu. When you begin typing, the Code Hint menu scrolls to find the first option that matches the characters you type.

7. Type `Mou`, then scroll through the Code Hint menu to find MouseEvent. With MouseEvent highlighted, press Return/Enter to add it to the script.

This function needs to be called on the click of the mouse, so you use MouseEvent as the argument.

Typing scrolls the Code Hint menu to the first item that matches what you type.

Pressing Return/Enter adds the highlighted item to the script.

A new Line 1 is added to the top of your script; this import statement imports the definition of the MouseEvent class. Basically, the line tells the file to import a predefined package that defines all of the methods and properties that can be applied to a MouseEvent.

```
1  import flash.events.MouseEvent;
2
3  lila_mc.addEventListener(MouseEvent.CLICK, raceStart);
4
5  function raceStart (event:MouseEvent) {
6  }
7
```

When you write and use ActionScript entirely within Flash, import statements that begin with the words "flash" are not typically necessary because Flash manages this issue for you in most cases. However, if you write ActionScript code outside of Flash, store ActionScript in a separate linked file (using the ".as" extension), or export the code to be used in some other application, the import statements must be included so the code can accurately process and compile the classes that are used.

Note:

You can also double-click a Code Hint to add the item to the script.

Note:

If you accidentally add the Mouse event instead of the MouseEvent event from the Code Hint list, the wrong import statement will be added to Line 1. You would have to delete the incorrect code and then redo Steps 6–7 to add the correct import statement.

Note:

Although this import statement is not necessary in the case of this project, it doesn't hurt anything to leave it there.

8. **Type a colon outside the closing parenthesis, and then choose void from the Code Hint menu.**

   The last part of a function specifies the return type for the function. If a function does not return a value, the keyword "void" is used (referred to as voiding the return type).

   ```
   1  import flash.events.MouseEvent;
   2
   3  lila_mc.addEventListener(MouseEvent.CLICK, raceStart);
   4
   5  function raceStart (event:MouseEvent):vo {
   6  }
   7
   ```
   XMLUI – adobe.utils
   Zoom – fl.transitions
   adobe
   fl
   flash
   flashx
   mx
   void

Note:

Every function has a return type that specifies the type of data (integer, string, etc.) that returns from the function. Void is used if the function returns no data.

9. **Click after the opening brace and press Return/Enter to add a line between the function braces.**

   Although not required, coding convention typically places each action of a function on a separate line in the script.

10. **Type** `this.gotoAndPlay(2);`.

    Because you are adding the function on the main timeline, **this** refers to the main movie timeline.

    The body of this function is going to address a number of objects on the Stage. Although the **this** reference is not strictly necessary, it can be useful to include (especially for anyone new to coding) so that each line in the function body includes an object path before the applied method.

    ```
    1  import flash.events.MouseEvent;
    2
    3  lila_mc.addEventListener(MouseEvent.CLICK, raceStart);
    4
    5  function raceStart (event:MouseEvent):void {
    6      this.gotoAndPlay(2);
    7  }
    8
    ```
    Click here and press Return/Enter to add a line to the body of the function.

Note:

*The code between the braces represents the **function body** — the actions that will be performed when the function is called.*

11. **Press Return/Enter to move to the next line in the script.**

12. **Add another statement to send the racetrack_mc timeline to Frame 2:**

    `racetrack_mc.gotoAndPlay(2);`

    Moving the playhead to Frame 2 of the Racetrack symbol bypasses the stop function that you defined on Frame 1, which allows the movie clip's timeline to play.

    ```
    1  import flash.events.MouseEvent;
    2
    3  lila_mc.addEventListener(MouseEvent.CLICK, raceStart);
    4
    5  function raceStart (event:MouseEvent):void {
    6      this.gotoAndPlay(2);
    7      racetrack_mc.gotoAndPlay(2);
    8  }
    9
    ```

13. Press Return/Enter to add a new line to the script. Add a statement that sends the lila_mc timeline to the frame labeled "fire":

```
lila_mc.gotoAndPlay("fire");
```

Don't forget to enclose the frame label in quotation marks.

You used frame labels in the characters' symbol timelines, which means you do not need to remember which frame number should be included as the method argument. (Of course, you do need to remember the labels that you defined for various frames in the symbol timeline.)

When a user clicks the Lila instance, the Lila movie clip timeline moves to Frame 21 (which you labeled "fire"), where Lila stops looking at her watch and fires the starting pistol.

```
1   import flash.events.MouseEvent;
2
3   lila_mc.addEventListener(MouseEvent.CLICK, raceStart);
4
5   function raceStart (event:MouseEvent):void {
6       this.gotoAndPlay(2);
7       racetrack_mc.gotoAndPlay(2);
8       lila_mc.gotoAndPlay("fire");
9   }
```

14. Press Return/Enter to move to the next line. Add additional commands to the function body to move the remaining characters to the "set" position. Use the Actions toolbox or simply type the necessary code:

```
char1_mc.gotoAndPlay("set");
```

```
char2_mc.gotoAndPlay("set");
```

```
char3_mc.gotoAndPlay("set");
```

```
char4_mc.gotoAndPlay("set");
```

```
1    import flash.events.MouseEvent;
2
3    lila_mc.addEventListener(MouseEvent.CLICK, raceStart);
4
5    function raceStart (event:MouseEvent):void {
6        this.gotoAndPlay(2);
7        racetrack_mc.gotoAndPlay(2);
8        lila_mc.gotoAndPlay("fire");
9        char1_mc.gotoAndPlay("set");
10       char2_mc.gotoAndPlay("set");
11       char3_mc.gotoAndPlay("set");
12       char4_mc.gotoAndPlay("set");
13   }
14
```

15. Press Command-Return/Control-Enter to test the movie. In the Player window, click the Lila movie clip.

When you click the starter, the racetrack and characters all move as they should:

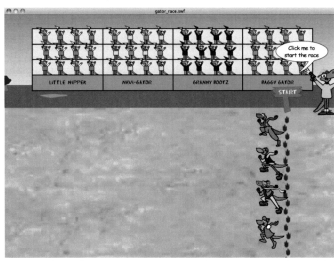

- Lila raises the gun and fires.
- The four racers get into the set position, and then start running.
- The racetrack moves to the right.
- After the main timeline reaches Frame 105 (approximately 9 seconds), the Restart button appears.

You should notice, however, that the banner_mc remains visible throughout the entire movie.

16. Close the Player window and return to Flash.

17. With Frame 1 of the Event Handlers layer selected, click in the Script pane to place the insertion point after the last statement in the raceStart function body.

18. Press Return/Enter to add a new line to the function body. Add another statement to the function body to hide the banner_mc instance after a user clicks the Lila instance:

> **banner_mc.visible=false;**

Note:

You still need to program the Restart button, but the necessary functions won't make sense until you program the rest of the game's functionality. You will create the second event handler and function later in this project.

```
1    import flash.events.MouseEvent;
2
3    lila_mc.addEventListener(MouseEvent.CLICK, raceStart);
4
5    function raceStart (event:MouseEvent):void {
6        this.gotoAndPlay(2);
7        racetrack_mc.gotoAndPlay(2);
8        lila_mc.gotoAndPlay("fire");
9        char1_mc.gotoAndPlay("set");
10       char2_mc.gotoAndPlay("set");
11       char3_mc.gotoAndPlay("set");
12       char4_mc.gotoAndPlay("set");
13       banner_mc.visible=false;
14   }
15
```

Add this line of code to the function to hide the banner when a user clicks Lila.

19. Save the file and continue to the next stage of the project.

Stage 4 Working with Variables and Arrays

When you last tested your movie, you saw that clicking the Lila instance correctly sets all the pieces of the movie into motion. You might have noticed, however, that the racers keep running after the race is over, and all four runners are in an even tie throughout the race.

Programming the remaining game functionality requires two additional blocks of code: one to move a different character at specific intervals throughout the race, and another to determine which character wins the race. To create the first half of this functionality, you will use variables and randomly generated numbers.

Variables are named spaces in the memory that point or refer to values. A variable exists as soon as you declare it; in other words, a variable exists because you say so.

Data types define the type of information to which a particular variable can refer. ActionScript 3 defines several primitive data types:

Note:

A variable is simply a container that holds information needed in other parts of the code.

- The **Number** data type can hold any number.

- The **int** (integer) data type can hold any whole number (a number with no decimal point).

- The **uint** (unsigned integer) data type can hold any whole, non-negative number.

- The **String** data type holds the actual characters within a set of quotation marks.

- The **Boolean** data type can have only one of two values: true or false.

To declare a variable, you need to use the var statement.

> **var myAge : uint;**

In this statement, you have declared a variable named *myAge*. The second part of the statement determines that the variable value must be a non-negative whole number (as ages typically are, unless you refer to yourself as "something and a half").

When you declare a variable, you can also define an initial value at the same time by using the equal sign (called an **assignment operator**):

> **var myAge : uint = 40;**

Note:

You can declare a variable without declaring its type, but the best practice is to declare a variable with its data type.

SCRIPT WITH VARIABLES AND RANDOM NUMBERS

At several times throughout the race, one of the racers will move to the left, basically "taking the lead." To accomplish this movement, you need a mechanism to randomly move one character instance to the left. ActionScript 3 includes the tools you need to accomplish this task.

Here's a plain-English explanation of the script you will write in this exercise:

- The first part of the script will generate a random number between 0 and just less than 1 (0.9̄9̄).

- That number will be multiplied by 4 so it can be associated with one of the characters' numbers (1–4). Since the characters are numbered 1, 2, 3, and 4 — whole numbers — you also have to round up the randomly generated number to match one of the character numbers.

- The second part of the script will add strings of text to the generated number to create the instance name of the character that will move when the script runs.

This script will repeat a number of times throughout the race. Every time the code runs, a different character moves or "takes the lead" in the race. (Because the process is random, the same racer might move more than once.)

1. **With `gator_race.fla` open, add a keyframe at Frame 24 of the Actions layer on the main timeline.**

2. **With Frame 24 selected, click in the Script pane of the Actions panel and type `var temp`.**

   The word *temp* is the name of the variable you are defining. As with instances of symbols, you have to name variables if you want to use them in ActionScript.

Note:

You can declare a variable using the Language Elements> Statements, Keywords & Directives>Definition Keyword class in the Actions toolbox.

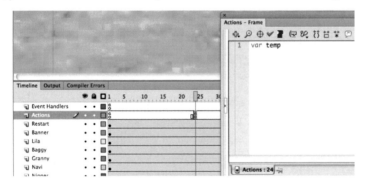

3. **Type a colon after the variable name in the script, and then choose Number from the Code Hint menu.**

   This command declares one variable (*temp*) of a Number data type. You defined this variable to hold a number that will be generated through the use of random numbers.

4. **Type a semicolon at the end of the line and press Return/Enter twice to move to Line 3 in the Script pane.**

5. **Type `temp=` in Line 3 of the Script pane.**

Remember, *temp* is the name of the variable you just declared. The next step is to assign a value to this variable.

6. **In the Actions toolbox, expand Top Level>Math>Methods. Double-click random in the list to insert it into the script.**

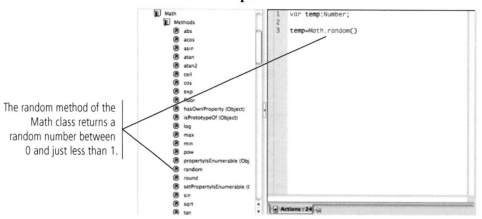

The random method of the Math class returns a random number between 0 and just less than 1.

7. **Type `*4;` after the Math.random function parentheses, and add a semicolon at the end of the line.**

Because there are four participants in this race, you also have to multiply the random number by 4. Every time the value is generated, it is converted into a value ranging from 0 to 4. (Don't worry if this sounds a bit confusing; it will make more sense shortly.)

8. **On the next line in the Script pane, type `temp=`.**

9. **In the Actions toolbox, double-click ceil (in Top Level>Math>Methods).**

10. **Type `temp` in the parentheses of the Math.ceil function, and then add a semicolon at the end of the line.**

The ceil method converts a number — in this case, the value in the *temp* variable — to the nearest integer that is either greater than or equal to that number; in other words, it rounds up to the next whole number.

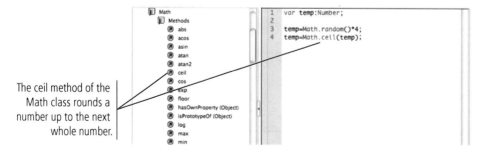

The ceil method of the Math class rounds a number up to the next whole number.

Together, these two lines of code generate a random number (between 0 and 1), multiply that number by 4, and round the number up to the nearest whole number. The result is stored as the value of the temp variable.

11. Press Return/Enter twice to move to Line 6 of the Script pane, and then type

```
this["char"+temp+"_mc"]
```

This statement calls the particular racer whose value is generated by the *temp* variable. (Remember, you previously assigned instance names char1, char2, char3, and char4 to the different character instances.)

```
1  var temp:Number;
2
3  temp=Math.random()*4;
4  temp=Math.ceil(temp);
5
6  this["char"+temp+"_mc"]
```

The brackets after the word "**this**" contain a single piece of code that defines the instance name. Within the brackets:

- **"char"** and **"_mc"** (including the quotation marks) are strings; the actual characters inside the quotation marks will be used.

- **+** is the addition or concatenation operator. When used between two numbers, the operator causes the two numbers to be added together. If used between text strings, the two strings are combined or concatenated into a single string.

  If the + operator is placed between a string and another data type, such as a number, the number is treated as a string and concatenated — as you are doing in this exercise.

- *temp* is the variable you defined earlier; it refers to a whole number between 1 and 4.

Note that the variable name is not surrounded by quotation marks because you want to use the value of the variable. If you surrounded the variable name with quotes, the function would treat the variable name as a text string.

12. Outside the closing bracket, type .x=.

As we stated earlier, you are writing this code to move one character to the left. By appending the x property (including the required dot) to the statement, this statement now refers to the horizontal (x) position of the instance identified by the randomly generated variable.

The x property determines an object's horizontal position (across) on the Stage. The y property determines the object's vertical position (up or down) on the Stage. To move something left on the Stage, you have to subtract from the object's x position.

13. After the "=" in the Script pane, type

```
this["char"+temp+"_mc"].x-50;
```

Remember, the right side of the statement is assigned to the left side of the statement. This line of code moves the selected character 50 pixels left from its previous position.

```
1  var temp:Number;
2
3  temp=Math.random()*4;
4  temp=Math.ceil(temp);
5
6  this["char"+temp+"_mc"].x=this["char"+temp+"_mc"].x-50;
```

14. Select Lines 3–6 in the Script pane, and then press Command/Control-C to copy the selected code.

Note:

In this case, you do not need to add the dot between the word "this" and the statement that defines the appropriate character name.

Note:

The different operators in ActionScript 3 can be accessed in the Language Elements>Operators class in the Actions toolbox.

Note:

*The equal sign (**assignment operator**) assigns values from right to left and not vice-versa.*

Note:

Add to the x property to move an object right.

Subtract from the x property to move an object left.

Add to the y property to move an object down.

Subtract from the y property to move an object up.

15. **In the timeline, add keyframes at Frames 36, 48, 60, 72, 84, and 96 of the Actions layer.**

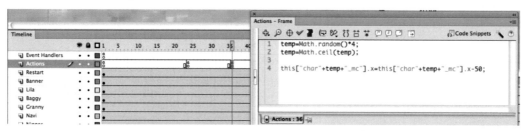

16. **Select Frame 36 and paste (Command/Control-V) the copied code into the Script pane.**

You don't need to declare the variable every time you want to use it, so you didn't need to copy Line 1 of the script in Frame 24.

When the main timeline reaches Frame 36, the temp variable gets a new value, which maps to a (possibly) different character. The character identified by the new variable value will then move 50 pixels to the left.

```
1   temp=Math.random()*4;
2   temp=Math.ceil(temp);
3
4   this["char"+temp+"_mc"].x=this["char"+temp+"_mc"].x-50;
```

17. **Repeat Step 16 to paste the copied code into Frames 48, 60, 72, 84, and 96 of the Actions layer.**

One character will move left every second of the race.

This code is pasted into the keyframes at Frame 36, 48, 60, 72, 84, and 96 to move one character every second.

```
1   temp=Math.random()*4;
2   temp=Math.ceil(temp);
3
4   this["char"+temp+"_mc"].x=this["char"+temp+"_mc"].x-50;
```

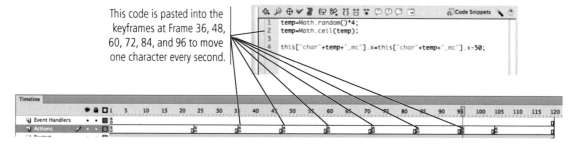

18. **Save the file and continue to the next exercise.**

Translating ActionScript to Plain English

FLASH FOUNDATIONS

Have you ever heard the terms "left-brain thinker" and "right-brain thinker"? Right-brain thinkers are typically more visually oriented, while left-brain thinkers are typically more logically oriented.

This delineation is why ActionScript sometimes causes problems for graphic designers, who are frequently right-brain thinkers. To put it simply, ActionScript is computer programming. Being successful at computer programming requires a good understanding of logic and quantitative reasoning — which can be a challenge for many right-brain thinkers.

Using examples is the best way to explain the code you are creating in this project. The table to the right shows examples of what is happening with the *temp* variable.

Random number	*4	Ceiled	Calls character
0.1	0.4	1	char1_mc
0.2	0.8	1	char1_mc
0.3	1.2	2	char2_mc
0.4	1.6	2	char2_mc
0.5	2.0	2	char2_mc
0.6	2.4	3	char3_mc
0.7	2.8	3	char3_mc
0.8	3.2	4	char4_mc
0.9	3.6	4	char4_mc

CREATE AN ARRAY TO STORE OBJECT POSITIONS

To determine which character wins the race, you need to write a script that captures the positions of each racer instance after the race is over. An **array** is essentially a type of variable that stores a group of values instead of a single value. A simple array declaration uses the following syntax:

var classmates:Array = new Array();

This statement declares an array variable named *classmates* of unlimited size, so you can enter any number of elements into this particular array.

You can also create an array with a pre-defined length by typing a number between the array object parentheses (in this case, the value of each array item is yet undefined):

var classmates:Array = new Array(3);

You can also directly define elements of the array by typing the values inside the array object parentheses:

var classmates:Array = new Array("Linus", "Sally", "Charlie");

If you don't directly define the array elements, you can define them in a list based on the array **index value**, which is simply the numbered position of specific elements in the array, beginning with 0. Array indexes always begin with 0, not 1.

classmates[0]="Linus"

classmates[1]="Sally"

classmates[2]="Charlie"

1. **With `gator_race.fla` open, insert a keyframe at Frame 100 of the Actions layer on the main timeline.**

    The final change of character position occurs at Frame 96, and the racetrack tween stops moving at Frame 100. You define the race winner as soon as the race ends — at Frame 100 on the main timeline.

2. **With Frame 100 selected, click in the Actions panel Script pane and type**

    `var arrPosition:`

3. **Choose Array from the Code Hint menu and press Return/Enter.**

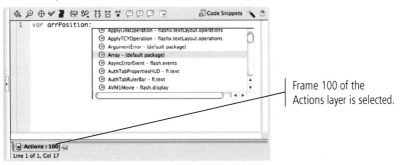

Frame 100 of the Actions layer is selected.

4. Type =new Array(4); at the end of the line.

There are only four participants in this race, so you are creating an array object that has four elements. The next step is to store the positions of each racer in the different elements of the array.

```
1  var arrPosition:Array=new Array(4);
```

5. On the next line of the Script pane, type:

```
arrPosition[0]=char1_mc.x;
```

Each array element is identified by an index value in brackets; the first element of the array always has an index value of 0, the second element has an index value of 1, and so on.

The left half of this statement says that you are defining the first element (**arrPosition[0]**) in the array. The right half of the statement (after the equal sign) assigns a value to that array element.

In plain English, this line says that the first value in the array (**arrPosition[0]**) equals the current x position of char1.

```
1  var arrPosition:Array=new Array(4);
2  arrPosition[0]=char1_mc.x;
```

6. Beginning on Line 3, add the following three lines of code to store the x positions of the other race participants in the different array positions:

```
arrPosition[1]=char2_mc.x;
arrPosition[2]=char3_mc.x;
arrPosition[3]=char4_mc.x;
```

Make sure each line of code has a different index number in the array position brackets (e.g., **arrPosition[0]**, **arrPosition[1]**, etc.), as well as a different character number in the instance names.

```
1  var arrPosition:Array=new Array(4);
2  arrPosition[0]=char1_mc.x;
3  arrPosition[1]=char2_mc.x;
4  arrPosition[2]=char3_mc.x;
5  arrPosition[3]=char4_mc.x;
```

7. Save the file and continue to the next stage of the project.

Note:

This array stores the positions of each racer at Frame 100 (after the position changes that occur throughout the race). The current x position of char1 is stored in the first array element, the position of char2 is stored in the second element, and so on.

Stage 5 Scripting Loops

In the previous exercise, you defined an array to store the position of each racer. The next step is to write code that compares the different values stored in the array to determine which value is lowest. Whichever character is farthest to the left (the lowest x value) "wins" the race.

To successfully evaluate each value in the array, you need to run the comparison script for each element in the array; in the case of the *arrPosition* array, you run the script four times. ActionScript loops are useful whenever you need something to happen more than once. There are five basic types of loops in ActionScript 3.

A **for loop** takes three arguments between the parentheses on the first line: the initialization value for a variable (**i=0**), a statement that determines under what conditions the loop continues to run (**i<5**), and the increment statement that increases or decreases the value of the variable after the loop is run (**i++**). All actions you want to execute are contained in the braces.

```
for (i=0; i<5; i++) {
    char1.x = char1.x–50;
}
```

In plain English, this code says:

> Start with the variable *i* equal to zero. Perform the actions within the braces (i.e., move the instance named char1 left 50 pixels), and then add one to variable *i*. As long as variable *i* is less than five, repeat these actions.

A **while loop** has only one argument — the condition in which the loop runs. As soon as the condition is false, the loop is terminated. In the example shown here, the variable is incremented at the end of the loop action.

```
while (i<5) {
    char1.x = char1.x–30;
    i++;
}
```

A **do while loop** is basically the same as a while loop, except the condition is placed at the end of the loop, which allows the loop to run at least once even if the condition is false. For example:

```
do {
    char1.x = char1.x–30;
} while (i<5)
```

If the variable *i* has a value greater than 5 before the program reaches this loop, the char1 instance will still move one time before the loop condition is determined to be false and the loop is exited.

The **for...in loop** evaluates the elements of an array. The for...in loop uses the following syntax:

```
for (var i in array) {
    perform some action;
}
```

The first part of the condition (*variable* in *array*) declares a variable that represents the current index value in the array. The second part of the condition identifies the specific array being evaluated.

Note:

*The ++ notation, called an **increment operator**, adds 1 to the value of the variable.*

*To reduce a variable by one, you can use the **decrement operator** (--).*

Note:

*Each time a loop runs is called an **iteration**.*

A for…in loop automatically executes the braced code for each element stored in the array. If an array stores four elements, for example, the loop will execute four times. The first time the loop runs, the variable value is 0 (the first index value). The second time the loop runs, the variable value is 1 (the second index value). This continues until the loop has evaluated all elements in the array, at which point the loop is exited.

The **for each…in loop** is similar to the for…in loop, except the variable in the array condition is the actual value in the array elements. For example:

> **var sample:Array=new Array ("Patty", "Lucy", "Schroeder");**

> **for each (var i in sample) {**
>
> > **perform some action;**
>
> **}**

Using a for…in loop, the value of *i* increments to the next array position every time the loop iterates, so:

First pass	**i=0**
Second pass	**i=1**
Third pass	**i=2**

Using the for each…in loop, the variable in the loop condition equals the actual value of the array element, so:

First pass	**i="Patty"**
Second pass	**i="Lucy"**
Third pass	**i="Schroeder"**

DEFINE A FOR…IN LOOP TO EVALUATE ARRAY ELEMENTS

At this point in the race movie, you have stored each racer's position in an array named *arrPosition*. The next task is to add a mechanism to evaluate the different values stored in the array to determine which character's movie clip instance is farthest left on the Stage, effectively "winning" the race. As we explained, a for…in loop is specifically designed to cycle through all of the values in an array.

Several variables are required to compare the different array values:

lowest	This variable will be initialized as the x position of the first character, creating a basis of comparison. The loop will then compare this value to the next value in the array to determine which is lower.
pos	This variable will hold the index value of the lowest array value (0, 1, 2, or 3) as determined by the comparison.
i	This variable will represent the current index position in the array evaluated in the for…in loop.

1. With **gator_race.fla** open, select Frame 100 of the Actions layer in the main timeline.

2. Place the insertion point at the end of the existing script (in the Actions panel Script pane) and press Return/Enter twice to move to Line 7.

3. At the current insertion point, type the following code to declare a variable named "lowest" of the integer data type.

   var lowest:int

4. After the variable declaration, add the following code to initialize the lowest variable to equal the first value stored in the arrPosition array.

   =arrPosition[0];

   By initializing the variable as the first value in the array, you establish a basis for comparing the other values in the array.

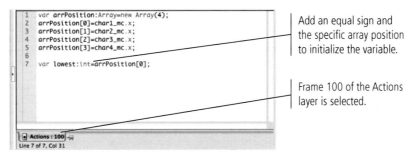

Add an equal sign and the specific array position to initialize the variable.

Frame 100 of the Actions layer is selected.

5. Press Return/Enter to move to the next line in the Script pane.

6. On Line 8 of the Script pane, type:

   var pos:int=0;

   This declares a variable named *pos* of the integer data type with an initial value of 0. This variable will store the array position associated with the lowest value in the array. (If you do not initialize the variable to 0, running the game multiple times in the Flash Player can cause problems in determining the true winner of the race.)

```
1   var arrPosition:Array=new Array(4);
2   arrPosition[0]=char1_mc.x;
3   arrPosition[1]=char2_mc.x;
4   arrPosition[2]=char3_mc.x;
5   arrPosition[3]=char4_mc.x;
6
7   var lowest:int=arrPosition[0];
8   var pos:int=0;
```

7. Press Return/Enter twice after the second variable declaration to move to Line 10 in the Script pane.

8. In the Actions toolbox, expand Language Elements>Statements, Keywords & Directives>statement and double-click for..in.

   As with other pieces of code, using the Actions toolbox adds the necessary syntax for a specific command. We recommend using the Actions toolbox for loops and conditions until you are familiar and comfortable with the syntax rules.

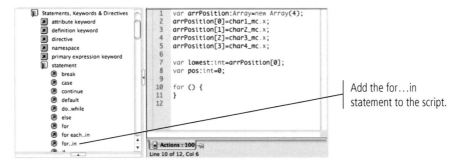

Add the for…in statement to the script.

9. **Between the parentheses of the for statement on Line 10, type:**

   var i in arrPosition

   In plain English, this line means, "while the value of i is within the length of the array named arrPosition." In other words, this loop will execute while the value of i is less than or equal to the number of values stored in the *arrPosition* array.

   The first time the loop runs, i is 0 (the first index value in the array). Each time the loop **iterates** (repeats), the value of i becomes the next index value in the array; so the second time the loop runs, i is 1 (the second index value in the array), and so on. The number of elements in the array is four, so this loop will continue until the value of i reaches 4.

   ```
   1  var arrPosition:Array=new Array(4);
   2  arrPosition[0]=char1_mc.x;
   3  arrPosition[1]=char2_mc.x;
   4  arrPosition[2]=char3_mc.x;
   5  arrPosition[3]=char4_mc.x;
   6
   7  var lowest:int=arrPosition[0];
   8  var pos:int=0;
   9
   10 for (var i in arrPosition) {
   11 }
   12
   ```

10. **Place the cursor after the opening brace and press Return/Enter.**

    Like the custom function you defined earlier, any code you want to execute during the loop must be placed within the braces.

    Note:

    When you create a for…in loop, you can declare the variable inside the loop condition. It is not necessary to declare the variable before entering the loop.

11. **In the Actions toolbox, double-click if (in Language Elements>Statements, Keywords & Directives>statement) to add the condition to the script.**

    The if statement is a called a **conditional statement**. In other words, you are now writing code to determine what happens *if a certain condition is true.*

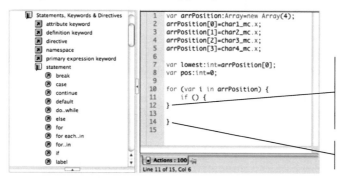

The if statement has its own parentheses (where you define the condition to check) and braces (where you define the action to perform if the condition is true).

This closing brace closes the entire for statement.

12. **Between the parentheses of the if statement on Line 11, type:**

    arrPosition[i]<lowest

    Remember that in Step 4, you initialized the variable lowest to equal the first value stored in the array. In this conditional statement, you compare the value of the current array position to the value of the variable *lowest.*

    The variable i represents the current array position based on the for…in loop.

    ```
    1  var arrPosition:Array=new Array(4);
    2  arrPosition[0]=char1_mc.x;
    3  arrPosition[1]=char2_mc.x;
    4  arrPosition[2]=char3_mc.x;
    5  arrPosition[3]=char4_mc.x;
    6
    7  var lowest:int=arrPosition[0];
    8  var pos:int=0;
    9
    10 for (var i in arrPosition) {
    11     if (arrPosition[i]<lowest) {
    12 }
    13
    14 }
    15
    ```

    - The first time the loop runs, $i = 0$. The conditional if statement compares the value in arrPosition[0] to the value of the variable *lowest.*

    - The second time the loop runs, $i = 1$. The conditional if statement compares the value in arrPosition[1] to the value of the variable *lowest.*

    - The third time the loop runs, $i = 2$. The conditional if statement compares the value in arrPosition[2] to the value of the variable *lowest.*

    - The fourth time the loop runs, $i = 3$. The conditional if statement compares the value in arrPosition[3] to the value of the variable *lowest.*

13. After the opening brace of the if statement, press Return/Enter. On Lines 12 and 13 of the script, type:

> lowest=arrPosition[i];
>
> pos=i;

```
 5    arrPosition[3]=char4_mc.x;
 6
 7    var lowest:int=arrPosition[0];
 8    var pos:int=0;
 9
10    for (var i in arrPosition) {
11        if (arrPosition[i]<lowest) {
12            lowest=arrPosition[i];
13            pos=i;
14    }
15
16    }
17
```

The statements between the conditional if braces determine what happens if the condition is true. In other words:

If the value of arrPosition[i] is less than the value of the *lowest* variable,

then the value of arrPosition[i] becomes the new value of the *lowest* variable

and the value of *i* becomes the new value of the *pos* variable.

Since the x-axis origin ("0") is the left edge of the screen, smaller numbers are closer to the left edge and thus "in the lead" — ultimately, the character with the lowest x value will win the race.

This loop compares each value in the array to find the smallest number — the winner. When the comparison finds a value that is less than the current value of the *lowest* variable, that value becomes the new value of the *lowest* variable. The associated array position (i) is then stored in the *pos* variable to identify the location of the winning character.

14. Save the file and continue to the next exercise.

FLASH FOUNDATIONS

Translating ActionScript to Plain English, Part 2

Suppose the racers' x positions at the end of the race are:

char1_mc (Baggy)	25	so	arrPosition[0] = 25
char2_mc (Granny)	23	so	arrPosition[1] = 23
char3_mc (Navi)	20	so	arrPosition[2] = 20
char4_mc (Nipper)	21	so	arrPosition[3] = 21

Before the program enters the for...in loop, the variable *lowest* contains the value of arrPosition[0], or 25. When the program reaches the for...in loop:

First run of the loop:	i = 0	arrPosition[0] < 25 ?	(Is 25 less than 25?)	False
Second run of the loop:	i = 1	arrPosition[1] < 25 ?	(Is 23 less than 25?)	True, so
				lowest = arrPosition[1] = 23
				pos = 1
Third run of the loop:	i = 2	arrPosition[2] < 23 ?	(Is 20 less than 23?)	True, so
				lowest = arrPosition[2] = 20
				pos = 2
Fourth run of the loop:	i = 3	arrPosition[3] < 20 ?	(Is 21 less than 20?)	False
Fifth run of the loop:	i = 4, which is not a position in the arrPosition array			

The loop is exited.

When the loop is finished, the *lowest* variable is the leftmost x position of the four characters — the winning character (in this example, Navi). The *pos* variable equals the index value of the array position in the *lowest* variable.

In this example, arrPosition[2] holds the lowest value, so the *pos* variable equals 2.

Use an If Else Condition to Call Movie Clip Frames

In the previous exercise, you defined a script that identifies the lowest value in the arrPosition array. Next, you need to write a script that maps the index value of the lowest array value (which you stored in the *pos* variable) to the appropriate character instance on the Stage — effectively identifying the "winner" of the race.

You also need write a script to tell each character what to do, based on the results of evaluation. Because you have four characters, you need to execute the script four times; a for loop is ideally suited for running a set of actions a specific number of times.

For each character (each iteration of the loop), you will write a conditional if else statement that tells the character what to do. If the character wins, the script should call the "cheering" frame of the movie clip's timeline. If the character does not win, the script should call the "crying" frame of the movie clip's timeline.

1. **With `gator_race.fla` open, select Frame 100 of the Actions layer on the main timeline. Place the insertion point at the end of the code in the Actions panel Script pane. Press Return/Enter twice to add white space.**

2. **Declare a variable named "chk" that will hold an integer value; initialize the value of this variable as "pos+1" by typing:**

   ```
   var chk:int=pos+1;
   ```

   Remember, the *pos* variable holds the array position of the lowest value in the array. Because the array positions are numbered from 0–3 and the characters are numbered from 1–4, you have to add 1 to the array position of the lowest value to determine which character wins the race.

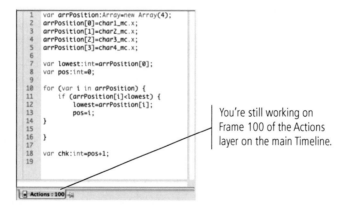

You're still working on Frame 100 of the Actions layer on the main Timeline.

3. **On the next line of the script, declare a variable named "a" that will hold an integer value by typing:**

   ```
   var a:int;
   ```

   This variable will be incremented in a loop and compared to the variable *chk* to define the action of each character instance.

4. **Press Return/Enter twice to move to Line 21. In the Actions toolbox, expand Language Elements>Statements, Keywords & Directives>statement and double-click for in the list.**

5. **Between the parentheses of the for statement, type:**

   a=1; a<=4; a++

   ```
   17
   18   var chk:int=pos+1;
   19   var a:int;
   20
   21   for (a=1; a<=4; a++) {
   22   }
   ```

   The first argument (**a=1**) initializes variable *a* to 1.

   The second argument (**a<=4**) adds a condition so the loop executes only while the value of *a* is less than or equal to 4.

   The third argument (**a++**) increments the value of *a* by 1 at the end of each pass through the loop.

6. **Place the cursor after the opening brace and press Return/Enter.**

7. **In the Actions toolbox, expand Language Elements>Statements, Keywords & Directives>statement and double-click if in the list.**

8. **Between the parentheses in the if condition, type:**

   a==chk

   The **equality operator** (**==**) evaluates the values on either side to determine if they are equal, and returns a Boolean value — true if they are equal or false if they are not equal. In plain English, this line of code checks to see if the value of variable *a* is equal to variable *chk*.

   Note:

   *Remember, **chk** = pos+1, or the number of the character instance with the lowest x position.*

   This brace closes the if statement.
   This brace closes the for statement.

   ```
   20
   21   for (a=1; a<=4; a++) {
   22       if (a==chk) {
   23       }
   24
   25   }
   ```

   This loop runs as long as *a* is less than or equal to 4; *a* is incremented each time the loop runs.

First run:	a = 1	The program enters the if condition, compares the value of *a* to the value of *chk*.
Second run:	a = 2	The program enters the if condition, compares the value of *a* to the value of *chk*.
Third run:	a = 3	The program enters the if condition, compares the value of *a* to the value of *chk*.
Fourth run:	a = 4	The program enters the if condition, compares the value of *a* to the value of *chk*.
Fifth run:	a = 5	The variable *a* is not less than or equal to 4, so the program exits the loop.

   You now need to tell the program what to do when *a* equals *chk* (the character wins the race), and what to do if *a* does not equal *chk* (the character does not win the race).

9. **After the opening brace of the if condition, press Return/Enter.**

10. **On Line 23 of the Script pane, type:**

    this["char"+a+"_mc"].gotoAndPlay("cheering");

    ```
    17
    18   var chk:int=pos+1;
    19   var a:int;
    20
    21   for (a=1; a<=4; a++) {
    22       if (a==chk) {
    23           this["char"+a+"_mc"].gotoAndPlay("cheering");
    24       }
    25
    26   }
    ```

    This statement applies the same technique you used earlier to move a character based on a randomly generated number. In other words, you are appending strings to the value of variable *a* to result in a character instance name (char1_mc, char2_mc, and so on).

    So if the value of *a* equals the value of *chk*, the character wins the race. The gotoAndPlay command calls the frame labeled "cheering" on that movie clip's timeline.

11. **Press Return/Enter to move to the next line in the Script pane.**

12. **In the Actions toolbox, double-click else (Language Elements>Statements, Keywords & Directives>statement).**

The previous line defines what happens if *a* is equal to *chk*. Using the else command, you define what happens if *a* is not equal to *chk*.

13. **In the empty line after the opening brace of the else statement (Line 25 in our example), type:**

```
this["char"+a+"_mc"].gotoAndPlay("crying");
```

Note:

In the Script pane, text strings (including the enclosing quotation marks) appear in green. If you forget to include a quotation mark in your code, the script will not work properly.

```
18    var chk:int=pos+1;
19    var a:int;
20
21    for (a=1; a<=4; a++) {
22        if (a==chk) {
23            this["char"+a+"_mc"].gotoAndPlay("cheering");
24        } else {
25            this["char"+a+"_mc"].gotoAndPlay("crying");
26    }
27
28    }
```

Double-clicking else in the Actions toolbox adds a closing brace for the **if** statement and an opening brace for the **else** statement.

This is now the closing brace for the **else** statement.

This is still the closing brace for the **for** statement.

If *a* is not equal to *chk*, this line of code calls the frame labeled "crying" on the character's movie clip (the beginning of the "losing" animation).

14. **Click the Auto Format button in the Actions panel to clean up the code.**

You have the correct code after the previous step, but it might be difficult to decipher visually. Although double-clicking an item in the Actions toolbox adds the necessary code (including punctuation), the code is not always formatted to be readable by humans. Clicking the Auto Format button makes the code easier to read.

If your script contains syntax errors, clicking the Auto Format button warns you of the errors; code with syntax errors is not reformatted.

Note:

You can customize the code formatting options in the ActionScript pane of the Preferences dialog box.

Auto Format button

```
1    var arrPosition:Array = new Array(4);
2    arrPosition[0] = char1_mc.x;
3    arrPosition[1] = char2_mc.x;
4    arrPosition[2] = char3_mc.x;
5    arrPosition[3] = char4_mc.x;
6
7    var lowest:int = arrPosition[0];
8    var pos:int = 0;
9
10   for (var i in arrPosition)
11   {
12       if (arrPosition[i] < lowest)
13       {
14           lowest = arrPosition[i];
15           pos = i;
16       }
17
18   }
19
20   var chk:int = pos + 1;
21   var a:int;
22
23   for (a=1; a<=4; a++)
24   {
25       if (a==chk)
26       {
27           this["char" + a + "_mc"].gotoAndPlay("cheering");
28       }
29       else
30       {
31           this["char" + a + "_mc"].gotoAndPlay("crying");
32       }
33
34   }
```

Using the Auto Format button indents the pieces of the else statement to show their positions within the main for statement.

15. **Save the file and continue to the next exercise.**

PROGRAM THE RESTART BUTTON FUNCTIONALITY

Now that you understand what is happening throughout the race, you can determine what needs to happen when a user clicks the Restart button and the game is reset:

- The "Click Me" banner needs to become visible.

- The main movie timeline needs to return to Frame 1.

- The Restart button instance needs to be hidden.

- The lila_mc instance needs to return to her original position and start tapping her foot.

- The racetrack_mc instance needs to return to its original position.

- The baggy_mc, granny_mc, navi_mc, and nipper_mc instances need to stop moving, and then return to their original positions.

Because all of these actions need to occur simultaneously when a user clicks the Restart button, the best way to accomplish these tasks is to use an event listener and a custom function.

1. **With `gator_race.fla` open, select Frame 1 of the Event Handlers layer on the main Stage.**

2. **Place the insertion point at the end of the existing script and press Return/Enter twice.**

3. **On Line 16 of the Script pane, add a reference to the restart_btn instance.**

4. **In the Actions toolbox, expand flash.events>EventDispatcher>Methods, and double-click addEventListener.**

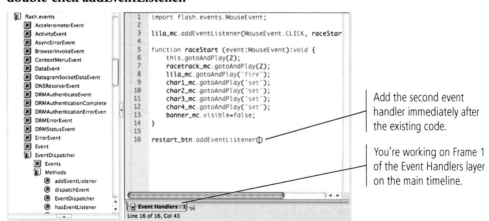

Add the second event handler immediately after the existing code.

You're working on Frame 1 of the Event Handlers layer on the main timeline.

5. **Inside the parentheses of the addEventListener function, type `MouseEvent.` and then choose CLICK from the Code Hint menu.**

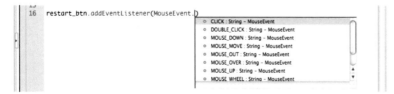

6. After the word CLICK, type `, raceReset` as the name of the function that will be called when the event is triggered.

7. Add a semicolon outside the closing parenthesis.

```
16    restart_btn.addEventListener(MouseEvent.CLICK, raceReset);
```

raceReset is the name of the function that will be called when the restart_btn instance is clicked. Next, you need to define the raceReset function.

8. Press Return/Enter twice to move the insertion point to Line 18 of the Script pane.

9. On Line 18 of the Script pane, type:

```
function raceReset () {}
```

```
16    restart_btn.addEventListener(MouseEvent.CLICK, raceReset);
17
18    function raceReset () {}
```

10. Place the insertion point inside the function parentheses and type `event:` and then choose MouseEvent from the resulting Code Hint menu.

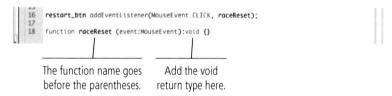

11. Type a colon outside the function parameter parentheses, and then choose void from the Code Hint menu.

Because this function returns no value, you have to use the void return type.

```
16    restart_btn.addEventListener(MouseEvent.CLICK, raceReset);
17
18    function raceReset (event:MouseEvent):void {}
```

The function name goes before the parentheses.　　Add the void return type here.

12. Place the insertion point between the function braces, and then press Return/Enter twice to add an empty line between the opening and closing braces.

13. On Line 19 (within the function braces), add the following code:

```
banner_mc.visible=true;
```

The banner_mc instance is hidden by the function called when a user clicks the lila_mc instance. This command shows the banner_mc instance again.

```
16    restart_btn.addEventListener(MouseEvent.CLICK, raceReset);
17
18    function raceReset (event:MouseEvent):void {
19        banner_mc.visible=true;
20    }
```

14. **Beginning on Line 20 of the Script pane (within the function braces), add lines of code to reset the main movie timeline and the Racetrack movie clip:**

    ```
    this.gotoAndStop(1);
    racetrack_mc.gotoAndStop(1);
    ```

    Frame 1 of the Actions layer on the main timeline includes a function to hide the Restart button instance, so you don't need to add a separate command to hide the Restart button here.

    Although the main timeline and the racetrack timeline include stop functions at Frame 1, the commands in this function override the stop commands. To prevent the main timeline and Racetrack movie clip timeline from automatically playing when the race resets, you have to use the gotoAndStop command instead of gotoAndPlay.

    ```
    16  restart_btn.addEventListener(MouseEvent.CLICK, raceReset);
    17
    18  function raceReset (event:MouseEvent):void {
    19      banner_mc.visible=true;
    20      this.gotoAndStop(1);
    21      racetrack_mc.gotoAndStop(1);
    22  }
    ```

Note:

You have already used most of the commands that are needed in this function. You can simply type the necessary code or use the Actions toolbox to add the various elements.

15. **On Line 22, add another line to the script that resets the Lila movie clip to its original state:**

    ```
    lila_mc.gotoAndPlay("waiting");
    ```

    You want the Lila character to tap her foot when the race is reset, so you have to use the gotoAndPlay command instead of gotoAndStop.

    Remember, you defined frame labels of "waiting" in all the character symbols (including Lila). Rather than using a frame number, you should use the defined frame labels.

    ```
    16  restart_btn.addEventListener(MouseEvent.CLICK, raceReset);
    17
    18  function raceReset (event:MouseEvent):void {
    19      banner_mc.visible=true;
    20      this.gotoAndStop(1);
    21      racetrack_mc.gotoAndStop(1);
    22      lila_mc.gotoAndPlay("waiting");
    23  }
    ```

Note:

The gotoAndStop command is similar to the gotoAndPlay command, except the playhead does not automatically move from the referenced frame.

Note:

You can access the gotoAndStop command in the Actions toolbox by expanding flash.display> MovieClip>Methods.

16. **Add four more lines to the raceReset function to reset the timelines of each racer's movie clip:**

    ```
    char1_mc.gotoAndStop("waiting");
    char2_mc.gotoAndStop("waiting");
    char3_mc.gotoAndStop("waiting");
    char4_mc.gotoAndStop("waiting");
    ```

    These lines of code move each racer's movie clip timeline to its original position, where the racers wait for a user to click the Lila movie clip.

    ```
    16  restart_btn.addEventListener(MouseEvent.CLICK, raceReset);
    17
    18  function raceReset (event:MouseEvent):void {
    19      banner_mc.visible=true;
    20      this.gotoAndStop(1);
    21      racetrack_mc.gotoAndStop(1);
    22      lila_mc.gotoAndPlay("waiting");
    23      char1_mc.gotoAndStop("waiting");
    24      char2_mc.gotoAndStop("waiting");
    25      char3_mc.gotoAndStop("waiting");
    26      char4_mc.gotoAndStop("waiting");
    27  }
    ```

17. **Using the Selection tool, click the Baggy instance on the Stage. Note the x position in the Properties panel.**

Because the characters move only as a result of code throughout the race, their positions on the development Stage never change. All four racers have the same horizontal (x) starting position.

Note:

Access the x property in the Actions toolbox by expanding flash.display> DisplayObject> Properties.

18. **Add four more lines to the raceReset function, changing the x property of each racer movie clip instance to the same x position you noted in Step 17:**

```
char1_mc.x=780;

char2_mc.x=780;

char3_mc.x=780;

char4_mc.x=780;
```

```
16   restart_btn.addEventListener(MouseEvent.CLICK, raceReset);
17
18   function raceReset (event:MouseEvent):void {
19       banner_mc.visible=true;
20       this.gotoAndStop(1);
21       racetrack_mc.gotoAndStop(1);
22       lila_mc.gotoAndPlay("waiting");
23       char1_mc.gotoAndStop("waiting");
24       char2_mc.gotoAndStop("waiting");
25       char3_mc.gotoAndStop("waiting");
26       char4_mc.gotoAndStop("waiting");
27       char1_mc.x=780;
28       char2_mc.x=780;
29       char3_mc.x=780;
30       char4_mc.x=780;
31   }
```

Note:

Because the motion in the Background and Lila movie clips is based on tweens, moving the playhead for those movie clips restores them to their original positions.

The racers' movement, however, is based on random numbers generated in ActionScript code; to return the racers to their original positions, you have to use ActionScript to change the x properties of those movie clip instances.

19. **Save the file and continue to the next exercise.**

TEST THE MOVIE AND CHANGE CURSOR STYLES

Most programmers have a motto that says code once, test often (or something to that effect). When you are writing code — regardless of how basic or complex — it's a good idea to test your scripts as often as possible. Even the best planner can overlook a critical design element during initial development. Problems or issues only become obvious when you watch the actual movie as it will appear when exported.

1. **With `gator_race.fla` open, press Command-Return/Control-Enter to test the movie.**

2. **In the Player window, click Lila to play the race.**

3. **After the race is over, click the Restart button to verify that everything resets properly. The race should play again.**

There is one remaining issue that you should address. Users typically expect to see a pointing-hand cursor when the mouse hovers over a button. The Restart button is an instance of a button symbol, so the cursor changes to a pointing hand when the mouse is over the button instance. The Lila instance, which starts the race, functions as a button, but it was created as a movie clip because it contains a number of frame animations. To meet users' expectations, you should change the cursor style for the lila_mc instance.

4. **Close the Player window and return to Flash.**

5. **Select Frame 1 of the Event Handlers layer on the main timeline.**

6. **Place the insertion point at the beginning of Line 3 in the Script pane. Press Return/Enter to move the existing script to Line 4.**

7. **On Line 3 of the Script pane, type:**

```
lila_mc.buttonMode=true;
```

This code allows the movie clip instance to appear to behave as a button without being created as a formal button symbol. When a user moves the mouse over the movie clip in the movie, the cursor will change to a pointing-hand icon that indicates a clickable object.

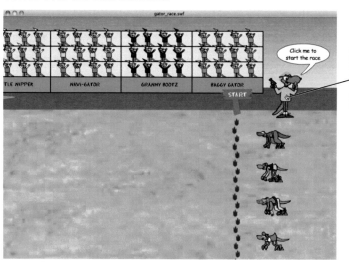

```
1  import flash.events.MouseEvent;
2
3  lila_mc.buttonMode=true;
4  lila_mc.addEventListener(MouseEvent.CLICK, raceStart);
5
6  function raceStart (event:MouseEvent):void {
7      this.gotoAndPlay(2);
8      racetrack_mc.gotoAndPlay(2);
9      lila_mc.gotoAndPlay("fire");
10     char1_mc.gotoAndPlay("set");
11     char2_mc.gotoAndPlay("set");
12     char3_mc.gotoAndPlay("set");
```
Event Handlers : 1
Line 3 of 32, Col 25

Note:

The buttonMode command can be accessed in the Actions toolbox by expanding flash.display> MovieClip>Properties.

8. **Press Command-Return/Control-Enter to test the movie. Move your mouse cursor over the Lila instance.**

You set the buttonMode property of the movie clip instance to true, so the cursor changes to a pointing hand.

9. **Close the Player window and return to Flash.**

10. **Save the file and continue to the final stage of the project.**

Stage 6 **Adding Sound with ActionScript**

When you last tested the movie, all elements moved correctly. To add life into the race game, you need to add sounds to two of the graphic elements — a gun shot when Lila starts the race, and the cheers from the crowd as they watch the race.

In Project 2: Talking Kiosk Interface, you learned how to import sound files into Flash, and to control sound files on the timeline. The sound files for this project have already been imported into the Library panel, and the running feet sounds are attached at appropriate spots in the racer characters' symbol timelines. In this stage of the project, you are going to use ActionScript to add the remaining sounds at appropriate points in the movie.

CREATE ACCESS FOR SOUND FILES IN THE LIBRARY

ActionScript 3 contains a built-in Sound class in the flash.media package that contains the events, methods, and properties related to controlling sound files.

To access sound files from the Library panel, you have to create a linkage for each sound file — defining new custom classes that are essentially sub-groupings of the main sound class (referred to as "extending the sound class").

If this seems confusing, don't worry; it will make more sense as you complete the following exercises.

Note:

This embeds the sound files into the final movie, even though they are not placed on the Stage at any point.

1. **With `gator_race.fla` open, expand the Sounds folder in the Library panel.**

2. **Control/right-click the gun_shot.wav file and choose Properties from the contextual menu.**

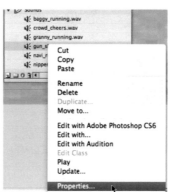

3. **In the resulting Sound Properties dialog box, click the ActionScript tab to display those options.**

4. **In the Linkage area, check the Export for ActionScript option.**

   The first step in adding sound through ActionScript requires making those sounds available without placing them on a timeline. This is called **linkage**. You must create a new class for each sound file; those new classes extend the base sound class.

   By default, the Export in Frame 1 option is also selected. The Base Class field shows that the file is recognized as part of the flash.media.Sound class.

5. **Change the Class field to ShotSound.**

   This is the name of the new class you are creating to make this file accessible in ActionScript.

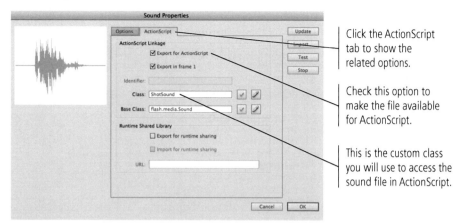

Click the ActionScript tab to show the related options.

Check this option to make the file available for ActionScript.

This is the custom class you will use to access the sound file in ActionScript.

6. **Click OK.**

   Flash displays a warning that a definition for the class does not exist, so the class will be defined for you when the file is exported.

The class name appears in the Linkage column of the Library panel.

7. **Click OK to close the warning dialog box.**

8. **Repeat Steps 2–7 for the crowd_cheers.wav sound file, defining CrowdSound as the custom class name.**

   The other files are attached to frames in the specific characters' timelines, so you do not need to define classes for those.

9. **Save the file and continue to the final exercise.**

CREATE SCRIPTS TO CALL SOUNDS

Now that the two library sounds are accessible to ActionScript, you have to write the code to play the different files at the appropriate points in the movie.

- The gun shot sound should play when Lila fires the starting pistol.

- The crowd cheering sound should play when the crowd is visible on the Stage, based on the motion tween in the Racetrack movie clip.

1. **With gator_race.fla open, select Frame 1 of the Event Handlers layer on the main timeline.**

2. **In the Actions panel, click at the end of the last line in the raceStart function body and press Return/Enter twice.**

   The gun shot sound should play when the race starts, so you are adding a command as part of the raceStart custom function.

3. **On Line 16 of the script, type:**

   ```
   var audioGun:ShotSound=new ShotSound();
   ```

   Creating an instance of an object (often referred to as *instantiating* the object) requires the **new** operator with a constructor function. In this example, "ShotSound( )" is the constructor function.

   The first half of this statement declares a variable named "audioGun" of the class ShotSound. That variable stores the instance created by the second half of the statement.

Note:

The gun_shot.wav file is the sound that an instance of the ShotSound class represents.

```
 5
 6   function raceStart (event:MouseEvent):void {
 7       this.gotoAndPlay(2);
 8       racetrack_mc.gotoAndPlay(2);
 9       lila_mc.gotoAndPlay("fire");
10       char1_mc.gotoAndPlay("set");
11       char2_mc.gotoAndPlay("set");
12       char3_mc.gotoAndPlay("set");
13       char4_mc.gotoAndPlay("set");
14       banner_mc.visible=false;
15
16       var audioGun:ShotSound=new ShotSound();
17   }
18
19   restart_btn.addEventListener(MouseEvent.CLICK, raceReset);
```

You are adding the sound to the raceStart function body.

Frame 1 of the Event Handlers layer is selected.

Event Handlers : 1
Line 16 of 34, Col 41

To better understand what is happening in this code, it helps if you look at the statement backwards:

Create an object of the ShotSound class (i.e., the gun_shot.wav file), and store it in the variable named audioGun.

4. **Press Return/Enter to move to Line 17 of the Script pane, and then type:**

   ```
   audioGun.play();
   ```

Note:

You can access the play function in the Actions toolbox by expanding flash.media> Sound>Methods.

```
 9       lila_mc.gotoAndPlay("fire");
10       char1_mc.gotoAndPlay("set");
11       char2_mc.gotoAndPlay("set");
12       char3_mc.gotoAndPlay("set");
13       char4_mc.gotoAndPlay("set");
14       banner_mc.visible=false;
15
16       var audioGun:ShotSound=new ShotSound();
17       audioGun.play();
18   }
19
```

Event Handlers : 1
Line 17 of 35, Col 18

5. **Choose View>Magnification>Fit In Window to make the entire Stage area visible.**

6. **Double-click the racetrack_mc instance on the Stage to edit the symbol in place.**

7. **Click the playhead and drag right until the second crowd (near the finish line) moves into the Stage area (past the guide that marks the left edge of the Stage).**

   Our timeline shows the crowd appearing at Frame 74.

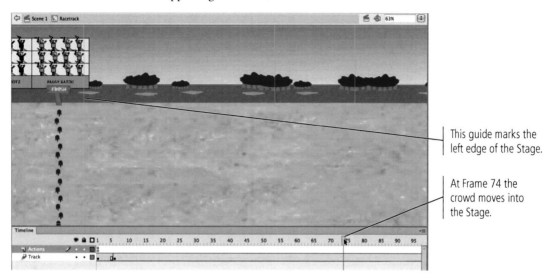

This guide marks the left edge of the Stage.

At Frame 74 the crowd moves into the Stage.

8. **Click Scene 1 in the Edit bar to return tot he main Stage.**

9. **Add a new layer named Sound at the top of the main timeline. Insert keyframes at Frames 2 and 74 of the Sound layer.**

   The existing layer contains a tween, so you are using a separate layer to contain the sound actions. Layers are not only useful for organizing visual elements, they also help in systematic development of a project that contains sounds. By inserting the sound file in a separate layer, you can more easily modify or delete that particular file (if necessary).

10. **Select Frame 2 of the Sound layer and click in the Actions panel Script pane.**

11. **Declare a variable named "audioCrowd" of the "CrowdSound" class by typing:**

    `var audioCrowd:CrowdSound=new CrowdSound();`

12. **On the next line of the Script pane, add the play command for the audioCrowd variable by typing:**

    `audioCrowd.play();`

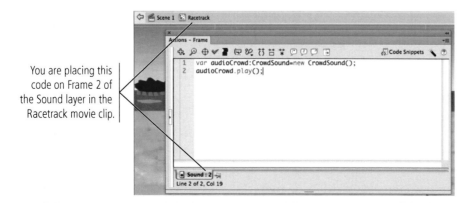

You are placing this code on Frame 2 of the Sound layer in the Racetrack movie clip.

13. **Select Frame 74 of the Sound layer. In the Script pane of the Actions panel, add the play command for the audioCrowd variable by typing:**

```
audioCrowd.play();
```

Note:

You don't need to declare the variable every time you want to use it.

14. **Press Command-Return/Control-Enter to test the completed game.**

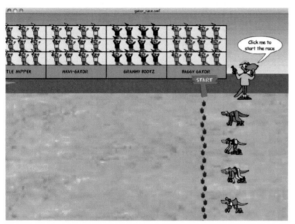 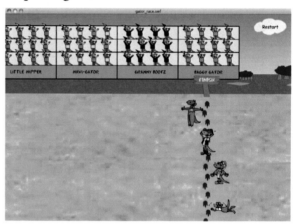

15. **Close the Player window and return to Flash.**

16. **Save and close the completed Flash file.**

fill in the blank

1. _____ are required to target specific symbol instances using ActionScript.

2. A(n) _____ relates a trigger object and event to a defined function.

3. Scripting in ActionScript 3 uses _____ to apply script commands to specific objects.

4. The _____ contains all of the functions of the ActionScript language, sorted into classes.

5. In the Actions panel, the _____ button lets you choose which object you want to affect with a function or event.

6. In the Actions panel, _____ show different options depending on the context of the code you are typing.

7. The code between braces represents the _____, or the actions that will be performed when the function is called.

8. _____ is also known as the assignment operator.

9. A(n) _____ is used to define actual characters (inside of quotation marks) in a script.

10. _____ is the numbered position of specific elements in an array.

short answer

1. Briefly define "variable." Provide two examples of how variables can be useful.

2. Briefly define "array." Provide two examples of how arrays can be useful.

3. Briefly define "loop." Include a short explanation of at least three different types of loops.

Portfolio Builder Project

Use what you learned in this project to complete the following freeform exercise.
Carefully read the art director and client comments, then create your own design to meet the needs of the project.
Use the space below to sketch ideas; when finished, write a brief explanation of your reasoning behind your final design.

art director comments

To help promote the new adventure book, *Up and Away*, your client has asked you to create a game similar to the gator race.

To complete this project, you should:

❏ Download the client's supplied files in the **FL6_PB_Project5.zip** archive on the Student Files Web page.

❏ Review the client's supplied artwork. If you use this artwork, import the file into Flash and create the necessary movie clips from the different elements.

❏ If you prefer, find or create other artwork as appropriate for a hot air balloon race.

❏ Add animations to the different elements of the race.

❏ Program the game functionality using ActionScript.

client comments

When we thought of this game, we were thinking about the carnival games where kids shoot water guns at a lever, and the balloons rise up a pipe based on how much water hits the trigger mechanism. We don't need something that complicated; we only need a Start button so kids can set the balloons flying. (And don't forget to add a Restart button so the kids can play the game more than once.)

We've supplied you with an Illustrator file that's being kicked around as the back-cover artwork, but you can use a different background or balloon artwork if you have something else.

We do want you to add some kind of frame animation to the balloons so they do more than just rise when the game is started. You also need to come up with some way to differentiate between the winner and the losers. We thought about the losers deflating and falling, but again, we're open to suggestions.

project justification

Project Summary

This project was designed as an introduction to ActionScript. By completing these exercises, you learned how to use many foundational programming techniques — event handlers, variables, operators, loops, and conditional statements — to control the elements in a Flash movie. You also learned about the terms related to writing code, and you discovered how functions relate to a Flash movie.

Many designers devote their entire careers to building applications with ActionScript. To enhance your marketability as a professional Flash developer, we encourage you to continue your ActionScript education beyond this project.

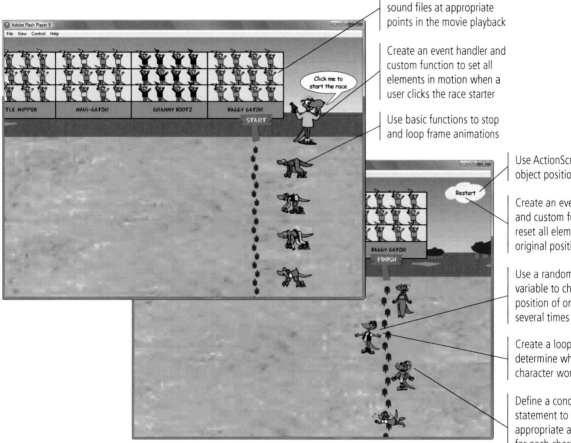

Use ActionScript to add sound files at appropriate points in the movie playback

Create an event handler and custom function to set all elements in motion when a user clicks the race starter

Use basic functions to stop and loop frame animations

Use ActionScript to control object position and visibility

Create an event handler and custom function to reset all elements to their original positions

Use a random-number variable to change the position of one character several times

Create a loop to determine which character won the race

Define a conditional statement to call the appropriate animation for each character at the end of the game

Gopher Golf Game

Your client operates a family entertainment complex in suburban Pittsburgh. The facility includes a nationally recognized professional golf course, as well as a miniature golf course to lure families looking for inexpensive entertainment. You have been hired to develop an interactive game for the client's Web site that appeals to both children and adults.

This project incorporates the following skills:

❏ Importing and analyzing artwork from an Adobe Illustrator file

❏ Creating various types of symbols based on imported artwork

❏ Editing symbol artwork to meet the unique needs of the job

❏ Using a mask to hide parts of a tween animation

❏ Writing scripts to animate different symbols in a random sequence

❏ Defining a variable to control game duration

❏ Using dynamic text areas to track user actions

❏ Adding script to show specific symbols based on the user's performance

❏ Creating an app for Android devices

client comments

I was at the amusement park with my niece last weekend and she played Whac-A-Mole for nearly an hour! I swear, it was contagious, because I played right along with her the whole time! That would be just the thing for our site.

To appeal to both adults and children, we also thought it would be fun to use gophers instead of moles — call it Gopher Golf. The kids will like the silly cartoon animals, and we think most adults will understand the gopher-on-the-golf-course reference.

art director comments

I had our staff illustrator build all of the necessary artwork components in Adobe Illustrator, so you should be able to import the file and get straight to work on the animation and programming. The artist is very good, but some organization will be part of your job simply because you have a better idea of the game structure than the illustrator does.

This game is going to require some random-number generation to make different gophers pop up around the golf course. You also need to use variables to control the game's duration and keep track of the user's hits and misses.

Finally, make sure the game can be played more than once. The Replay button should reset the counting variables and make gophers randomly pop up again.

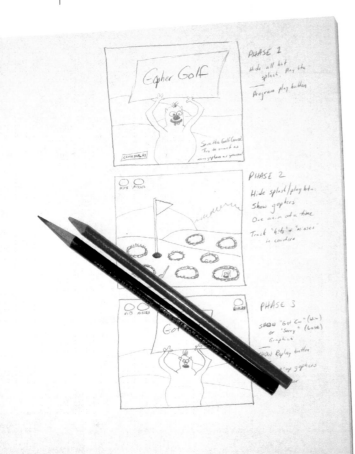

project objectives

To complete this project, you will:

- ❏ Import provided artwork
- ❏ Define symbols and instances
- ❏ Organize layers and library folders
- ❏ Map timeline frames and keyframes
- ❏ Change button states
- ❏ Add dynamic text areas
- ❏ Animate the pop-up gopher movie clip
- ❏ Mask a motion tween
- ❏ Add timeline controls
- ❏ Program basic game functionality
- ❏ Use the trace function and code comments
- ❏ Track user actions
- ❏ Pass values to the parent timeline
- ❏ Program the game results
- ❏ Create an app for Android devices

Stage 1 Preparing Game Artwork

The best way to start a complicated project such as this one is to develop a clear plan of what needs to take place. Consider the sketches from the original project-planning meeting; as you can see, you need to complete a number of tasks to create the final game.

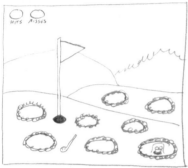

 IMPORT PROVIDED ARTWORK

The first step in most Flash projects is creating a new file. Then you can import the provided artwork and review the content.

1. **Download FL6_RF_Project6.zip from the Student Files Web page.**

2. **Expand the ZIP archive in your WIP folder (Macintosh) or copy the archive contents into your WIP folder (Windows).**

    This results in a folder named **Gophers**, which contains the files you need for this project. You should also use this folder to save the files you create in this project.

3. **In Flash, create a new Flash file for ActionScript 3.**

    For now, you will develop this game as a typical Flash movie. After you complete the game, you will adapt the file for export to an Android device.

4. **Using the Properties panel, change the file frame rate to 15 fps.**

    Don't worry about the Stage size. You're going to change that shortly, based on the size of imported artwork.

5. **Choose File>Import>Import to Stage. Navigate to the file gophers.ai in the WIP>Gophers folder and click Open.**

6. **In the Import to Stage dialog box, review the primary level of items that will be imported.**

7. **At the bottom of the dialog box, make sure the Flash Layers option is selected in the Convert Layers To menu.**

8. **Check the Place Objects at Original Position and Set Stage Size options.**

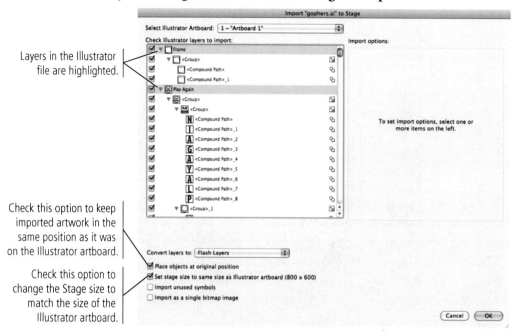

Layers in the Illustrator file are highlighted.

Check this option to keep imported artwork in the same position as it was on the Illustrator artboard.

Check this option to change the Stage size to match the size of the Illustrator artboard.

9. **Click OK to import the file.**

10. **Click in the gray area outside the Stage area to deselect all imported objects. If necessary, use the Fit In Window magnification option to show the entire Stage in the document window.**

The Stage is now 800 × 600 pixels; the imported artwork fits entirely inside the new Stage area.

The imported artwork is contained on eleven layers.

11. **Save the file as gophers.fla in your WIP>Gophers folder and continue to the next exercise.**

DEFINE SYMBOLS AND INSTANCES

When you work with provided artwork, you never know what you are going to get. Some designers simply illustrate with no concern for how the file will be used. Others understand the requirements of artwork that will be used in a Flash movie, and organize files as much as possible to ensure an almost seamless transition from an Illustrator artboard to the Flash Stage. The artwork for this project — like the majority of artwork you'll receive from external sources — falls somewhere in between these two extremes.

1. **With gophers.fla open, Option/Alt-click the visibility icon for the top layer (Frame).**

   This method hides all layers but the one you Option/Alt-click.

   The Frame layer has no functional purpose other than (as the name suggests) framing the overall artwork. You don't need to address this object in any way, so you don't need to convert it to a symbol.

Note:

Many Flash projects involve artwork created in Adobe Illustrator. The amount of required setup work depends on the complexity of the project, but also on the quality of the artwork.

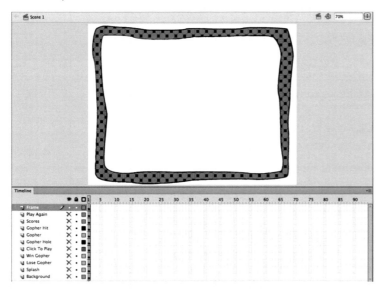

2. **Option/Alt-click the visibility icon for the next layer in the stack (Play Again).**

   The Play Again layer contains the artwork for a button that users can click to replay the game.

3. **Control/right-click inside the selected object on the Play Again layer and choose Convert to Symbol from the contextual menu.**

Note:

If you Option/Alt-click the visibility icon for a layer that is already hidden, it becomes visible; all other layers in the file are hidden.

4. **In the resulting dialog box, name the new symbol Replay, choose Button from the Type menu, and choose the center registration point. Click OK to create the new symbol.**

You are using the Button type for this symbol because you are going to use the special button frames to change the text color when a user clicks the button.

Note:

If the Convert to Symbol dialog box appears in Advanced mode, click the button to collapse it to Basic mode.

5. **With the symbol instance selected on the Stage, type replay_btn in the Instance Name field of the Properties panel.**

Name the instance on the Stage "replay_btn".

Note:

Keep in mind that ActionScript is case sensitive. When you write script to address this instance, you must use the lowercase instance name that you defined in this step.

6. **Show the next layer in the stack (Scores) and hide all other layers.**

This layer contains two groups, which will track the users' "hits" and "misses" as they play the game. These don't need to be clickable, but you do need to address them with ActionScript — and that means you must convert each group to a movie clip symbol.

7. **Using the Selection tool, click away from the selected objects to deselect them, then click the Hits group to select it.**

If you clicked the Scores layer name in the timeline after showing the layer, both objects on the layer will be automatically selected. If this is the case, click away from the two groups to deselect them so you can select only the Hits group.

8. **Control/right-click the Hits group and choose Convert to Symbol in the contextual menu. Name the new symbol Hits, choose Movie Clip in the Type menu, and choose the center registration point. Click OK to return to the Stage.**

The original artist grouped related objects in the Illustrator file, so the entire "Hits" golf ball artwork is a single group.

9. **With the Hits instance selected on the Stage, type hits_mc in the Instance Name field.**

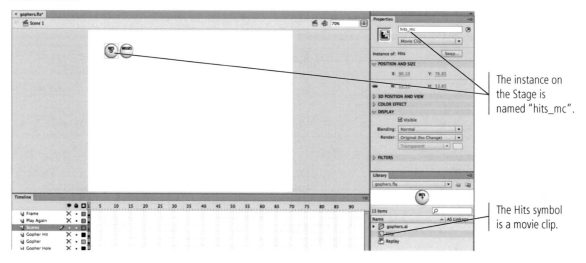

The instance on the Stage is named "hits_mc".

The Hits symbol is a movie clip.

10. Repeat Steps 8–9 for the Misses golf ball, naming the instance `misses_mc`.

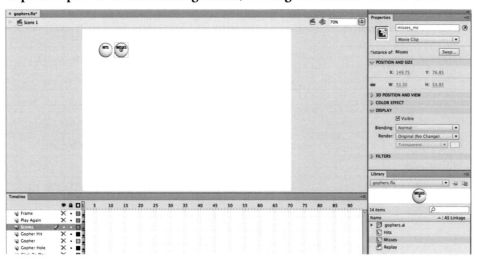

11. Show the next layer in the stack (Gopher Hit) and hide all other layers.

This layer contains artwork that will be part of the pop-up gopher animation, which you will create later. You are going to create the animation on a motion tween, which means you need to convert this artwork to a movie clip symbol.

12. Click the visible object on the Stage to select the group. Using the same method from Step 8, create a new movie clip symbol named `Gopher_Hit`, using the bottom-center registration point.

13. After you create the symbol, delete the selected instance from the Stage.

This symbol will be used inside another movie clip; it does not appear by itself in the game, so you can delete the instance from the Stage.

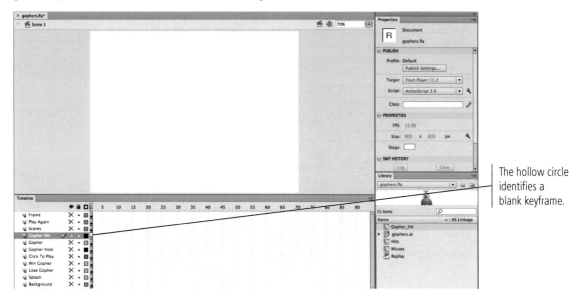

The hollow circle identifies a blank keyframe.

14. **Using the same process outlined in the previous steps, review the contents of each layer and create the necessary symbols. Use the following information to create each movie clip and instance:**

Layer	Symbol Name	Symbol Type	Registration Point	Action to Take on the Stage
Gopher	Gopher	Movie clip	bottom-center	Delete the instance from the Stage
Gopher Hole	Gopher_Hole	Movie clip	bottom-center	Delete the instance from the Stage
Click To Play	Play	Button	center	Name the instance **play_btn**
Win Gopher	Win_Gopher	Movie clip	bottom-center	Name the instance **win_mc**
Lose Gopher	Lose_Gopher	Movie clip	bottom-center	Name the instance **lose_mc**
Splash	Splash_Gopher	Movie clip	bottom-center	Name the instance **splash_mc**
For the Splash layer, the artist did not group all the objects that make up the layer. You must first select all of the pieces on the layer before creating the movie clip for this layer's contents.				
Background		None		

Although this process can be time consuming, reviewing the provided artwork and creating the necessary components is a vital part of any Flash project. (To minimize the amount of required set-up work for this project, we optimized the original artwork significantly. When you work on your own projects, you never know what you will get until you start clicking in the imported artwork.)

Note:

Symbol and instance names should reflect the content and purpose. Descriptive names will help significantly when you create the required ActionScripts.

15. **Save the file and continue to the next exercise.**

 ## Organize File Layers

Before you build the animations and scripts that are required for this game, you need to complete one more organizational step. To more easily manage the large number of components throughout the project, you should remove unnecessary layers and add any new layers that you know you will need.

Of course, you can add layers at any time in a project, but taking the time to properly set up a project at the beginning of the process can make your work easier and more efficient. This process also helps you to get into the habit of analyzing a project's needs before you create a single animation or script — a vital step in successful Flash development.

1. **With gophers.fla open, show all layers on the timeline.**

   To help you more easily manage the various components at different points in time, the final game will include seven basic layers:

   - The Background and Frame layers, which already contain the necessary artwork
   - A Splash layer to manage the objects that appear when the movie first opens
   - A Content layer that manages the pop-up gopher animations and the counters
   - Two results layers to contain the "Win" and "Lose" movie clips, respectively
   - An Actions layer to manage the various scripts

2. **Lock the Background and Frame layers.**

Note:

You will use ActionScript to control the visibility of the "Win" and "Lose" movie clips, so you could use a single layer for the two results movie clips. However, using separate layers makes it easier to manage the separate objects during development.

3. **Select the three layers with no content and delete them.**

   You can identify empty layers by the blank keyframe icon (the open circle).

The blank keyframes identify the empty layers.

Click this button to delete selected layers.

4. **Click the Click To Play layer name to select the objects on that layer. Choose Edit>Cut.**

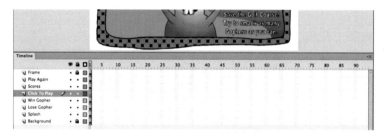

5. **Click the Splash layer name to select that layer, and then choose Edit> Paste in Place.**

   The Paste in Place command makes it easy to move objects from one layer to another.

Paste in Place puts the cut object in the same spot on the Splash layer.

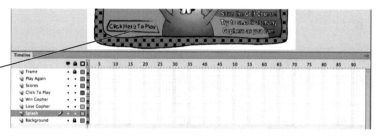

Note:

Press Command/Control-Shift-V to paste the copied object in place.

6. **Delete the now-empty Click To Play layer.**

   The artist who created these projects used Illustrator layers to manage the various pieces of the game. The layers in provided artwork will not always match the layers you need or want in your Flash file, so this type of clean-up and reorganization might be required whenever you import artwork from a client-supplied file.

7. **Double-click the Scores layer name and change it to Content.**

   It's always a good idea to use descriptive names for assets, including layers.

8. **Repeat Steps 4–6 to move the replay_btn instance from the Play Again layer to the Win Gopher layer and remove the Play Again layer.**

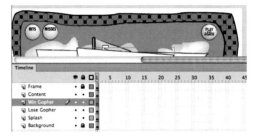

9. **Select the Frame layer, and then click the New Layer button. Name the new layer Actions.**

By selecting the top layer before adding the layer, the new layer automatically appears at the top of the stack in the timeline.

10. **In the timeline, drag the Splash layer above the Content layer.**

Although not strictly necessary, it can be easier to understand complex processes if you arrange layers in a linear (top-to-bottom) manner, in the order their content appears when the movie plays.

11. **Save the file and continue to the next exercise.**

MAP TIMELINE FRAMES AND KEYFRAMES

In this exercise, you add keyframes and manipulate regular frames to define the basic timeline structure for the three stages of the final game:

- The splash screen and Play button appear when the movie first opens.

- The pop-up gophers and score clips appear when a user clicks the Play button.

- The results and "Play Again" button appear after the game is finished. The score clips should remain visible in the third stage.

1. **With `gophers.fla` open, click the keyframe on the Content layer to select it. Click the selected frame again, and then drag it to Frame 2.**

   You can't simply click and immediately drag a frame. You have to first select the frame(s), release the mouse button, and then click again to drag the selected frame(s).

   When the movie opens, only the splash objects and Play button should be visible. By moving the Content layer keyframe from Frame 1 to Frame 2, you effectively hide the Content layer objects until a user clicks the Play button. (Later in this project, you will add a script to prevent the timeline from automatically playing when the game first opens.)

Note:

This is not a double-click; there must be a pause between the clicks. Flash indicates that you can drag a frame with a small square in the cursor icon.

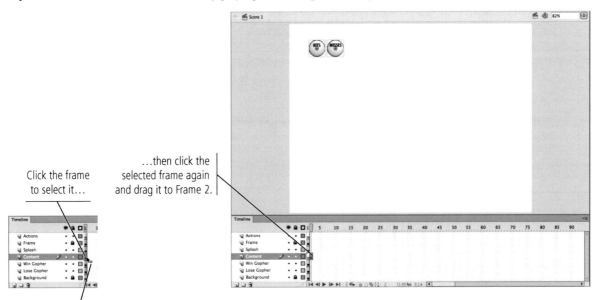

Click the frame to select it...

...then click the selected frame again and drag it to Frame 2.

The square in the cursor icon indicates that you can click and drag to move the selected frames.

2. **Click the keyframe on the Win Gopher layer, then Shift-click to select the keyframe on the Lose Gopher layer. Click either selected frame and drag them both to Frame 3.**

The third phase of the game shows the user's results. The Play Again button and the "Win" or "Lose" movie clip should not be visible until the third phase — which you accomplish by moving these layers to Frame 3.

Shift-click to select
contiguous frames...

...then click either
selected frame and
drag to move all
selected frames.

3. **Click Frame 3 of the Actions layer to select it. Press Command/Control and then click Frame 3 of the Frame, Splash, Content, and Background layers to add them to the selection.**

Command/Control-clicking selects individual noncontiguous frames. Shift-clicking selects all contiguous frames between your first and second selections.

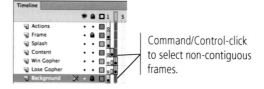

Command/Control-click
to select non-contiguous
frames.

4. **Press the F5 key to add new regular frames to all selected frames.**

All layers now extend to Frame 3, and the background and frame graphics are visible on all frames of the movie.

5. **Click Frame 3 of the Actions layer, and then Shift-click Frame 3 of the Background layer to select Frame 3 on all layers.**

6. **Click any of the selected frames and drag to Frame 22.**

The pop-up gopher animation requires 20 frames. Although the animation will occur within the timeline of its own movie clip, you have to allow enough time for the movie clip timeline to finish before the main timeline reaches the next process.

Because the content layer begins on Frame 2, allowing 20 frames for the animation means the main content should end on Frame 21. The third phase (the results) must then begin on Frame 22.

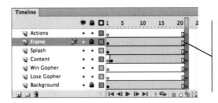

When you drag frames, keyframes are automatically added to those frames. You can remove any of these that are not necessary.

Note:

Be patient; it might take a few seconds for Flash to add the necessary frames to all the layers.

7. **On the Splash layer, select Frame 2, and then Shift-click Frame 22 to select all but Frame 1 on that layer. Control/right-click any of the selected frames and choose Remove Frames from the contextual menu.**

Shift-click to select all but Frame 1 on the Splash layer...

...then use the Remove Frames command to delete the selected frames.

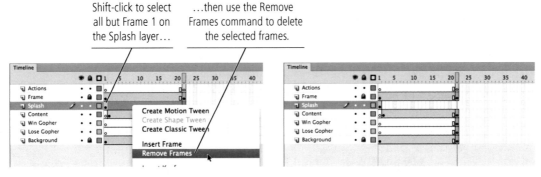

The objects on this layer should only be visible when the game first opens, so it is not necessary to extend the layer to the end of the movie. Rather than simply deleting the unnecessary keyframe, you are deleting all unnecessary frames from the layer.

Note:

You can't simply press Delete to remove frames; you have to choose the Remove Frames command.

8. **Control/right-click Frame 22 of the Frame layer and choose Clear Keyframe from the contextual menu.**

If you choose Remove Frame, the keyframe is deleted but so is the regular frame — which would cause the layer to have only 21 frames instead of the necessary 22.

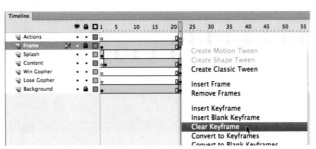

9. **Repeat Step 8 for the Background layer.**

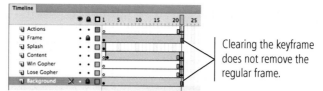

Clearing the keyframe does not remove the regular frame.

10. **Save the file and continue to the next exercise.**

 CHANGE BUTTON STATES

The game requires two buttons (Play and Replay), which should change in appearance when the user's mouse rolls over them. In Project 2: Talking Kiosk Interface, you learned about the special frames that are created in a button symbol. In this exercise, you change the Over frame of the two button symbols.

1. **With gophers.fla open, double-click the Play button symbol icon in the Library panel to enter into the symbol Stage.**

2. **Add a new keyframe to the Over frame in the button symbol timeline.**

3. **With the Over frame selected, double-click the artwork on the Stage to enter into the group.**

4. **Double-click the blue area in the background to enter into the nested group.**

5. **Double-click the blue area again to access the blue drawing object.**

   Depending on how the imported artwork was created, you might have to drill down several levels to access the individual drawing objects in client-supplied artwork.

Note:

Press F6, or choose Insert>Timeline> Keyframe, to insert a keyframe at the selected frame on the timeline.

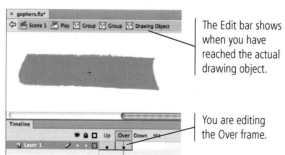

The Edit bar shows when you have reached the actual drawing object.

You are editing the Over frame.

6. **Click the drawing object to select it, and then use the Color or Properties panel to change the fill color to a bright green.**

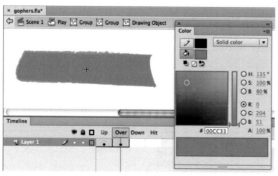

7. **In the Library panel, double-click the Replay button symbol icon to enter into the symbol Stage.**

   It is not necessary to return to the main Stage before you enter a different symbol Stage.

8. **Add a new keyframe to the Over frame of the timeline.**

9. **With the Over frame selected, double-click the artwork on the Stage to enter into the group, and then double-click any of the letters to enter the group containing just the words.**

10. **With the individual letter objects selected, use the Color or Properties panel to change the fill color to the same green you used in the Play button symbol.**

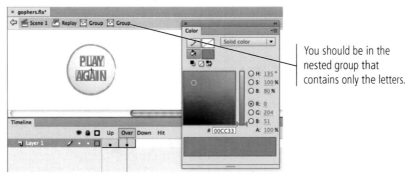

You should be in the nested group that contains only the letters.

11. **Click Scene 1 in the Edit bar to return to the main Stage.**

12. **Save the file and continue to the next exercise.**

 ## ADD CLASSIC DYNAMIC TEXT AREAS

Throughout a game, the "Hit" and "Miss" golf balls in the top-left corner should reflect the user's score. To accomplish this, you need to add dynamic text fields, which can contain variable text. (You will write the scripts for these fields later.)

TLF Text areas, which you used in Project 3: Animated Internet Ads, allow a large number of formatting options that are not available for Classic Text areas. The text areas you need for this project, however, will contain nothing more than basic numbers with no special formatting; Classic Text is suitable for this type of application.

1. **With gophers.fla open, double-click the Hits movie clip icon in the Library panel to enter into the symbol Stage.**

2. **Choose the Text tool in the Tools panel.**

3. **In the Properties panel, choose Classic Text in the Text Engine menu and choose Dynamic Text in the secondary Text Type menu.**

4. **In the Character options, choose a sans-serif font such as Arial or Helvetica. Choose Bold in the Style menu, set the size to 12 pt, and choose a bright green color.**

5. **In the Paragraph options area, click the Align Center button.**

By defining character and paragraph properties before creating a text area, the properties will be applied to new text areas you create. Although the dynamic text area has no content of its own, the formatting will be applied to text that is scripted into the area.

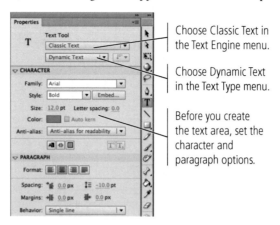

Choose Classic Text in the Text Engine menu.

Choose Dynamic Text in the Text Type menu.

Before you create the text area, set the character and paragraph options.

Note:

When you create a Classic Text object using the Static Text type, the object height is determined by the formatting that is applied to text in the area. You can drag the area handles to change its width, but you can't define a specific object height.

When you choose the Dynamic Text option, however, you can change both the height and width of the text object.

6. **Click and drag over the ball shape to create a new text area. Use the handles to resize the text area to fit within the ball shape, below the word "HITS."**

7. **With the text area active on the Stage, type `hit_count` in the Instance Name field of the Properties panel.**

Later in the project, you will use ActionScript to dynamically change the value in this text area — hence, the name "dynamic text area." Like symbol instances, text areas must be named before they can be addressed in ActionScript.

As with symbols, dynamic text areas must be named to be addressed in ActionScript.

8. **In the Properties panel, click the Embed button in the Character area.**

9. **In the Options area, check the Numerals box. Leave all other options at their defaults, and click OK.**

Because these areas can only display numbers, you only need to embed the font's numeral characters.

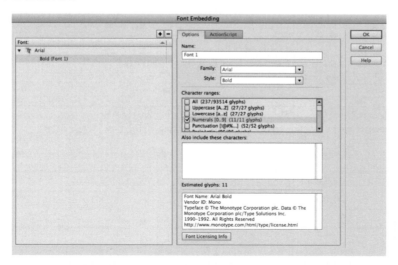

10. **Repeat this process to add a dynamic text area named `miss_count` in the Misses movie clip symbol. Use red as the text color for this area.**

    Make sure you add this text area inside the symbol Stage, and not on top of the placed instance on the main Stage.

    You can skip Steps 8–9, because the font is already embedded (assuming you use the same font for both text areas).

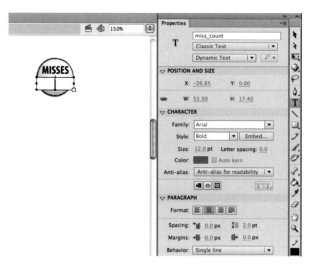

11. **Click Scene 1 in the Edit bar to return to the main Stage.**

12. **Save the file and continue to the next exercise.**

ANIMATE THE POP-UP GOPHER MOVIE CLIP

The final step in preparing this file for scripting is animating the pop-up gophers. When you imported the artwork, you created three separate movie clip symbols, which you need to combine into an animated movie clip. The animation is a simple up-and-down motion tween, with a few extra steps required to create the desired effect.

1. **With `gophers.fla` open, choose Insert>New Symbol. Name the new symbol `Gopher_Popup`, choose Movie Clip in the Type menu, and click OK.**

2. **On the new symbol Stage, change the name of Layer 1 to `Hole`.**

3. **Drag an instance of the Gopher_Hole movie clip onto the Stage. Using the Align panel, align the bottom center of the symbol instance to the symbol Stage's registration point.**

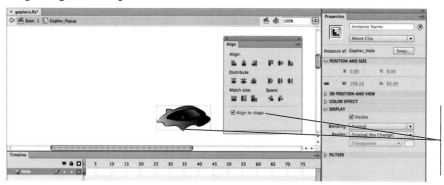

When the Align To Stage option is active, the Align buttons position the object relative to the symbol Stage's registration point.

4. **Add a new layer named Gopher to the top of the layer stack in the timeline.**

5. **With the new Gopher layer active in the timeline, drag an instance of the Gopher movie clip symbol onto the Stage. Using the Properties panel, define the position of the symbol instance as X: 0, Y: 50.**

   Remember, the symbol's X/Y position is based on the symbol's registration point.

6. **With the Gopher instance selected on the Stage, type gopher_tracked in the Instance Name field of the Properties panel.**

   You are going to write a script that tracks the number of times the user hits this symbol. To be addressed by ActionScript, the instance must be named.

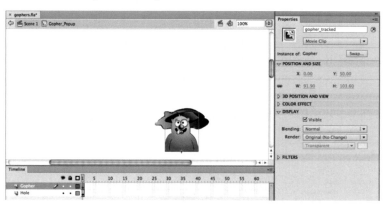

7. **Select Frame 20 of both layers and insert new regular frames to extend the timeline on both layers.**

8. **Control/right-click anywhere between Frame 1 and Frame 20 on the Gopher layer and choose Create Motion Tween.**

Note:

Press F5 to insert regular frames at the selected frames on the timeline.

9. **Select Frame 10 of the Gopher layer. Click the Gopher instance on the Stage using the Selection tool, and then change the Y position of the instance to –15.**

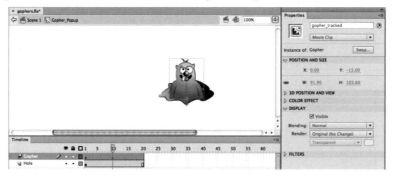

10. **Select Frame 20 of the Gopher layer. Click the Gopher instance on the Stage to select it, and then change the Y position back to 50.**

By changing only the Y position, the motion tween moves the gopher straight up and then straight down over the 20 frames in the tween.

11. **Control/right-click Frame 21 of the Gopher layer and choose Insert Blank Keyframe.**

If the user clicks ("hits") the gopher in Frames 1–20, a different gopher will move back down into the hole. Frames 21–30 will include the "pop-down" animation for the Gopher_Hit movie clip.

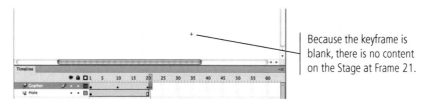

Because the keyframe is blank, there is no content on the Stage at Frame 21.

12. **Drag an instance of the Gopher_Hit movie clip onto the Stage, and position it at X: 0, Y: –15.**

This is the same position as the regular gopher at the height of the up-and-down tween (on Frames 1–20).

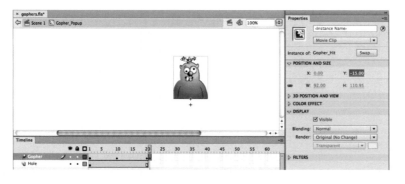

Note:

Later you will add ActionScript to determine whether a user clicks the pop-up gopher instance (named "gopher_tracked" from Step 6) in the tween on Frames 1–20. The instance on Frame 21 will only appear if the user clicks the gopher_tracked instance, based on the code you will later create. You don't need to name the Gopher_Hit instance on Frame 21 because you are not going to directly address that instance.

13. **Add a new frame at Frame 30 of both layers to extend the timeline of both layers.**

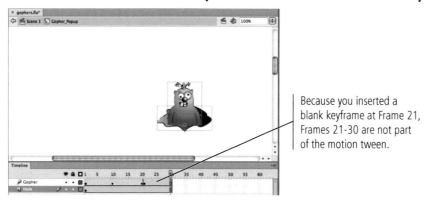

Because you inserted a blank keyframe at Frame 21, Frames 21-30 are not part of the motion tween.

14. **Create a motion tween between Frame 21 and Frame 30 on the Gopher layer.**

You are adding a second, separate tween to the Gopher layer.

15. **Select the Gopher_Hit instance on Frame 30 and change its position to Y: 50.**

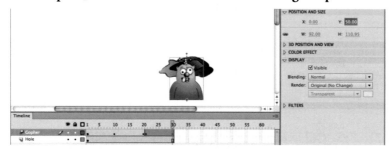

16. **Press Return/Enter to play the timeline on the Stage.**

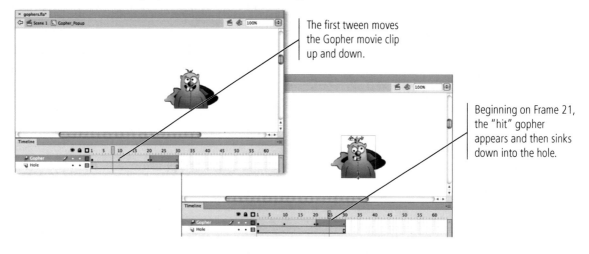

The first tween moves the Gopher movie clip up and down.

Beginning on Frame 21, the "hit" gopher appears and then sinks down into the hole.

17. **Save the file and continue to the next exercise.**

CREATE A MASK

Your current animation has two problems. First, the gopher pops up and sinks down, then jumps to show the "hit" gopher, which also goes back down into the hole. You will use ActionScript to solve this problem later.

Second, the gopher animation appears entirely in front of the hole artwork because the Gopher layer is on top of the Hole layer. If you move the Gopher layer down in the stacking order, however, the entire tween will fall behind the hole graphic. To achieve the desired effect of the gopher rising up from inside the hole, you need to create a mask to show only a specific part of the gopher animation.

In most graphics applications, a **mask** is used to isolate specific parts of an image or object, such as removing a photographed object from its background (often referred to as silhouetting). Anything outside the area of the mask will not be visible.

In Flash, masks are created by converting a specific layer to a mask. You can use a shape created with one of the Flash drawing tools (such as the Square or Oval tool), or you can use other objects (such as a bitmap image) in the mask layer. Objects on a mask layer are not visible in the final file; they are only used to define the visible area of underlying (masked) layers.

1. **With gophers.fla open, make sure you are editing the Gopher_Popup symbol, and make the rulers visible (View>Rulers).**

2. **Click Frame 10 to select it.**

   This frame shows the gopher at the highest point in the animation. You need to know this position so you can define the area for the mask.

3. **Click the vertical ruler on the left side of the document window and drag right to create a guide that marks the left edge of the Gopher instance.**

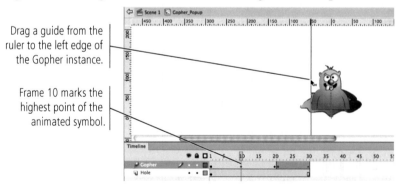

Drag a guide from the ruler to the left edge of the Gopher instance.

Frame 10 marks the highest point of the animated symbol.

4. **Drag a second guide from the vertical ruler to mark the right edge of the Gopher instance.**

5. **Click the horizontal ruler at the top of the document window and drag down to create a guide that marks the top edge of the Gopher instance.**

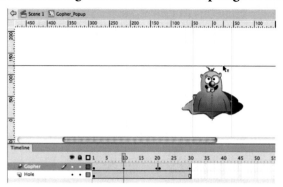

6. **Create a new layer named Mask above the Gopher layer, and then hide the Gopher layer.**

A mask in Flash requires a separate layer to hold the masking object.

7. **Select Frame 1 of the Mask layer. Using any method you prefer, draw a shape with a bright-colored fill that covers the entire area of the gopher. Use the guides as references for the top, left, and right sides of the shape. Adjust the bottom edge of the shape to match the front of the hole (as shown here).**

A mask object is not visible in the final movie; it simply defines the area that will be visible on the masked layer; it doesn't matter what color you use.

Note:

It is a good idea to use a solid-colored fill and no stroke on the mask object.

Note:

Because the mask shape covers the elements you want to reveal, it can be difficult to finetune the shape and placement of your mask object. It can be helpful to show the mask layer in Outline mode while you refine the mask shape.

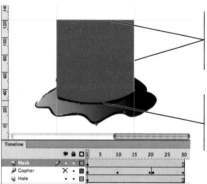

Use the guides as references to position the top, left, and right edges of the mask shape.

Because the Gopher layer is hidden, you can see the line that marks the "front" of the hole.

8. **Show the Gopher layer, then Control/right-click the Mask layer name and choose Mask from the contextual menu.**

Both layers are automatically locked. The layer immediately below is automatically masked by the Mask layer, and now you can see only the top of the gopher's head; the fill attributes of the actual mask shape are hidden.

Note:

If more than one object exists on the mask layer, only the bottommost object on the layer will define the mask shape.

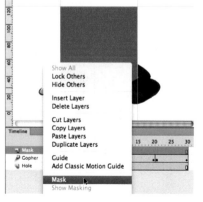
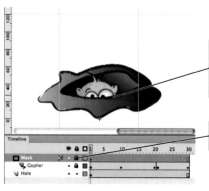

After creating the mask, only the area of the Gopher layer that is inside the mask shape is visible.

The masking and masked layers are both locked.

9. **Click Frame 1 of any layer to select it, and then press Return/Enter to test the animation.**

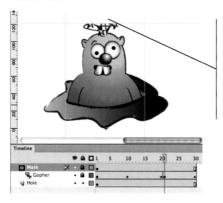

When the timeline reaches Frame 21, you can see that the "hit" gopher artwork is higher than the regular gopher. This tells you that some of the artwork falls outside the mask area.

Note:

To reveal the mask shape for editing, simply unlock the Mask layer. When you are finished editing the shape, relock the Mask layer to reapply the mask.

10. **In the timeline, unlock the Mask layer.**

11. **Click Frame 21 to show the "hit" gopher at its highest point.**

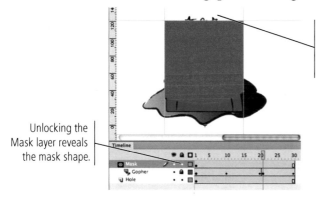

With Frame 21 selected, you can see the top (unmasked) part of the Gopher_Hit symbol instance.

Unlocking the Mask layer reveals the mask shape.

12. **Use any method you prefer to move the top edge of the mask shape up so it extends beyond the top edge of the "hit" gopher artwork.**

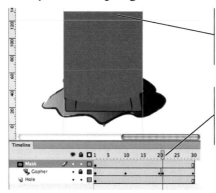

We used the Subselection tool to move up the anchor points on the top corners of the mask shape.

Because Frame 21 is selected, you can see where you should place the top edge of the mask shape.

13. Lock the Mask layer.

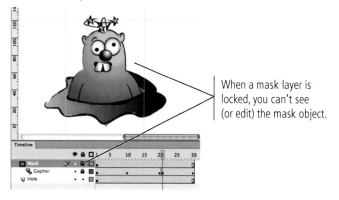

When a mask layer is locked, you can't see (or edit) the mask object.

14. Click Scene 1 on the Edit bar to return to the main Stage.

15. Select Frame 2 of the Content layer. Drag eight instances of the Gopher_Popup movie clip onto the Stage, and arrange the instances around the golf course.

Space the instances around the Stage to leave enough room above each instance so the gophers do not pop up on top of other instances. (Use the image after Step 16 as a rough guide for placing the instances.)

16. Select each Gopher_Popup instance and assign instance names using the gopher1_mc naming convention.

Each instance name should contain the word "gopher", a sequential number (1–8), then "_mc". For the ActionScript that you will write later to work properly, you must follow this naming convention exactly.

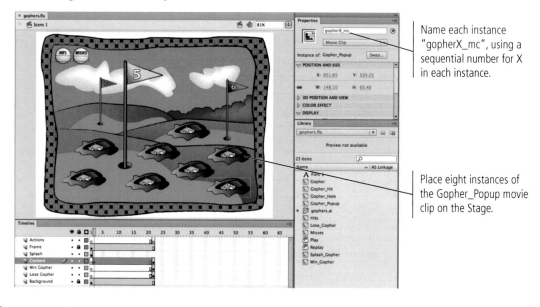

Name each instance "gopherX_mc", using a sequential number for X in each instance.

Place eight instances of the Gopher_Popup movie clip on the Stage.

17. Save the file and continue to the next stage of the project.

Stage 2 Programming Interactivity

After considerable setup, all the pieces of your game are now in place. Now you are ready to begin scripting. The skills you learned in Project 5: Gator Race Game are foundational building blocks for programming in ActionScript; some of those same methods are useful in developing this game, as they will be in many Flash projects. Others — specifically the interactive elements — require different methods.

To complete this project, you will script the following functionality:

- Program button functionality for the Play and Replay buttons.

- Pop up only one gopher at a time, repeating 10 times.

- Based on hits and misses, show the correct gopher during the "down" half of the animation.

- Track a user's hits and misses throughout the game.

- Calculate a win or loss based on the number of hits, and then show the appropriate movie clip after the game is complete.

ADD TIMELINE CONTROLS

Rather than beginning as soon as the file opens, you need to stop the timeline until a user clicks the Play button. In this exercise, you add basic functions to control the playback of the main and gopher animation timelines.

1. **With gophers.fla open, press Command-Return/Control-Enter to test the movie.**

    By default, the timeline plays from beginning to end and then loops (starts over). All eight gophers pop up at the same time when the main timeline reaches Frame 2. The first task, then, is to stop the animations from playing as soon as the file opens.

2. **Close the Flash Player window and return to Flash.**

3. **Select Frame 1 of the Actions layer on the main timeline and open the Actions panel (Window>Actions).**

4. **Click in the Actions panel Script pane to place the insertion point, and type stop();.**

As you learned in Project 5: Gator Race Game, this statement stops the active timeline.

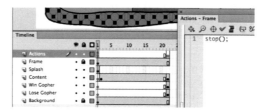

5. **Select the keyframe on Frame 22 of the Actions layer. In the Actions panel Script pane, type stop();.**

If you don't add the stop function to the final frame, the timeline automatically returns to Frame 1 after it plays the whole way through. This would show the splash screen graphics again (which is incorrect).

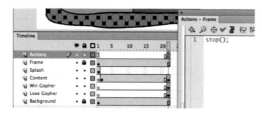

6. **Double-click the Gopher_Popup symbol icon in the Library panel to enter into the symbol Stage.**

7. **Add a new layer named Actions at the top of the timeline.**

8. **Select Frame 1 of the Actions layer. In the Script pane, type stop();.**

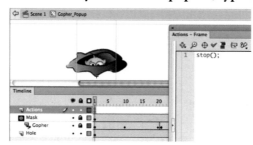

9. **Insert a new keyframe on Frame 20 of the Actions layer. In the Script pane, type stop();.**

In this movie clip, two sections of the Gopher layer represent the "down" half of the animation. This stop function prevents the movie clip timeline from showing both "down" tweens.

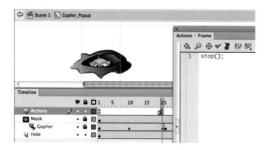

10. **Click Scene 1 on the Edit bar to return to the main Stage.**

11. Press Command-Return/Control-Enter to test the movie.

Because you added the stop functions to the main timeline and the gopher animation timeline, nothing happens in the movie.

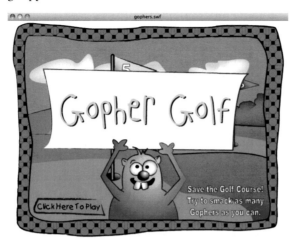

12. Close the Flash Player window and return to Flash. Save the file and continue to the next exercise.

 ## PROGRAM BASIC GAME FUNCTIONALITY

When a user clicks the Play button, the game should begin. The Gopher_Popup instances should randomly animate ten times, and the user's mouse action should be tracked to determine "hits" and "misses." This functionality requires a number of variables, loops, and conditional statements, which you learned about in Project 5: Gator Race Game.

1. With gophers.fla open, select the Frame 1 keyframe on the Actions layer of the main timeline.

2. Type the following code beginning on Line 3 of the Actions panel Script pane:

```
play_btn.addEventListener(MouseEvent.CLICK, playGame);

function playGame (event:MouseEvent):void {
    play();
```

Note:

Remember, the actions of a function must be contained within the braces of the function.

When you type the opening brace of the function and press Return/Enter, Flash automatically adds the required closing brace for you.

Note:

The import statement is added on Line 1 as soon as you define the argument of the function (event:MouseEvent); this moves the code you are typing so it begins on Line 5.

This is the same basic technique you used in Project 5: Gator Race Game to make the gators start running when a user clicked the race starter. The first line attaches an event listener to the play_btn instance on the Stage. When a user clicks the button (**MouseEvent.CLICK**), the function **playGame** is called. The second half of the script is a custom function named "playGame", which simply plays the current timeline.

The next step is to create code that makes something happen when the main timeline reaches Frame 2.

3. **Insert a new keyframe on Frame 2 of the Actions layer. On Line 1 of the Script pane, insert code to declare a random-number integer-type variable, multiplied by 8 and rounded to the next highest whole number (ceiled):**

```
var randomGopher:int = Math.ceil(Math.random()*8);
```

In Project 5: Gator Race Game, you did the same basic thing in two lines: one line to generate the random number (and multiply it by the required whole number), and one line to ceil that number. In this case you are combining the two lines, generating the random number as the argument of the ceil function. The two methods accomplish the same goal.

Note:

All code in this exercise was explained in Project 5: Gator Race Game. Refer to that chapter for detailed discussion.

4. **On Line 2 of the Script pane, add the gotoAndPlay function, using the random-number variable to call one of the eight Gopher instances:**

```
this["gopher"+randomGopher+"_mc"].gotoAndPlay(2);
```

5. **Press Command-Return/Control-Enter to test the movie. Click the Play button to test the animation.**

Because you used the random-number generator, a single gopher pops up when the main timeline reaches Frame 2. You haven't yet defined a loop to show more than one gopher. When the timeline reaches Frame 22, the movie stops.

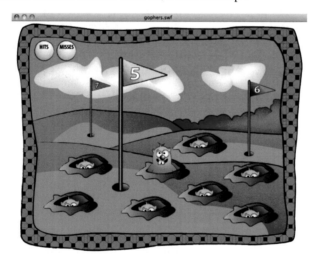

6. **Close the Flash Player window and return to Flash.**

The final game requires ten random gophers to pop up. To accomplish this task, you will increment a variable and use a conditional statement to make the timeline loop ten times.

7. **Select Frame 1 of the Actions layer. On Line 11 of the Script pane, define a number-type variable named `counter`, initialized to 0:**

```
var counter:Number=0;
```

This variable will be used in a conditional statement to determine how many times the timeline loops.

```
1   import flash.events.MouseEvent;
2
3   stop();
4
5   play_btn.addEventListener(MouseEvent.CLICK, playGame);
6
7   function playGame (event:MouseEvent):void {
8       play();
9   }
10
11  var counter:Number=0;
```

Actions : 1
Line 11 of 11, Col 22

8. **Select Frame 2 of the Actions layer. Place the insertion point at the beginning of the existing script and press Return/Enter twice to add space above the script.**

9. **On Line 1 of the Script pane, add code to increment the counter variable:**

```
counter++;
```

```
1   counter++;
2
3   var randomGopher:int=Math.ceil(Math.random()*8);
4   this["gopher"+randomGopher+"_mc"].gotoAndPlay(2);
```

Actions : 2
Line 1 of 4, Col 11

10. **Insert a new keyframe at Frame 21 of the Actions layer. With the Frame 21 keyframe selected, add a conditional statement that evaluates the value of the counter variable, and then directs the timeline to the correct frame (based on that value):**

```
if (counter<10) {
    gotoAndPlay(2);
} else {
    nextFrame();
}
```

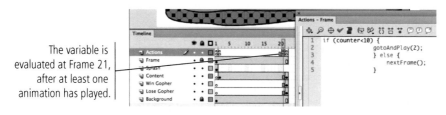

The variable is evaluated at Frame 21, after at least one animation has played.

The nextFrame and prevFrame commands control the timeline of the defined movie clip, moving the playhead forward or backward one frame from the current position. (Because these commands control movie clip timelines, they are part of the flash.display package, in the Methods folder of the MovieClip class.)

11. **Press Command-Return/Control-Enter to test the movie.**

12. **Click the Play button in the Flash Player window to test the randomness of the animated gophers. Watch the pop-up gophers to see if any overlap with other instances.**

   After 10 gophers appear, the main timeline moves to Frame 22 (accomplished by the nextFrame statement that appears on Frame 21).

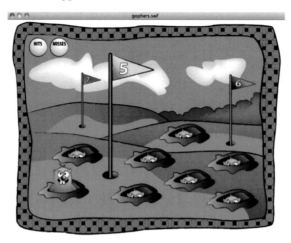

13. **Close the Flash Player window and return to Flash.**

14. **If any pop-up gophers overlapped with other instances when the movie played, rearrange the instances on the Stage as necessary.**

15. **Save the file and continue to the next exercise.**

 ## USE THE TRACE FUNCTION AND CODE COMMENTS

When you work with variables that control internal functionality, it can be difficult to understand exactly what is happening behind the scenes. The trace function can be very useful — especially during development — in understanding how variables work in relation to your scripts.

1. **With gophers.fla open, select the Frame 2 keyframe on the Actions layer of the main timeline.**

2. **Place the insertion point at the end of the existing script and press Return/ Enter twice.**

3. **Type trace(counter); at the current location in the script.**

   This line tells the script to show (trace) the value of the counter variable in the Flash Output panel.

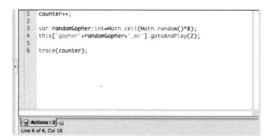

4. **Press Command-Return/Control-Enter to test the movie.**

5. **Move the Flash Player window on your screen so you can see both the Player window and the timeline.**

You should be able to see the Timeline panel behind the Flash Player window.

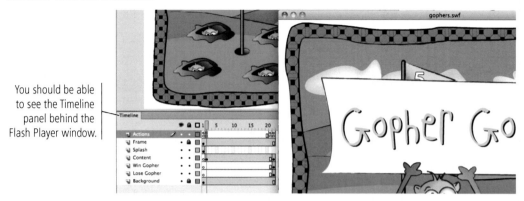

6. **Click the Play button in the test movie.**

7. **While the movie plays, watch the timeline in the Flash application.**

When the trace function occurs in a script, the Flash Output panel automatically appears in the same position as it was last used. If you have never used the Output panel, it is docked with the timeline by default.

Because you added the trace function to show the value of the counter variable, you can better understand how the conditional loop makes the correct number of gophers pop up throughout the game:

- Clicking the Play button plays the current timeline, so the timeline moves to Frame 2.
- The counter variable is incremented when the timeline reaches Frame 2.
- Each time the timeline reaches Frame 21, the if condition evaluates the value of the counter variable.
- If the value of the counter variable is less than 10, the timeline loops back to Frame 2, where the counter variable is incremented again.
- When the timeline has looped enough times to increment the counter variable to 10, the timeline continues and shows one more instance of the pop-up animation (the tenth gopher).
- When the timeline reaches Frame 21 after the counter has been incremented to 10, the if condition determines that 10 (the value of the counter variable) is not less than 10 (the defined condition) and sends the timeline to the next frame (the else condition).

The Output panel shows each time the counter variable changes.

Each time the Timeline reaches Frame 2, the counter variable is incremented.

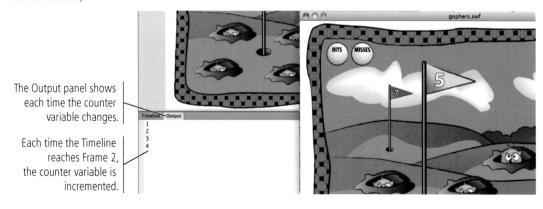

8. **Close the Player window and return to Flash.**

9. **Place the insertion point before the trace function in the Script pane and type //.**

Two forward slashes identify a **comment**, which is simply a non-functioning line in the Script pane; comments are ignored when the movie plays. Using comments is an easy way to temporarily deactivate specific lines of code during development.

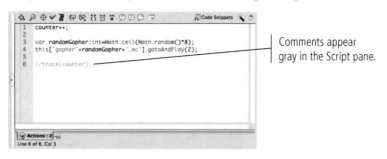

Comments appear gray in the Script pane.

Note:

You can also convert multiple lines of code to a comment by typing /* at the beginning of the first line and */ at the end of the last line you want to comment. This is called a **block comment**.

10. **Place the insertion point at the beginning of Line 3 in the Script pane (where you declare the randomGopher variable).**

11. **Press Return/Enter twice, and then move the insertion point back to Line 3 in the script. Type:**

    ```
    //Generate a random number that calls one of the gopher animation instances.
    ```

    Comments are also useful for creating text-based notes to yourself or other programmers.

Note:

Clicking the Remove Comment button removes all comment notations and converts commented lines to regular code. The commented lines remain in the script.

12. **If the entire comment does not fit into the Script pane, open the panel Options menu and choose Word Wrap.**

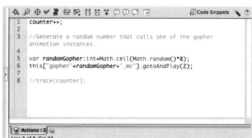

13. **Save the file and continue to the next exercise.**

TRACK USER ACTIONS

The next element of this game's interactivity requires some method of tracking a user's mouse clicks. You first need to determine whether a user clicked the pop-up Gopher instance; then you can call the correct animation based on the user's action; and finally, you can use variables to count the number of hits and misses in a single game.

1. With **gophers.fla** open, double-click the **Gopher_Popup** symbol icon in the Library panel to enter the symbol Stage.

2. Add a new keyframe at Frame 2 of the Actions layer.

3. With the Frame 2 keyframe selected, define an event listener and custom function so that, if a user clicks the gopher, the timeline moves to Frame 21 (the "hit" gopher going back into the hole):

   ```
   gopher_tracked.addEventListener(MouseEvent.CLICK, clicked);

   function clicked(event:MouseEvent):void{
        gotoAndPlay(21);
   }
   ```

   Remember, gopher_tracked is the instance name you defined for the Gopher symbol instance when you created the pop-up gopher animation. This is a movie clip symbol, so you don't have to define a specific hit-frame area; any click within the movie clip area will trigger the function (score a "hit").

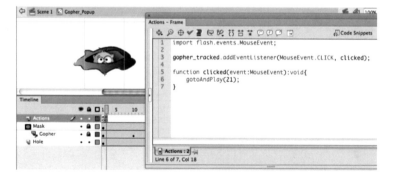

Note:

The import statement is added on Line 1 as soon as you define the argument of the function (event:MouseEvent).

4. Click Scene 1 on the Edit bar to return to the main Stage.

5. Select the Frame 1 keyframe on the Actions layer.

6. At the end of the existing script, define two new number-type variables to store the number of hits and misses, and initialize both values to 0:

   ```
   var hitcount:Number=0;
   var misscount:Number=0;
   ```

7. **Select the Frame 2 keyframe on the Actions layer. Place the insertion point immediately after the counter increment (Line 1) and press Return/Enter twice.**

8. **With the insertion point on the new empty Line 3, type:**

   `hits_mc.hit_count`

Note:

You can always use the Insert Target Path button if you forget the exact names you defined for specific symbol instances.

   The first two pieces of this code define the sepcific object you want to address — the **hit_count** dynamic text object, which is nested inside the **hits_mc** movie clip instance. Whenever you want to address an object that is nested inside another, you have to define the entire path to the object you want to target; each object in the path must be separated by a dot.

9. **After the end of the path, type:**

   `.text=hitcount;`

   This statement changes the text property (**.text**) of the hit_count dynamic text area to the value of the hitcount variable (**=hitcount**).

10. **Add another statement on the next line that maps the misscount variable to the miss_count dynamic text area:**

    `misses_mc.miss_count.text=misscount;`

11. **Save the file and continue to the next exercise.**

PASS VALUES TO THE PARENT TIMELINE

You still need some way to increment the hitcount and misscount variables. Because different frames in the animation represent "hits" and "misses," the easiest way to accomplish this is to create simple statements that are triggered when the playhead reaches specific frames.

1. **With gophers.fla open, double-click the Gopher_Popup symbol icon (in the Library panel) to enter the symbol Stage.**

2. **Select the Frame 20 keyframe on the Actions layer.**

3. **After the existing script, type the following code:**

```
MovieClip(parent).misscount++;
```

The **MovieClip(parent)** statement refers to the object that contains the current movie clip (its parent). In this case, the gopher instances are placed on the main Stage, so the **MovieClip(parent)** statement refers to the main timeline. The **misscount++** statement increments the value of the misscount variable, which you declared on Frame 1 of the main timeline.

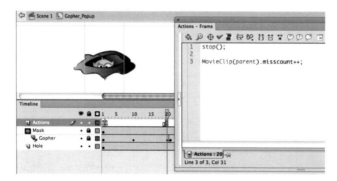

MovieClip(parent) — referred to as a **constructor function** — tells Flash that the parent object is a MovieClip. This allows you to access the methods and properties of the MovieClip class.

Without this constructor function, Flash would only know that the parent has to be a DisplayObjectContainer, which does not have all the properties and methods of the MovieClip class. The DisplayObjectContainer class also does not allow you to reference variables or functions that it does not recognize. If you tried to access properties of the MovieClip class that are not part of a DisplayObjectContainer, or if you tried to access properties that you added (like the misscount variable), Flash would produce an error when you try to create an .swf file

4. **Add a new keyframe on Frame 21 of the Actions layer.**

5. **Add a statement to Frame 21 that increments the hitcount variable on the main timeline:**

```
MovieClip(parent).hitcount++;
```

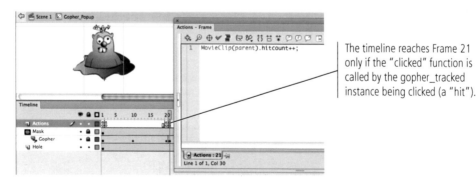

The timeline reaches Frame 21 only if the "clicked" function is called by the gopher_tracked instance being clicked (a "hit").

6. **Click Scene 1 on the Edit bar to return to the main Stage.**

7. **On Frame 2 of the Actions layer, select and copy the code on Lines 3–4.**

8. **Select the Frame 22 keyframe. Place the insertion point at the beginning of the existing script and press Return/Enter two times.**

9. **Move the insertion point to Line 1 and paste the code you copied in Step 7.**

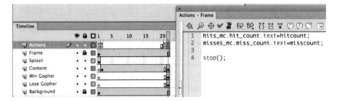

When the timeline reaches Frame 21 and **counter=10**, the main timeline moves to Frame 22; the timeline on the gopher movie clip might not have triggered the final hitcount or misscount increment, so the final score only reflects the first nine user results.

If a user clicks the gopher at Frame 15 of the pop-up animation, that action triggers the 10-frame pop-down animation of the "hit" gopher. The entire revolution takes 25 frames to complete, which means the main timeline will have reached Frame 22 before the final hit or miss was mapped to the dynamic text areas.

If you don't include these statements on Frame 22, the final "hit" or "miss" (from the tenth run) might not reflect in the counters.

10. **Press Command-Return/Control-Enter to test the movie. Play the game and review the tracking functionality in the dynamic text objects.**

11. **Close the Flash Player window and return to Flash.**

12. **Save the file and continue to the next exercise.**

PROGRAM THE RESULT

The final tasks are to show the correct "results" movie clip when the game is finished, and to program the Replay button. Like the first exercise in Stage 2 of this project, both of these tasks require code that you learned about in Project 5: Gator Race Game.

1. **With gophers.fla open, select the Frame 22 keyframe on the Actions layer of the main timeline.**

2. **After the existing script, define a conditional "if…else" statement that shows the lose_mc instance if the hitcount value is less than 10, and the win_mc instance if the hitcount value is 10:**

```
if (hitcount<10) {
      lose_mc.visible=true;
      win_mc.visible=false;
} else {
      win_mc.visible=true;
      lose_mc.visible=false;
}
```

Notice that each condition defines the visibility property of both movie clips. Also notice that the else statement does not define a specific condition; the code assumes only two possibilities: *if* the condition is true, or *else* it is not.

The condition in the if statement (hitcount<10) requires a user to hit all 10 gophers in order to win the game. You could set a lesser threshold for victory by reducing the number in the condition, or even use both variables in the condition to define a victory as any game with more hits than misses (misscount<hitcount).

3. **Define a new event listener and function to reset the game when a user clicks the Replay button:**

```
replay_btn.addEventListener(MouseEvent.CLICK, replay);

function replay (event:MouseEvent):void {
      counter=0;
      hitcount=0;
      misscount=0;
      gotoAndPlay(2);
}
```

The variables that control the timeline loop and increment the counters are initialized to 0 in Frame 1. However, this function resets the game to Frame 2, so the variables need to be reset to 0 within the function body.

4. **Press Command/Control-Enter to test the movie. Play the game at least twice to test the Replay functionality.**

5. **Close the Player window and return to Flash.**

6. **Save the Flash file, then continue to the next stage of the project.**

FLASH FOUNDATIONS

Using the Compiler Errors Panel

When you test a file that includes ActionScript, your code must be processed or compiled before you can preview the file in the Flash Player window. If your script contains errors when you try to test the movie, Flash shows a list of the problems in the Compiler Errors panel (grouped with the Timeline panel by default).

Go To Previous

Go To Source   Go To Next   Clear

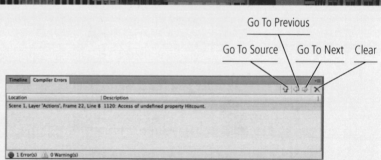

In addition to simply reviewing the list of problems, you can use the Compiler Errors panel to navigate to the specific problem. If you select an item in the list and click the Go To Source button (or simply double-click the problem in the list), the Actions panel displays the relevant line code.

In the example shown here, Flash could not find a property named "Hitcount". The source of the problem shows that the programmer was trying to reference the variable named "hitcount", but accidentally typed a capital H instead of lowercase — a very common error, especially for new programmers.

ActionScript is case-sensitive; if you define a variable with a lowercase name, you must always use the lowercase name when referring to that variable. Although certainly not the only potential error in a script, capitalization is a good place to begin searching for the root cause of compiler errors.

Stage 3 Creating a Mobile App

Now that your game is complete, you need to convert it to an app that can be deployed on smartphone and tablet devices. Flash supports creating apps for both Android and Apple iOS (iPads and iPhones). However, the development process for iOS requires a number of steps outside the scope of the Flash interface — including paying a fee to Apple for membership in the iOS Developer Program. In this project, you will create the app for the Android operating system, which does not require any resources outside of the Flash toolset.

 DEFINE PUBLISH SETTINGS FOR ANDROID

When you create a new file, you can use the New Document dialog box to target a specific output type, such as ActionScript 3, AIR for Android, etc. In this project, for example, you started with a basic Flash file that targeted ActionScript 3. Once a file is created, you can change the target in the Properties panel so that you can define settings specific to the device for which you are publishing.

1. **With gophers.fla open, click the gray area outside the Stage to deselect everything in the file.**

2. **In the Properties panel, open the Target menu and choose AIR 3.2 for Android.**

   AIR stands for the Adobe Integrated Runtime. AIR allows you to deploy Flash content as though it's a native application on a number of different devices, including desktop and mobile environments.

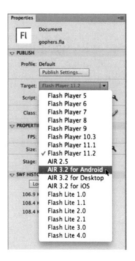

3. **Click the Edit Application Settings button (the wrench) to the right of the Target menu.**

Edit Application Settings button

4. Review the options in the resulting AIR for Android Settings dialog box.

The **Output File** is the actual file that will be created for your final game. The ".apk" extension is used for an Android app.

The **App Name** is the name that will appear under the app icon on the Android device.

The **AppID** is a path used by Flash to keep the various assets associated with each AIR release separated into unique packages.

The **Version** is used by the installer if there is already an app of the same AppID installed on that device. If you make changes to create a new version of your app, you should change this number to reflect the new version.

The **Version Label** allows you to add a specific label to a version, such as "bug fix".

The **Aspect Ratio** refers to the app's orientation on the device.

The **Full Screen** check box, when active, allows the app to scale up or down to take up the most available space on the screen of whatever device is being used to run the app.

The **Auto Orientation** check box, when active, allows the app to automatically reorient itself between portrait and landscape mode based on the device's physical orientation.

The **Render Mode** menu determines whether you want the device to render graphics using the central processing unit (CPU) or a separate graphics processing unit (GPU). GPU is faster, but won't always work correctly. Direct facilitates 3D rendering and some other advanced effects.

Included files are the files that will be wrapped up into your app file. The .swf file is the movie itself, and the .xml file is a document (called an application descriptor file) that instructs the app how to deploy that .swf file. The application descriptor is written automatically for you, consisting largely of the choices you make in the AIR for Android Settings dialog.

5. Change the following settings (leave all others at the default values):

App name:	**Gopher Golf**
AppID:	**gophers.android**
Aspect ratio:	**Landscape**
Full screen:	**Checked**
Render mode:	**Auto**

6. **Click the Deployment tab at the top of the dialog box to display those options.**

7. **Click the Create button to the right of the Certificate button.**

   When you develop an app for distribution, the people who install your app need to trust that it will not harm their devices. A verified certificate basically says that you are a legitimate developer; if something goes wrong, users will be able to find you.

   Certificates are available through a number of verification companies such as VeriSign. Although acquiring a verified certificate is beyond the scope of this book, you still need a certificate to be able to test your app.

   When you develop for the Android OS, Flash allows you to create your own self-signed certificate — basically, like writing a permission slip for yourself. This certificate is not trustworthy to the app-using community, but it is good enough for you to test your apps.

8. **In the resulting dialog box, fill in the following fields with your own information:**

   > **Publisher name**
   >
   > **Organization unit**
   >
   > **Organization name**
   >
   > **Country**
   >
   > **Password**
   >
   > **Confirm password**

   In general, these values can be whatever you want them to be. Just make sure that the password is easy for you to remember, and that the file is saved in a convenient location.

9. **Leave the Type and Validity Period values at the default options.**

10. **Click the folder icon/Browse button to the right of the Save As field. In the Select Certificate Destination dialog box, navigate to and select your WIP>Gophers folder. Change the Save As/File Name field to `testing.p12`, then click Save.**

    Here you are defining the location and filename of the certificate (using the extension .p12), not the actual game app. Because this is a certificate created only for testing purposes, the generic name is preferred over the specific (default) name for the game you are creating.

11. Click OK to create the certificate, then click OK to dismiss the resulting message.

12. In the Deployment options, make sure the Certificate field displays the target path to the certificate you just created.

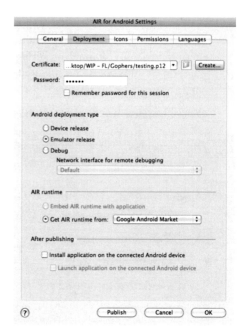

13. Type your password in the field.

This is the password you defined when you created your certificate.

If you are not working on a shared computer, you can check the Remember Password option so you don't have to re-enter your password every time you publish the file.

14. Select Emulator Release in the Android Deployment Type options.

The Emulator Release option allows you to test the actual game app on your computer, before the app is installed on a separate Android device.

If you have an Android device running Android OS 2.2 or higher, you can publish the app directly on the device (explained in the next exercise).

15. Click the Icons tab at the top of the dialog box to display those options.

App icons, which are provided for you for this project, must be in the ".png" format and match the listed sizes. Multiple icon sizes are included so that Android devices with different screen resolutions can use whichever icon is best suited for that device.

16. Select icon 36×36 item in the list at the top of the dialog box.

17. Click the folder icon next to the text field that is below the list. Navigate to WIP>Gophers>icons>icon36.png and click Open.

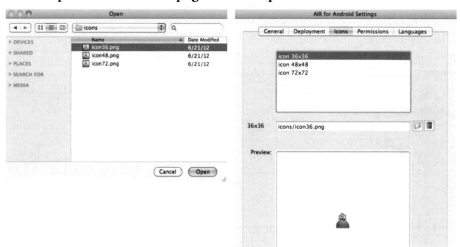

Note:

The AIR Runtime options tell the device where to get the software it needs to run your app. You can embed that software into your app, or allow the app to access the runtime software through the Google Android Market (now Google Play) or the Amazon Appstore. If you choose the first option, users will not be required to install the AIR Runtime separately.

18. Repeat Steps 16–17 for the 48×48- and 72×72-pixel icons.

19. **Click the Permissions tab at the top of the dialog box, and check the Internet option.**

The Permissions options allow your app access to various device capabilities, such as the camera or microphone. Internet is the only capability you need to access for the game you designed in this project.

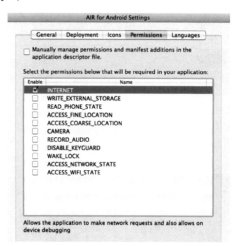

20. **Click the Languages tab at the top of the dialog box and review the options.**

If you do not choose any specific languages in this list, users of Android localized for any language will be able to see and install your application. If you want to restrict your app to users of one or more languages, you can select the appropriate languages in this list.

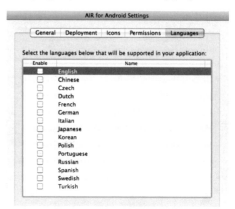

21. **Click OK to close the AIR for Android Settings dialog box.**

22. **Press Command-Return/Control-Enter to test your movie.**

The Simulator window that appears allows you to simulate motion, geolocation, and multi-touch gestures which would be impossible with a desktop or laptop device. However, these services are beyond the scope of this project.

Note:

The ADL appears as a separate app in the Dock (Macintosh) or in the Taskbar (Windows). If you switch away to another application from the test movie window, you have to choose the ADL option to navigate back to the test movie.

Macintosh

Windows

23. **Close the test window and return to Flash.**

When you test a movie that is being deployed for Android, it opens in the AIR Debug Launcher. There is no visual difference between the ADL and the Flash Player window that appears when you test a regular SWF file.

24. **Save the gophers.fla file.**

If you do not have an Android device, you will not be able to complete the final exercise. You can close the file.

Publishing an App for iOS

FLASH FOUNDATIONS

Developing an app for the iPhone or iPad is similar to developing one for an Android device. In the AIR for iOS Settings dialog box, the General, Icons, and Languages options have the same general options as in the AIR for Android Settings dialog box. (There is no Permissions tab.)

If you develop an app for iOS, you do not have the option of creating a self-signed certificate; instead, you must apply for a certificate from Apple. You also need to join the iOS Developer Program in order to acquire a Provisioning Profile. (An individual membership costs $99 annually.)

A Provisioning Profile is basically a permission slip from Apple that allows you to install your app onto an iOS device. It allows you to bypass the App Store during the development process, and allows you access to the App Store once you have completed development.

In brief, the steps are as follows:

- Register as an Apple Developer
- Request a Signing Certificate
- Request a Development Certificate
- Register the iOS device(s) you want to use for testing
- Create a Development Provisioning Profile that links your AppID to your iOS testing device
- Install the Provisioning Profile
- Sync your app to your device

You can find out more about Apple's Developer Program, and about the entire process of developing, testing, and distributing apps for iOS at Apple's iOS Developer Program page at https://developer.apple.com/programs/ios/

TEST THE APP ON AN ANDROID DEVICE

If you have an Android device running Android OS 2.2 or higher, you can install your app directly onto the device through a USB cable.

If you are working on Windows, you will also need to make sure that you have the developer USB drivers for your Android device installed on your computer. You can find drivers for your device and OS at http://developer.android.com/tools/extras/oem-usb.html.

Macintosh computers will install the device's drivers automatically.

Note:

To complete this exercise, you must have an Android device running Android OS 2.2 or higher.

1. **Locate the Developer Options in your Android device's Settings.**

   Depending on the version of Android you are running, this might also be called Development or Applications>Development.

2. **Make sure the USB Debugging is active.**

The USB Debugging option should be checked.

This option should be available in the Developer Options settings.

3. **Connect your device to your computer using a USB cable.**

4. **On your computer, make sure gophers.fla is open in Flash.**

5. **In the Properties panel, click the Edit Application Settings button (the wrench icon) next to the Target menu.**

Edit Application Settings button

6. **In the AIR for Android Settings dialog box, click the Deployment tab to show those options.**

7. **If you did not check the option to remember the password for this session, or if you quit and restarted since the last exercise, enter your password in the Password field.**

   Remember, this is the password you defined when you created your certificate.

8. **Choose Device Release in the Android Deployment Type options.**

9. **In the After Publishing area, check the Install Application... and Launch Application... options.**

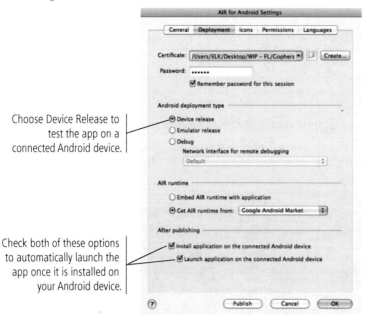

Choose Device Release to test the app on a connected Android device.

Check both of these options to automatically launch the app once it is installed on your Android device.

10. **With your Android device still connected to your computer via USB, click Publish.**

   After several seconds, the Gopher Golf game should automatically launch on your Android device.

11. **On your computer, save the Flash file (gophers.fla) and close it.**

fill in the blank

1. The _____ command makes it easy to move objects from one layer to another without changing the objects' X,Y position.

2. _____ to delete selected frames from the timeline.

3. Press _____ to select noncontiguous frames in the timeline.

4. When named, a _____ can be addressed by ActionScript to display variable values in the exported movie.

5. The _____ operator is used to increase the value of a variable by 1.

6. The _____ function can be used during development to show specific results in the Output panel.

7. Type _____ at the beginning of a line of code to convert that line to a comment.

8. A _____ appears between /* and */ in the Script pane.

9. True or false: The variable names "Hitcount" and "hitcount" are interchangeable in ActionScript. _____

10. The _____ code is used to send values from a movie clip's timeline to the timeline on which the symbol instance is located.

short answer

1. Briefly explain how dragging frames on the timeline affects objects on the Stage.

2. Briefly explain how the Output panel can aid in development.

3. Briefly explain two advantages of dynamic text.

Portfolio Builder Project

Use what you learned in this project to complete the following freeform exercise.
Carefully read the art director and client comments, then create your own design to meet the needs of the project.
Use the space below to sketch ideas; when finished, write a brief explanation of your reasoning behind your final design.

art director comments

The assistant program director for the Field Museum in Chicago has hired you to create an interactive map that can be accessed on kiosks placed throughout the facility, as well as using hand-held digital devices.

❏ Download the museum's floor plans (www.fieldmuseum.org/visit/floor_maps).

❏ Create symbols and buttons to identify the different facilities and services (stairs, restrooms, elevators, etc.) throughout the museum.

❏ Add artwork, images, and color however you prefer to identify the different exhibit halls and other areas of the museum.

client comments

We currently have PDF maps that people can download and print from the Internet, but we are in the process of building an interactive electronic self-guided tour that can be downloaded onto tablets or smartphones. We think the interactive map will be an excellent added service for this program.

Can you use the PDF maps we have now to create new interactive versions? We hunted through our archives — we don't have the original art files of the floor plans, so we're hoping you can create new line art for the floor plans as well.

You can add color or illustrations however you like to make the map more attractive. Just make sure the map is easily readable; don't let the artwork or imagery clutter the basic floor plan too much.

project justification

340

To complete this project, you worked with artwork imported from an Adobe Illustrator file. In this case, the artwork was prepared with a high degree of concern for integration into Flash. You saw, however, that even "good" supplied artwork will almost always require some additional setup once you import it into Flash. You also used Flash to create basic up-and-down tweens that, although very simple, still provide the underlying basis of the entire game. Different frames of the animation are called based on where the user clicks.

As you learned in this project, interactivity is far more than simply clicking a button. Using ActionScript 3, you can build complex games and other interactive tools that allow the user to control the behavior of every element on the screen.

Track user actions with dynamic text areas

Force a symbol instance to track the mouse movement

Randomly animate symbol instances throughout the game

Pass values from a movie clip timeline to the main timeline

Show different symbol instances based on user results

Travel Video Module

Your client is a travel company that offers a variety of customized and personalized tourism packages. As part of their promotional efforts, the client wants to include a series of video clips on their Web site to show examples of the tours they offer. They hired you to build a video-viewing module in Flash, including a series of thumbnail images that a user can click to watch the related videos.

This project incorporates the following skills:

❏ Using the Adobe Media Encoder application to encode video files for Flash

❏ Importing video files into Flash

❏ Defining video playback parameters

❏ Creating dynamic video captions

❏ Building thumbnail navigation

❏ Linking thumbnails to movies and captions

❏ Editing component skins and dynamic text styles

client comments

Our primary focus is building custom trips to meet any need or interest. We've taken people all over the world, and we can build packages for groups of any size — from two to two hundred.

We think the best way to promote our services to potential clients is to show them what we have done for other clients. Our Web site already includes testimonials from satisfied clients, but we also want to add actual trip videos on our home page.

We've taken hundreds of videos over the years, but we've narrowed down the field to eight movies that we want to show on the home page. Because the videos will be on our home page, we want each one to last only about 10 seconds. We might decide to expand the library later; if we do, we might add longer videos on other pages.

The module needs to include some way for users to select which video they want to see, and for each video, please include a caption that explains what the video is all about.

art director comments

Flash includes many built-in tools for working with video, but you have to start by encoding the client's files into a format that Flash can read. You can then import the video into Flash, build the rest of the module components, and tie everything together.

You could manually program the module functionality, but using the TileList component is the easiest way to create a thumbnail navigation system. This tool condenses several hours or days of work in ActionScript into a very simple, panel-based set of parameters.

Because there is only a single caption for each video, you can use a script to change the content in a dynamic text area based on which thumbnail is clicked. You'll use the same scripting methodology to change the specific video that displays in the playback component.

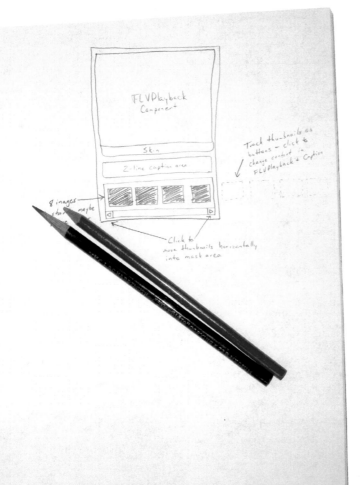

project objectives

To complete this project, you will:

❏ Use the Adobe Media Encoder to encode video files to a format appropriate for Flash

❏ Import video into Flash

❏ Define FLVPlayback component parameters

❏ Use the TextArea component for captions

❏ Add content to the TileList component

❏ Adjust TileList parameters

❏ Define a change event to link thumbnails to movies and captions

❏ Edit an interface component skin

❏ Format component text styles using ActionScript

❏ Export the video module for upload to a Web server

Stage 1 Encoding Video for Flash

Working with digital video is the topic of entire books, and most of the underlying mechanics are beyond the scope of this book. In this project, we focus on the common task of converting video to the proper format for placement in a Flash movie.

 ## CREATE A QUEUE IN THE ADOBE MEDIA ENCODER

A number of formats are used for video files on the Web. To work properly inside a Flash animation, however, video should be converted to the FLV (Flash Live Video) or F4V format. (The F4V format is a later variant of the format.)

Converting video to the proper format is easily accomplished using the stand-alone Adobe Media Encoder application, which is included with Flash Professional.

1. **Download FL6_RF_Project7.zip from the Student Files Web page.**

2. **Expand the ZIP archive in your WIP folder (Macintosh) or copy the archive contents into your WIP folder (Windows).**

   This results in a folder named **Tourism**, which contains the files you need for this project. You should also use this folder to save the files you create in this project.

3. **Launch the Adobe Media Encoder application.**

   On Macintosh, you will probably find this application in the Applications>Adobe Media Encoder CS6 folder on your hard drive.

   On Windows, click the Start menu and choose All Programs. Look for the Adobe Media Encoder application in the main Start menu.

4. **Open the Preferences dialog box for the Adobe Media Encoder.**

   As with any other application, preferences are accessed in the Application menu on Macintosh (in this case, Adobe Media Encoder) and the Edit menu on Windows.

5. **In the General options, check the Specify Output File Destination option.**

   If you don't define a target folder, the encoded files will be placed in the same folder as the source files.

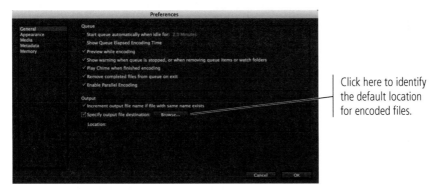

Click here to identify the default location for encoded files.

6. **Click Browse, and then navigate to the F4U folder that is in your WIP>Tourism folder. Click Choose/OK.**

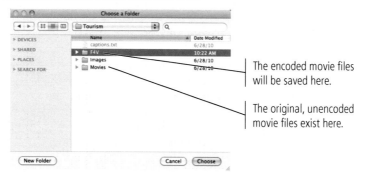

The encoded movie files will be saved here.

The original, unencoded movie files exist here.

You will save the primary Flash file for this project in the WIP>Tourism folder. The individual movie files will be exported to the F4V folder. References from the main Flash file to the various movie files will use the path from the main file to the individual movies in the F4V folder.

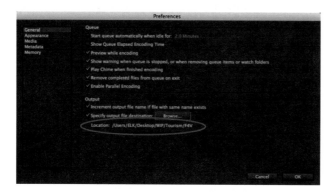

7. **Click OK to close the Preferences dialog box.**

8. **In the Adobe Media Encoder application, click the Add Source button below the Queue tab.**

The Queue pane lists all files that will be encoded when you click the Play button. You can use the + button to add files to the queue, or simply drag files from your desktop into the pane to add them to the list.

9. **Navigate to the WIP>Tourism>Movies folder. Shift-click to select all eight files in the Movies folder.**

Add Source button

Note:

Compression *is the process of reducing video file size relative to the original file size. There are two general types of compression:*

Spatial compression *removes redundant data from a file. Instead of defining each pixel, an area is specified using coordinates. Spatial compression is often called* **intraframe** *compression.*

Temporal compression *identifies differences between frames and stores only the differences. By doing so, a large portion of the file is deleted, and each frame is described based on the difference from the preceding frame. Temporal compression is often called* **interframe** *compression.*

Both spatial and temporal compression are considered **lossy** *methods, which means that redundant data is discarded during the compression process. The discarded data is not available when the file is decompressed for playback. (**Lossless compression** results in an exact copy of all original data when the file is decompressed.)*

10. **Click Open to add the selected files to the Media Encoder queue.**

The queue displays the Source Name, Format, Preset, and Output File location for each file in the list.

Click the hot text link to open the Export Settings dialog box for the selected format so you can define custom export settings.

Note:

Click and drag to reorder files in the queue.

11. **Continue to the next exercise.**

 ## DEFINE ENCODER SETTINGS

When you encode video files, you can control numerous export options in the various tabs of the Export Settings dialog box.

1. **In the Adobe Media Encoder application, click away from the added movie files to deselect them.**

2. **Click the F4V link for the first file in the Queue (video1.mov) to open the Export Settings dialog box for that file.**

   Clicking this link opens the Export Settings dialog box, where you can define custom output settings for the selected file.

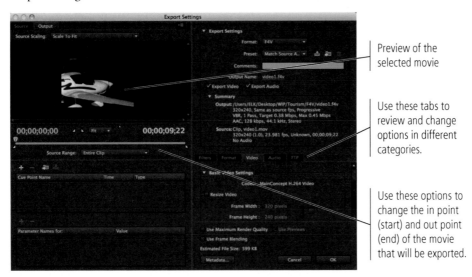

Preview of the selected movie

Use these tabs to review and change options in different categories.

Use these options to change the in point (start) and out point (end) of the movie that will be exported.

3. **In the top-left corner of the dialog box, click the Source tab (if it is not already active).**

4. **Click the Crop button, then set the Top field value to 10.**

Cropping a video means changing the physical area of the video that will be included in the output. Areas outside the crop area are not included in the exported video. When the Crop button is active, you can use the Left, Top, Right, and Bottom fields (above the preview image) to crop a specific number of pixels from each edge of the image. You can also use the handles in the preview to crop the file to a specific point.

Crop button

Change these values to define how many pixels to crop from each edge.

You can also click and drag these handles to change the crop area.

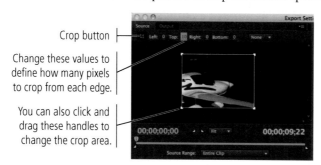

5. **Click the Crop button again to exit Crop mode.**

6. **Click the Out Point handle below the preview area and drag left until the time stamp below the preview shows approximately 8 seconds.**

These handles allow you to shorten the duration of a video clip. When you change the in point or out point of a video file, the new duration displays below the preview image.

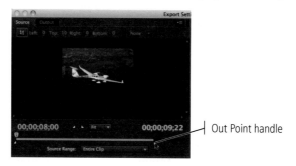

Out Point handle

7. **On the right side of the dialog box, click the Video tab (if necessary) and review the options.**

Although most of these settings are beyond the scope of this book, you should understand that you can use these options to resize the video (proportionally or non-proportionally) and define a different frame rate.

8. **Click OK to return to the main window of the Adobe Media Encoder.**

After you change the settings for a specific video, the Preset column in the queue changes to "Custom".

Note:

The word "codec" is the shortened form of enCOder/DECoder. This option refers to the software method used to compress video data on video frames, to reduce the size of the file and improve file transfer time.

Using the F4V format, video is compressed using the MainConcept H.264 codec.

Note:

The files for this project do not include audio information. If a video file includes audio information, you can use the options in the Audio tab to control the quality of audio included in the movie.

9. **In the main Adobe Media Encoder dialog box, click the Start Queue button.**

The Encoding panel (below the Queue panel) shows the progress of the encoding. As each file is completed, the Preset column in the Queue panel turns gray.

You can click the Stop Queue button to stop the encoding process midstream; you will see a warning dialog box asking if you want to finish encoding the current file.

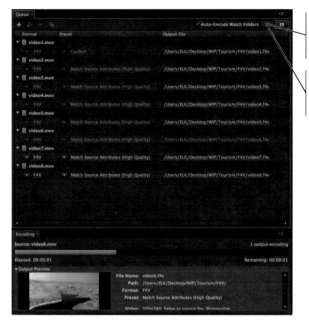

When the queue is not running, click here to start the process.

While the queue is running, click here to stop the process.

Note:

For the sake of this project, you are using the default settings for the remaining seven files.

10. **When all eight files have been encoded, quit the Media Encoder application.**

11. **Continue to the next exercise.**

 ## PREPARE THE CONTAINER FLASH FILE

The basic Flash file for this project will contain three primary elements: a video player, a caption area, and a scrolling area for numerous thumbnail images that will function as links. Now that the necessary video files have been converted to the F4V format, your next task is to create the Flash file itself.

1. **In Flash, create a new Flash file for ActionScript 3.0.**

2. **Using the Properties panel, change the document to 330 pixels wide and 425 pixels high. Click the Background Color swatch and choose #006666 from the Swatches panel.**

3. **Rename Layer 1 as** `Background`.

4. Choose File>Save As. Save the file as travel_videos.fla in your WIP>Tourism folder.

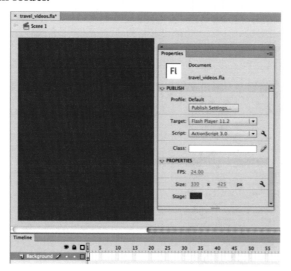

5. Continue to the next exercise.

 IMPORT VIDEO INTO FLASH

There are three different ways to deliver video using the Flash Player.

Embedded Video. Using this method, video content is placed on the Flash timeline and stored in the SWF file with other content. This method is popular because it is easier to visualize the interaction of the video with other elements on the timeline. There are, however, several disadvantages to embedding video on a timeline:

- Each frame in a video file will map to a single frame in the Flash timeline. This method can significantly reduce the video playback quality because the video will play back at the frame rate defined for the Flash file. (Video is typically captured at 24 fps or higher.)

- You might have problems synchronizing audio and video because the frame rates of the video and the Flash movie do not always match.

- Embedding video into a Flash file can significantly increase the file size and download time of the final SWF movie file.

Progressive Download. Using this method, the video file remains separate from the primary Flash file. When the video file is called by the SWF file, the video downloads from its source location on the Web server. As soon as enough information has downloaded, the video begins playing (although you can prevent automatic playback if you prefer). Because the video file remains external to the SWF file, this method offers the following benefits:

- You do not need to republish the SWF file to change content in the video file.

- The SWF file is considerably smaller than an embedded video file, so it takes less time to download the SWF file.

- Because the video file and the SWF file are separate, the video can maintain a higher frame rate; this also prevents problems with synchronization between the audio and video components of the video file.

Streaming Video. Using this method, each user (client) opens a connection to the streaming server; the streaming server streams the video bits to each individual client. These bits are used by the client, and then immediately discarded. (As with progressive download files, streaming video files remain external to other Flash content. Unlike progressive download video, which can be deployed with any Web server, streaming video requires streaming server software.) Streaming video offers the following advantages:

- With streaming video, the server can precisely control and deliver any portion of the stream to individual clients. Users can go to any point in the video and play it immediately.

- Streaming allows real-time broadcasting of live events, as well as the ability to control the playback of the content. Live video can be delivered from any connected video camera or Web cam.

- Streaming immediately serves the requested video; users do not need to wait until the entire video has downloaded.

The disadvantage to streaming video is that slow connection speed or bandwidth congestion might result in decreased quality.

In this exercise, you import the first client video into the Flash file. This video will be the default playback when a user opens the module. Later in the project, you will use ActionScript to change the video that appears in the module, based on a series of thumbnail images.

1. **With travel_videos.fla open, create a new layer named Video at the top of the layer stack.**

2. **Choose File>Import>Import Video.**

   The Import Video command works like a wizard, walking you through the necessary steps of the video import process. In the first screen, you define the location of the video file you want to import (whether on your local drive or already deployed on a Web server).

   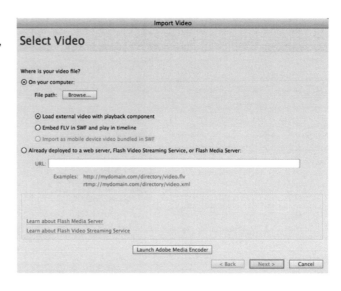

   The Import Video dialog box provides three import options:

   - **Load External Video with Playback Component.** If this option is selected, the video imports into an instance of the FLVPlayback component to control video playback. When you publish the Flash document as a SWF and upload it to your Web server, you must also upload the video file and configure the FLVPlayback component with the location of the uploaded video file.

   Note:

   The Import Video dialog box only mentions FLV files, but the options apply to F4V as well.

   - **Embed FLV in SWF and Play in Timeline.** You can use this option to embed an FLV/F4V into the Flash document. When you import video this way, the video is placed in the timeline; individual frames in the video correspond to individual frames on the Flash timeline.

   - **Import as Mobile Device Video Bundled in SWF.** Similar to embedding a video in a Flash document, this option allows you to bundle a video into a Flash Lite document for deployment on a mobile device.

3. **Make sure the Load External Video with Playback Component is selected, then click the Browse button.**

4. **Navigate to ᴠideo1.f4ᴠ in your WIP>Tourism>F4V folder and click Open to return to the Import Video dialog box.**

   Because the F4V folder is located in the same place as the Flash file, the Flash file can create an accurate relative path from the main file to the video file. If you change the relative location of either file — either on your local computer or when you upload files to the Web server — you will have to correct the video's link path.

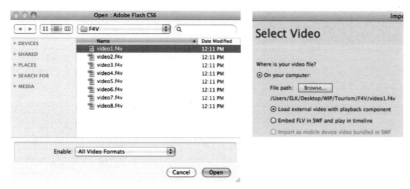

5. **Click Next to show the Skinning screen of the Import Video dialog box.**

   A **skin** is simply the superficial appearance of a component. The selected skin determines which controls are included in the video playback component created by the Import Video process.

6. **In the Skin menu, choose SkinUnderPlaySeekStop.swf.**

   The video files for this project do not include audio information, so you do not need to include volume and mute controls in the skin.

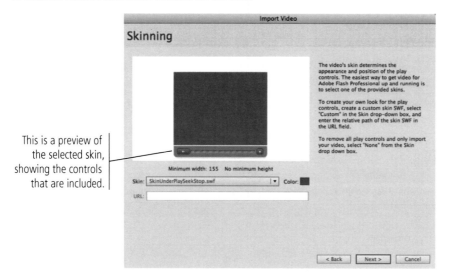

This is a preview of the selected skin, showing the controls that are included.

7. **Click Next to show the Finish Video Import screen of the Import Video dialog box.**

8. **Read the information in the Finish Video Import screen.**

As we mentioned earlier, the relative path from the exported SWF file (and the working Flash file) must be maintained. This screen also tells you that the skin you selected is actually a separate SWF file, which must also be available on the Web server for the video to work properly.

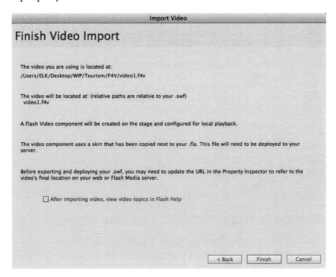

9. **Click Finish to add the video playback component to the Stage.**

When you import the video with the FLVPlayback component, the component is automatically centered on the Stage. (The skin is not considered in the object size for centering.) The playback component defaults to the same physical size as the video file you imported.

10. **Make sure the component is selected on the Stage. In the Properties panel, change the position of the component to X: 5, Y: 5.**

This defines the position of the component, which is the actual object on the Stage. Changing the position or size of the FLVPlayback component has no effect on the size of the video that plays inside the component.

This position provides 5 pixels of "breathing space" between the edges of the Stage and the left, top, and right sides of the component.

11. **Type player in the Instance Name field.**

Later in this project, you will use ActionScript to change the content in the playback component, based on which thumbnail is clicked. Like other elements on the Stage, the playback component instance must be named so it can be addressed in ActionScript.

The source parameter shows the file that will play in the component.

12. **Save the file, and then press Command-Return/Control-Enter to test the movie.**

13. **Close the Player window.**

14. **On your desktop, open your WIP>Tourism folder and review the contents.**

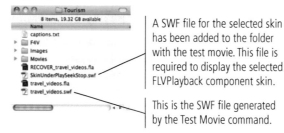

A SWF file for the selected skin has been added to the folder with the test movie. This file is required to display the selected FLVPlayback component skin.

This is the SWF file generated by the Test Movie command.

Note:

You might need to click away from the selected component before you can test the file.

15. **Return to Flash, and then continue to the next stage of the project.**

Stage 2 Working with Components

Flash CS6 includes a set of useful tools for building a variety of common interface elements — from scrolling panes to form fields to working video players with all of the necessary playback controls. These tools are very appealing to visually oriented designers because, in most cases, the tools can be used instead of writing complicated ActionScript — even when developing fairly complex interactive elements. (Many components still require some coding, but not as much as creating the components from scratch would require.)

User Interface Components

FLASH FOUNDATIONS

Two sets of components are available in the Components panel: User Interface and Video. User Interface components, described here, help you create common interactive elements.

Button. This component creates a rounded-corner button with a text label. You can change the text label on the button by changing the Label parameter for the component. One unique feature of the Button component (compared to regular button symbols) is the Toggle parameter; if you set this parameter to True, the button maintains the clicked state until a user clicks it a second time.

CheckBox. This component creates a standard toggle control with the label to the right of the check box shape.

ColorPicker. This component creates a color swatch that, when clicked, opens the Color Picker palette in the SWF file. You can use ActionScript to define the specific colors available in the palette.

ComboBox. This component creates a pop-up menu list. Double-clicking the dataProvider parameter for this component opens the Value dialog box, where you can define the labels and data for individual menu items.

DataGrid. This component creates a table to contain data in columnar format; the resulting grid is scrollable and sortable, based on column headings. The contents of the DataGrid component must be defined through ActionScript.

Label. This component creates a text area that is basically a dynamic text field. (There is no real reason to use this component; a standard dynamic text area serves the same purpose.)

List. This component creates a scrollable selection list. As with the ComboBox component, double-clicking the dataProvider parameter opens the Value dialog box, where you can define the necessary list items.

NumericStepper. This component creates a specialized field that contains only numbers. You can define the initial value in the field, minimum and maximum values, and the increments used by the Up Arrow and Down Arrow buttons attached to the field.

ProgressBar. This component creates a small animated bar that indicates a process is not yet finished.

RadioButton. This component creates a toggle control similar to a check box. The primary difference is that radio buttons are typically part of a group, from which a user can select only one option. You have to use ActionScript to define each radio button as part of the same group.

ScrollPane. This component creates an area intended to contain large images. You can define the size of the pane in the Properties panel, and then control the pane's content and scrolling behavior in the Properties panel.

Slider. This component creates a number-based slider bar. Like the NumericStepper component, you can define the initial, minimum, and maximum possible values, as well as the intervals at which the slider handle can stop.

TextArea. This component creates a stylized dynamic text area that can scroll both horizontally and vertically (if necessary). For smaller blocks of text that don't require scrolling, a regular dynamic text area might be easier to manage because you don't have to use ActionScript to control the appearance of text.

TextInput. This component creates a field in which users can input text.

TileList. This component creates a scrollable grid of images. You can double-click the dataProvider parameter to add items to the component, or you can use ActionScript to manually define the component contents. (If you use the dataProvider option, each item will include a label, which will be added to the selected thumbnail. If you don't want to include labels for the items, you should use ActionScript to manually define the items.)

UILoader. This component creates an invisible area you can use to easily load other files into a movie. You can use the Properties panel to define the file to load.

UIScrollBar. This component can be dragged onto a scrollable dynamic text field to easily add scrollbars to the text area. (To make a dynamic text area scrollable, Control/right-click a dynamic text area and choose Scrollable from the contextual menu.) The scrollbar component snaps to the closest edge of the text area, depending on where you drag the component.

DEFINE FLVPLAYBACK COMPONENT PARAMETERS

When you use the Import Video command and dialog box, the selected video is placed onto the Stage within an instance of the built-in FLVPlayback component. Although you can create custom players, this component includes skins with controls that are already designed to control the video inside the player. By selecting different skins and modifying the component parameters, you can customize the component to meet most video player needs without having to learn complex ActionScript.

1. **With travel_videos.fla open, select the FLVPlayback component on the Stage.**

    Components are controlled using built-in sets of parameters, which you can define in the Component Parameters section of the Properties panel.

2. **In the Component Parameters section of the Properties panel, uncheck the autoPlay option.**

3. **Click the skinBackgroundColor swatch and choose the light gray swatch (#CCCCCC) from the pop-up color palette.**

4. **Change the skinBackgroundAlpha value to 0.5 to increase the transparency of the skin.**

    By default, the FLVPlayback skin is partially transparent. The skinBackgroundAlpha parameter controls the transparency from 0 (fully transparent) to 1 (fully opaque).

Note:

Because you added the video using the Import Video command, the source path is already defined. You can also drag the FLVPlayback component from the Components panel, and then use the Source parameter to define which video to play.

Uncheck this option to prevent the movie from playing when the file first opens.

Change the skinBackgroundAlpha property to 0.5.

Click this swatch to change the skin color to #CCCCCC.

5. **Press Command-Return/Control-Enter to test the movie.**

    When the player module opens, the movie does not play until you click the Play button in the component skin. This follows the rule you defined using the autoPlay parameter.

Note:

The skinAutoHide value is set to false by default. If you change this parameter to true, the skin will only appear when a user rolls the mouse over the FLVPlayback component.

6. **Close the Player window and return to Flash.**

7. **Save the file and continue to the next exercise.**

FLASH FOUNDATIONS

FLVPlayback Component Parameters

You can customize the FLVPlayback component by changing many parameters in the Properties panel.

- **align.** This determines where the video is placed inside the FLVPlayback component if the video is physically smaller than the space available in the playback component. (When you place a video using the Import Video dialog box, the playback component automatically scales to match the physical size of the imported video.)

- **autoPlay.** This parameter has two options: true or false. Your choice determines whether the defined video plays as soon as the module opens.

- **cuePoints.** This parameter describes cue points for the selected video file. Cue points synchronize certain points in the video file with animation, text, or graphics.

- **isLive.** This parameter allows you to set a Boolean value that, if true, specifies that the video is streaming live from Flash Media Server (FMS). The default value is false.

- **preview.** This parameter has no function in Flash CS6; it is a remnant of CS4 (and earlier) technology, when the application could not display the actual video content on the Flash Stage.

- **scaleMode.** This determines whether the video will scale to fit the space available in the FLVPlayback component. You can choose maintainAspectRatio to scale the video proportionally (horizontal and vertical dimensions keep the same aspect ratio as the original), noScale (the video displays at its original size), and exactFit (the video scales to fit the available space, distorting the original aspect ratio if necessary).

- **skin.** This parameter enables you to customize the location of controls, as well as the specific options included in the component skin. Double-clicking this parameter opens the Select Skin dialog box, which includes the same options you saw in the Skinning screen of the Import Video dialog box.

- **skinAutoHide.** This parameter has two possible options: true or false. If you select true, the skin controls will only be visible when a user moves the mouse over the FLVPlayback component.

- **skinBackgroundAlpha.** This parameter controls the transparency of the component skin. Alpha can be set from 0 (fully transparent) to 1 (fully opaque).

- **skinBackgroundColor.** This parameter controls the color of the component skin.

- **source.** This parameter specifies the path to the video to be played. If a video file has already been uploaded to a Web server, you can specify the HTTP or RTMP (Real-Time Messaging Protocol) path in this parameter.

- **volume.** This parameter controls the default volume for the video in the component. Volume can be set from 0 (silent) to 1 (full volume).

USE THE TEXTAREA COMPONENT FOR CAPTIONS

The TextArea component is basically an enhanced dynamic text area that defaults to show a white background and a slightly beveled edge. (You can edit this appearance by editing the component skins.)

1. With **travel_videos.fla** open, create a new layer named **Caption** at the top of the layer stack.

2. Open the Components panel (Window>Components) and expand the User Interface folder. Drag an instance of the TextArea component onto the Stage.

3. Using the Properties panel, change the TextArea component dimensions to 320 pixels wide by 40 pixels high. Center the component horizontally on the Stage, and position it directly below the FLVPlayback instance.

   Make sure you turn off the Constrain option in the Properties panel so that you can change the component height without affecting its width (and vice versa).

4. **Type `caption` in the Instance Name field.**

Like all other elements, a TextArea component must be named before you can address it in ActionScript.

Component instance names, dimensions, and positions can be defined in the Properties panel.

5. **In the Component Parameters section of the Properties panel, click the text field to place the insertion point.**

6. **Type `Click a thumbnail to see video clips from recent tours.` and press Return/Enter to finalize the text in the field.**

This text will appear in the caption area when the module first opens (before a user clicks a specific thumbnail).

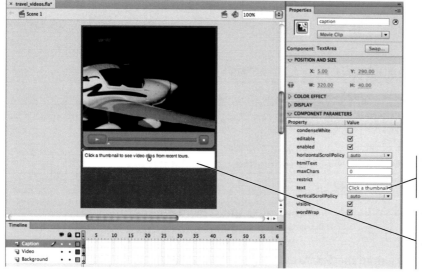

Click here and type the text that you want to appear in the component.

When you press Return/Enter after typing the caption, the text appears in the component.

7. **Save the file and continue to the next exercise.**

FLASH FOUNDATIONS

TextArea Component Parameters

A number of parameters are available for customizing TextArea components:

- **condenseWhite.** This option affects text areas set to export as HTML text. When this parameter is true, extra spaces and line breaks are removed from the text.

- **editable.** This option determines whether users can change the text in the area.

- **enabled.** This option sets a value that indicates whether the component can accept user interaction. A value of true indicates that the component can accept user interaction; a value of false indicates that it cannot. When the enabled property is set to false, the color of the container becomes dimmed and the user input is blocked (with the exception of the Label and ProgressBar components).

- **horizontalScrollPolicy.** This option can be set to true to allow horizontal scrolling.

- **htmlText.** If enabled, this option formats text in the area as HTML text.

- **maxChars.** This parameter limits the number of characters in the component.

- **restrict.** This option defines specific sets of characters that are allowed in the component (for example "0–9" allows only numerals).

- **text.** This field defines specific text that will appear in the component. (You can also use ActionScript to define and/or change the text content in a TextArea component.)

- **verticalScrollPolicy.** This option can be set to true to allow vertical scrolling.

- **visible.** This option can be set to true to allow the visibility of the text or caption added in the component.

- **wordWrap.** If this option is set to false, text in the text area does not automatically wrap, based on the component's defined width.

ADD THUMBNAIL NAVIGATION OBJECTS

Flash components are prebuilt elements that make it easy to create and control the behavior of common interface objects that would otherwise require complex design and coding. When you imported the video file into Flash, the video was added inside a predefined playback module. This module is actually an instance of the FLVPlayback component, which is available in the Video folder of the Components panel.

Other common interface elements are also available in the Components panel, including a scrolling area for holding multiple thumbnail images. Rather than building your own scrolling area from scratch, you can use the built-in TileList component to hold the video thumbnails that will act as links to the various video files.

1. With **travel_videos.fla open, create a new layer named Thumbnails at the top of the layer stack.**

2. **Drag an instance of the TileList component from the Components panel to the Stage.**

3. **Change the component width to 320 pixels and height to 85 pixels. Align the TileList component horizontally on the Stage, below the TextArea component.**

4. In the Properties panel, type thumblist in the Instance Name field.

As with other types of instances, component instances must be named before they can be addressed by ActionScript.

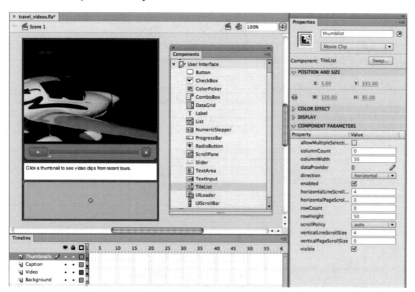

Note:

The TileList component is actually a type of scrolling area, but it is specifically designed to create a grid of images in a scrollable area. The ScrollPane component, on the other hand, is primarily intended to hold text-based content; you can't define columns or rows in the parameters of a ScrollPane component.

5. In the Component Parameters, choose on from the scrollPolicy menu.

By default, the scrolling TileList component scrolls only when necessary. By setting the scrollPolicy to on, the scrollbar appears within the component, even if the included content does not require scrolling.

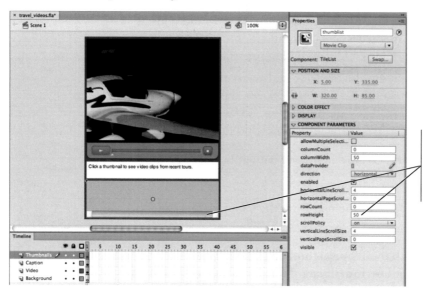

When the ScrollPolicy parameter is set to on, the scrollbar appears within the component on the Stage.

6. Save the file and continue to the next exercise.

ADD CONTENT TO THE TILELIST COMPONENT

If you use the dataProvider parameter to add content to the TileList component, each item will include a text label. You don't want labels on the thumbnails, and you will later add data to each item, so you are going to use ActionScript to manually define the items included in the TileList component.

The Flash file needs to be able to locate the thumbnail images you will script into the TileList component. The relative path that you define for the thumbnail images in the script must remain consistent when the exported SWF file is uploaded to a Web server.

1. With **travel_videos.fla** open, create a new layer named **Actions** at the top of the layer stack.

2. With Frame 1 of the Actions layer selected, open the Actions panel.

3. Click in the Script pane of the Actions panel to place the insertion point.

4. On Line 1 of the Script pane, type:

   thumblist.addItem({source:"Images/thumbnail1.jpg"});

   The first part of the script identifies the component (**thumblist**) where items will be added (using the **addItem** command). Inside the parentheses, you define the content that will be added to the TileList. Make sure you type the command exactly as we show here.

   • Capitalization counts; the folder containing the thumbnail images is named "Images" (capitalized).

   • The source statement must be enclosed within both parentheses and braces.

   • The location of the source parameter must be enclosed within quotation marks.

   • The statement must have a semicolon at the end of the line.

You are adding this statement to Frame 1 of the Actions layer.

5. Press Return/Enter to move to Line 2 of the Script pane.

6. Beginning on Line 2, add seven more lines to the script that will add each thumbnail image in the Images folder into the TileList component:

   thumblist.addItem({source:"Images/thumbnail2.jpg"});
   thumblist.addItem({source:"Images/thumbnail3.jpg"});
   thumblist.addItem({source:"Images/thumbnail4.jpg"});
   thumblist.addItem({source:"Images/thumbnail5.jpg"});
   thumblist.addItem({source:"Images/thumbnail6.jpg"});
   thumblist.addItem({source:"Images/thumbnail7.jpg"});
   thumblist.addItem({source:"Images/thumbnail8.jpg"});

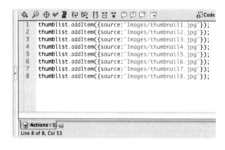

   If you copy and paste the first line of code, make sure you change the number for each thumbnail.

7. **Press Command-Return/Control-Enter to test the movie.**

The TileList scroll panel includes all eight thumbnail images. The thumbnails are proportionally scaled to fit into the grid size, which is defined by the default columnWidth and rowHeight parameters of the TileList component.

8. **Close the Player window and return to Flash.**

9. **Save the file and continue to the next exercise.**

TileList Component Parameters

The TileList component is a versatile tool for showing multiple thumbnails in a scrolling area. You can customize this component using a variety of parameters:

- **allowMultipleSelection.** This parameter has two possible values: true or false. False is the default value. If you select true, a user can select multiple thumbnails in the component. (This parameter should remain at the default value if you are using a TileList component to change other areas of the movie.)

- **columnCount.** This parameter limits the number of columns to a specific number. By default, a TileList displays as many rows and columns as possible, based on the component size and the defined columnWidth and rowHeight parameters.

- **columnWidth.** This parameter defines the horizontal size of cells in the TileList grid.

- **dataProvider.** This parameter defines the label and source for items in the TileList.

- **direction.** This parameter determines whether the TileList scrolls horizontally or vertically.

- **enabled.** This option sets a value that indicates whether the component can accept user interaction. A value of true indicates that the component can accept user interaction; a value of false indicates that it cannot. When the enabled property is set to false, the color of the container becomes dimmed and the user input is blocked.

- **horizontalLineScrollSize.** This parameter defines the scrolling increment for the Arrow button in a horizontal scrollbar.

- **horizontalPageScrollSize.** This parameter defines the scrolling increment when a user clicks the scrollbar track on a horizontal scrollbar.

- **rowCount.** This parameter limits the number of rows in the TileList grid.

- **rowHeight.** This parameter defines the vertical size of cells in the TileList grid.

- **scrollPolicy.** This parameter determines whether scrollbars are visible only when necessary (auto), always (on), or never (off).

- **verticalLineScrollSize.** This parameter defines the scrolling increment for the Arrow button in a vertical scrollbar.

- **verticalPageScrollSize.** This parameter defines the scrolling increment when a user clicks the scrollbar track on a vertical scrollbar.

- **visible.** This option can be set to true to allow the visibility of the text or caption added in the component.

FLASH FOUNDATIONS

ADJUST TILELIST PARAMETERS

By default, a TileList component shows small thumbnails (50 × 50 pixels) within the defined dimensions of the component. You can customize many parameters in a TileList, including the component's scroll behavior and the size of items in the component.

Note:

If the component is high enough to accommodate more than one row using the defined height, multiple rows will appear unless you use the rowCount parameter to restrict the number of rows to 1.

1. **On your desktop, click one of the thumbnail images in the WIP>Tourism> Images folder. If your system does not automatically show the physical dimensions of the selected file, Control/right-click a thumbnail file and choose Get Info/Properties from the contextual menu.**

   On Macintosh, the file dimensions are listed in the More Info section of the Info dialog box.

   On Windows, the dimensions are listed in the Summary tab of the Properties dialog box.

Each thumbnail is 90 pixels wide by 61 pixels high.

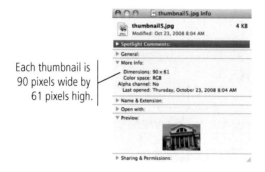

2. **With travel_videos.fla open in Flash, select the TileList component instance on the Stage.**

3. **In the Properties panel, change the columnWidth parameter value to 90.**

4. **Change the rowHeight parameter value to 70.**

   Remember, you defined this component's height as 85 pixels; the scrollbar takes up 15 pixels of that height. Although the thumbnails are only 61 pixels high, you are defining the rowHeight as 70 so that the row occupies the entire vertical space in the component.

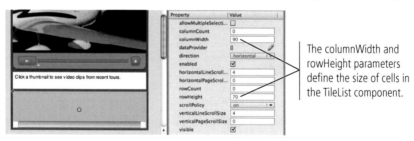

The columnWidth and rowHeight parameters define the size of cells in the TileList component.

5. **Change the horizontalLineScrollSize parameter value to 90.**

   This value is the same as the columnWidth value, so clicking the scroll buttons will move to the next or previous tile in the component.

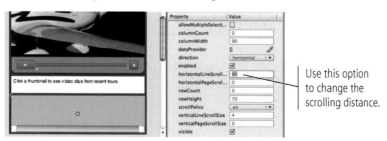

Use this option to change the scrolling distance.

6. **Press Command-Return/Control-Enter to test the movie. Click the scroll buttons in the TileList component to test their functionality.**

Clicking the scroll buttons moves the area by 90 pixels.

7. **Close the player window and return to Flash.**

8. **Save the file and continue to the next exercise.**

USE A CHANGE EVENT TO LINK IMAGES, MOVIES, AND CAPTIONS

Two things need to happen when a user clicks a thumbnail in the TileList component: first, the related video needs to appear in the FLVPlayback component; and second, the appropriate caption needs to appear in the caption dynamic text area.

To accomplish these goals, you first need to define the caption and video file related to each thumbnail. Then, you need to define an event listener and function to change the content in the video player and caption area.

1. **With trauel_uideos.fla open, select Frame 1 of the Actions layer.**

2. **Open the Actions panel Options menu and make sure the Word Wrap option is checked.**

   When this option is active (toggled on), each line of the script wraps to fit inside the Script pane. This is helpful if your panel is not wide enough to show an entire line of code in one line.

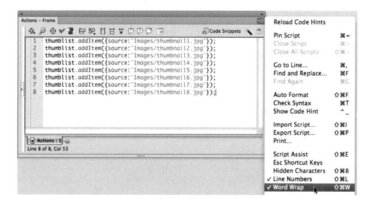

3. **In Line 1 of the Actions panel Script pane, place the insertion point after the closing quote but before the closing brace.**

Place the insertion point here.

4. **Type the following:**

```
, data1:"F4U/video1.f4u", data2:"caption"
```

For each thumbnail, **data1** will contain the path to the related movie, and **data2** will contain the caption for that movie.

```
1  thumblist.addItem({source:"Images/thumbnail1.jpg", data1:"F4V/video1.f4v", data2:"caption"});
2  thumblist.addItem({source:"Images/thumbnail2.jpg"});
3  thumblist.addItem({source:"Images/thumbnail3.jpg"});
4  thumblist.addItem({source:"Images/thumbnail4.jpg"});
5  thumblist.addItem({source:"Images/thumbnail5.jpg"});
6  thumblist.addItem({source:"Images/thumbnail6.jpg"});
7  thumblist.addItem({source:"Images/thumbnail7.jpg"});
8  thumblist.addItem({source:"Images/thumbnail8.jpg"});
```

5. **For each line of the script, add the same statements, making sure to change the number of the video in each line of the script.**

```
1  thumblist.addItem({source:"Images/thumbnail1.jpg", data1:"F4V/video1.f4v", data2:"caption"});
2  thumblist.addItem({source:"Images/thumbnail2.jpg", data1:"F4V/video2.f4v", data2:"caption"});
3  thumblist.addItem({source:"Images/thumbnail3.jpg", data1:"F4V/video3.f4v", data2:"caption"});
4  thumblist.addItem({source:"Images/thumbnail4.jpg", data1:"F4V/video4.f4v", data2:"caption"});
5  thumblist.addItem({source:"Images/thumbnail5.jpg", data1:"F4V/video5.f4v", data2:"caption"});
6  thumblist.addItem({source:"Images/thumbnail6.jpg", data1:"F4V/video6.f4v", data2:"caption"});
7  thumblist.addItem({source:"Images/thumbnail7.jpg", data1:"F4V/video7.f4v", data2:"caption"});
8  thumblist.addItem({source:"Images/thumbnail8.jpg", data1:"F4V/video8.f4v", data2:"caption"});
```

Each line should refer to a different number video.

6. **From the WIP>Tourism folder, open the file captions.txt in a text editor.**

7. **Select the first caption (excluding the opening number and space) and copy it.**

Don't select the opening number and space.

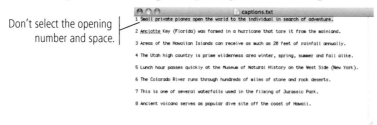

```
captions.txt
1 Small private planes open the world to the individual in search of adventure.

2 Anclotte Key (Florida) was formed in a hurricane that tore it from the mainland.

3 Areas of the Hawaiian Islands can receive as much as 20 feet of rainfall annually.

4 The Utah high country is prime wilderness area winter, spring, summer and fall alike.

5 Lunch hour passes quickly at the Museum of Natural History on the West Side (New York).

6 The Colorado River runs through hundreds of miles of stone and rock deserts.

7 This is one of several waterfalls used in the filming of Jurassic Park.

8 Ancient volcano serves as popular dive site off the coast of Hawaii.
```

8. **In Flash, select the word "caption" in Line 1 of the script and paste the copied caption.**

```
1  thumblist.addItem({source:"Images/thumbnail1.jpg", data1:"F4V/video1.f4v", data2:"Small private
   planes open the world to the individual in search of adventure."});
2  thumblist.addItem({source:"Images/thumbnail2.jpg", data1:"F4V/video2.f4v", data2:"caption"});
3  thumblist.addItem({source:"Images/thumbnail3.jpg", data1:"F4V/video3.f4v", data2:"caption"});
```

9. **For each line in the script, copy the appropriate caption from the text file and paste it in place of the word "caption" in the data2 value.**

```
1  thumblist.addItem({source:"Images/thumbnail1.jpg", data1:"F4V/video1.f4v", data2:"Small private
   planes open the world to the individual in search of adventure."});
2  thumblist.addItem({source:"Images/thumbnail2.jpg", data1:"F4V/video2.f4v", data2:"Anclotte Key
   (Florida) was formed in a hurricane that tore it from the mainland."});
3  thumblist.addItem({source:"Images/thumbnail3.jpg", data1:"F4V/video3.f4v", data2:"Areas of the
   Hawaiian Islands can receive as much as 20 feet of rainfall annually."});
4  thumblist.addItem({source:"Images/thumbnail4.jpg", data1:"F4V/video4.f4v", data2:"The Utah high
   country is prime wilderness area winter, spring, summer and fall alike."});
5  thumblist.addItem({source:"Images/thumbnail5.jpg", data1:"F4V/video5.f4v", data2:"Lunch hour
   passes quickly at the Museum of Natural History on the West Side (New York)."});
6  thumblist.addItem({source:"Images/thumbnail6.jpg", data1:"F4V/video6.f4v", data2:"The Colorado
   River runs through hundreds of miles of stone and rock deserts."});
7  thumblist.addItem({source:"Images/thumbnail7.jpg", data1:"F4V/video7.f4v", data2:"This is one of
   several waterfalls used in the filming of Jurassic Park."});
8  thumblist.addItem({source:"Images/thumbnail8.jpg", data1:"F4V/video8.f4v", data2:"Ancient volcano
   serves as popular dive site off the coast of Hawaii."});
```

10. **Click at the end of the existing script (in the Script pane), then press Return/Enter twice to move to Line 10 of the script.**

Note:

Remember, when you work in the Actions panel Script pane, you have to use the keyboard shortcuts:

Copy (Command/Control-C)

Cut (Command/Control-X)

Paste (Command/Control-V)

11. Type the following:

```
thumblist.addEventListener(Event.CHANGE, thumbclicked);
```

The images in the TileList component do not technically exist on the Stage, so you must add the event listener to the actual component (named **thumblist**).

In this case, you're going to change the contents of the FLVPlayback and TextArea components based on which thumbnail is clicked. The CHANGE event is used whenever you want to change something on the Stage.

12. Press Return/Enter twice, then type the following:

```
function thumbclicked(event:Event):void {
```

This creates the function that will be called by the event listener you just defined.

13. Press Return/Enter to move to the next line of the Script pane. On the empty line between the function braces, type the following line of code and then press Return/Enter.

```
player.source = event.target.selectedItem.data1;
```

This line of code says, "Change the source parameter of the player instance to the data1 value in the selected item (the clicked thumbnail)."

14. Inside the function body (on the empty line between the braces), type the following:

```
caption.text = event.target.selectedItem.data2;
```

This line of code says, "Change the text in the caption instance to the data2 value in the selected item (the clicked thumbnail)."

```
     country is prime wilderness area winter, spring, summer and fall alike."});
  7  thumblist.addItem({source:"Images/thumbnail5.jpg", data1:"F4V/video5.f4v", data2:"Lunch hour
     passes quickly at the Museum of Natural History on the West Side (New York)."});
  8  thumblist.addItem({source:"Images/thumbnail6.jpg", data1:"F4V/video6.f4v", data2:"The Colorado
     River runs through hundreds of miles of stone and rock deserts."});
  9  thumblist.addItem({source:"Images/thumbnail7.jpg", data1:"F4V/video7.f4v", data2:"This is one of
     several waterfalls used in the filming of Jurassic Park."});
 10  thumblist.addItem({source:"Images/thumbnail8.jpg", data1:"F4V/video8.f4v", data2:"Ancient
     volcano serves as popular dive site off the coast of Hawaii."});
 11
 12  thumblist.addEventListener(Event.CHANGE, thumbclicked);
 13
 14  function thumbclicked(event:Event):void {
 15      player.source = event.target.selectedItem.data1;
 16      caption.text = event.target.selectedItem.data2;
 17  }
```

15. Press Command-Return/Control-Enter to test the movie. Click the various thumbnails in the TileList component to test the link functionality.

16. Close the Player window and return to Flash.

17. Save the file and continue to the next exercise.

EDIT AN INTERFACE COMPONENT SKIN

As you might have noticed, each built-in component has a predefined default appearance; in most cases, this appearance is controlled by using skins. You can access and edit the pieces of skins by double-clicking a component instance on the Stage.

1. **With travel_videos.fla open, double-click the TileList component (the thumblist instance) on the Stage.**

   Double-clicking a component enters into the component. Unlike a standard symbol Stage, the component Stage shows the various elements that make up the overall component. You can double-click these elements to enter into and edit the appearance of those elements.

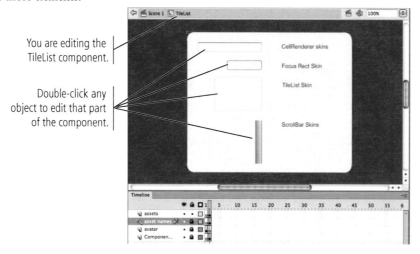

You are editing the TileList component.

Double-click any object to edit that part of the component.

2. **In the TileList component Stage, double-click the ScrollBar Skins instance to access that element on the Stage.**

3. **In the Layers panel, click Frame 2 of the Assets layer.**

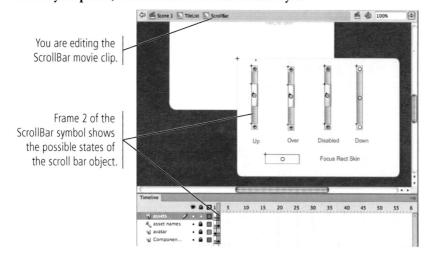

You are editing the ScrollBar movie clip.

Frame 2 of the ScrollBar symbol shows the possible states of the scroll bar object.

4. **Double-click the scrollbar track of the Down instance to edit that element.**

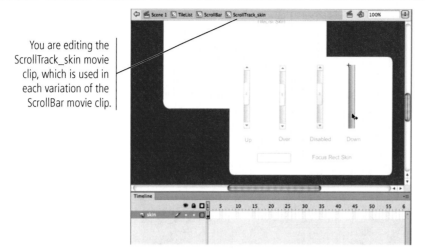

You are editing the ScrollTrack_skin movie clip, which is used in each variation of the ScrollBar movie clip.

5. **Double-click the only available object on the Stage to enter into the group, and then double-click again to access the drawing object — which is the scrollbar track.**

When you edit component skins, you might need to double-click multiple times to access the element you want. In this case, you have drilled five steps away from the main Stage to access the gradient in the scrollbar track.

Note:

When you edit the appearance of components, keep an eye on the Edit bar to see which element you are editing.

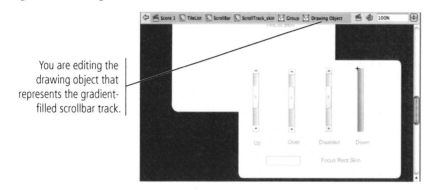

You are editing the drawing object that represents the gradient-filled scrollbar track.

6. **Click the scrollbar track with the Selection tool to select the object.**

7. **In the Color panel (Window>Color), double-click the left gradient stop to open the Color Picker for that stop. Choose #006666 as the new stop color.**

The Linear Gradient type is already applied to this object; you are simply changing the color of the existing gradient stop.

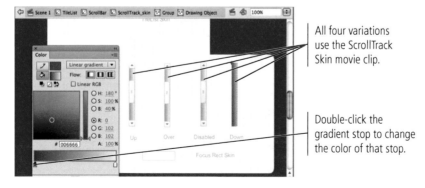

All four variations use the ScrollTrack Skin movie clip.

Double-click the gradient stop to change the color of that stop.

8. **Click Scene 1 in the Edit bar to return to the main Stage.**

   Unfortunately, the scroll track on the Stage does not show the modified color. This is a flaw in the application. Your changes will reflect in the exported file.

9. **Press Command-Return/Control-Enter to test the movie.**

   The scroll track shows the modified gradient color.

10. **Close the Player window and return to Flash.**

11. **Save the file and continue to the next exercise.**

Text Properties in ActionScript 3

FLASH FOUNDATIONS

The TextFormat class of ActionScript 3 controls character formatting. Properties are defined by typing the property, then an equal sign, and then the value (in the appropriate format). The properties of the TextFormat class include:

- The **align** property defines paragraph alignment, using a string as the value (e.g., align = "left").

- The **blockIndent** property defines the indent value for all lines in a paragraph (e.g., blockIndent = 0).

- The **bold** property determines whether text is bold, using a true-or-false value (e.g., bold = false).

- The **color** property uses the hexadecimal color value; the "#" in the color name is replaced in code with the character "0x" (e.g., color = 0x000000).

- The **font** property defines the typeface to use as a string (e.g., font = "Times New Roman"). The font may be substituted if the defined font is unavailable on a user's computer.

- The **indent** property defines the first-line distance from the left margin (e.g., indent = 0).

- The **italic** property determines whether text is italic, using a true-or-false value (e.g., italic = false).

- The **kerning** property determines whether text kerning is enabled, using a true-or-false value (e.g., kerning = false).

- The **leading** property defines the space between lines of text (e.g., leading = 0).

- The **leftMargin** property defines the size of the paragraph left margin (e.g., leftMargin = 10).

- The **letterSpacing** property defines an amount of space that is distributed across all characters (e.g., letterSpacing = 0).

- The **rightMargin** property defines the size of the paragraph right margin (e.g., rightMargin = 12).

- The **size** property defines the size of text (e.g., size = 10).

- The **underline** property determines whether text is underlined, using a true-or-false value (e.g., underline = false).

EDIT COMPONENT TEXT STYLES

Although a number of component types include text, component skins do not include predefined text areas — which means you can't use the Properties panel to control the appearance of text in a component. To control text styles for these components, you need to use ActionScript.

1. **With travel_videos.fla open, add a new layer named Format Code at the top of the layer stack.**

2. **With Frame 1 of the new layer selected, click in the Script pane of the Actions panel to place the insertion point.**

3. **On Line 1 of the script pane, type the following:**

   import fl.managers.StyleManager;

   The first step to formatting text in a component instance is to import the StyleManager class of commands. In most keyframe scripts (such as the ones you have built in this book), you don't need to import classes. The StyleManager class is an exception; you have to manually import this class into the script so the script will execute properly when you use StyleManager to change the properties of a component.

4. **On Line 3 of the Script pane, instantiate a new TextFormat object and store it in a variable named "capstyle":**

   var capstyle:TextFormat = new TextFormat();

   You must use the **new TextFormat()** constructor function to create or instantiate a TextFormat object before you can set its properties.

   When you define the TextFormat object in the script, Flash automatically adds the necessary

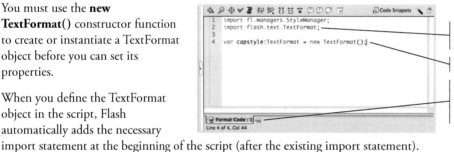

Flash added this line automatically.

You added this line manually.

You are adding this code to Frame 1 of the Format Code layer.

   import statement at the beginning of the script (after the existing import statement). This allows Flash to correctly interpret the code that you add in the rest of the script.

5. **Beginning on Line 5 of the Script pane, assign new property values to the capstyle variable:**

   capstyle.bold=true;
   capstyle.size=12;

   Each line defines a different text property of the capstyle TextFormat object.

6. **On Line 8 of the Script pane, use the StyleManager to set the style of text to the capstyle variable:**

```
StyleManager.setStyle("textFormat", capstyle);
```

The first parameter (**"textFormat"**) is the style you are setting; it refers to the overall text formatting of components. The second parameter (**capstyle**) is the specific object you are setting — in other words, the variable that contains the formatting instructions.

7. **Press Command-Return/Control-Enter to test the movie.**

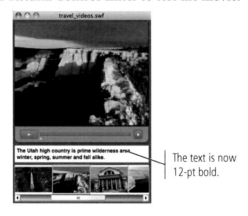

The text is now 12-pt bold.

8. **Close the Player window and return to Flash.**

9. **Close the Flash file, saving if prompted.**

10. **On your desktop, create a new folder named upload in your WIP>Tourism folder.**

11. **Move the following into the upload folder:**

Images folder	**F4V folder**
SkinUnderPlaySeekStop.swf	**travel_videos.fla**
travel_videos.swf	

As we mentioned earlier, the relative path to the encoded movie files and thumbnail image files must remain constant when files are uploaded. You are also moving the working Flash file into the upload folder so you can reopen the file and make changes later (if necessary) without changing the paths from travel_videos.fla to the movie and image files.

Project Review

fill in the blank

1. To work properly in a Flash animation, video should be converted to the _____ format.

2. When encoding video, the _____ option refers to the software method used to compress video data on video frames.

3. Using _____ video, video content is placed on the Flash timeline and stored in the SWF file with other content.

4. Using _____ video, the video file remains separate from the primary Flash file. When the video file is called by the SWF file, the video downloads from its source location on the Web server.

5. Using _____ video, each user opens a connection to the server, which sends video bits to individual users; these bits are used by the client, and then immediately discarded.

6. The _____ component is used to deliver video in a Flash movie.

7. The _____ panel is used to define the parameters of a user interface component.

8. The _____ component can be used to present a series of thumbnail images in a scrollable window.

9. True or false: You can change the appearance of text in a textArea component by double-clicking the object to enter into it. _____

10. A(n) _____ event can be used to define different contents for an object depending on which object a user clicks.

short answer

1. Briefly explain the three different ways to deliver video using the Flash Player.

2. Briefly explain the concept of a skin.

3. Briefly explain the concept of a component.

Use what you learned in this project to complete the following freeform exercise.
Carefully read the art director and client comments, then create your own design to meet the needs of the project.
Use the space below to sketch ideas; when finished, write a brief explanation of your reasoning behind your final design.

art director comments

Your client hosts several tours a year for extreme sports enthusiasts, ranging from skydiving over the Grand Canyon to wind surfing in the Grand Banks. They hired you to build a separate video module to feature movies from those trips.

To complete this project, you should:

❑ Download the client's supplied files in the **FL6_PB_Project7.zip** archive on the Student Files Web page.

❑ Develop a complete interface that shows the movie thumbnails, each linking to the appropriate movie in a FLVPlayback component.

client comments

Our extreme tourism business is very popular, so our site is going to include a whole different collection of related movies purely for those clients. We want you to create some kind of interface for that page to feature the movies.

This video module should be different than the one on our home page. We were thinking of an overall horizontal orientation, with a scrolling area on the left and the video on the right. We don't need to include captions, but there should be some kind of text that identifies the sport in each video.

The page title is 'Travel XTreme', which you can include in the module. We won't place it in the HTML page if it's already in the Flash file. We're open to any ideas you have for the page. Some kind of custom animation at the top or bottom might be a good addition.

Keep in mind the type of people who are interested in this kind of package: usually a younger demographic, people who like to be outdoors, and who look for an adrenaline rush.

project justification

_____
_____
_____
_____
_____
_____
_____
_____
_____
_____
_____
_____
_____
_____

Video is one type of multimedia file that adds depth and dimension to a Web page. Video provides the photographic realism that is not typically part of Flash animation. As an isolated task, importing video into Flash is not a particularly difficult process. You have to convert the video into the required format, but Flash walks you through this process and includes default settings that create acceptable quality for many applications.

To complete this project, you learned how to use several built-in components that simplify complex processes required for common interface tools. Rather than spending hours to manually build a scrolling area, for example, you learned how to use the TileList component to build a thumbnail navigation object. You also learned how to define parameters for, edit the visual appearance of, write ActionScript to interact with, and in general, customize built-in Flash components.

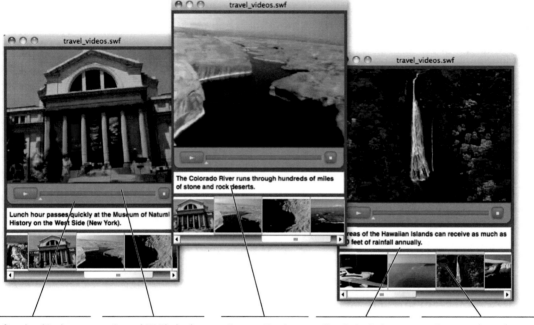

Define the skin that appears in the FLVPlayback component when you import video into Flash

Control FLVPlayback component parameters to customize the player module

Create a TextArea component to hold dynamic captions

Use ActionScript to format text in a component instance

Create and populate a TileList component to hold thumbnail navigation items

Photo Portfolio Interface

Your client is the owner of Crowe Photography, an art studio in central California near San Francisco. You have been hired to create a Flash-based portfolio site.

This project incorporates the following skills:

❑ Creating a site navigation structure

❑ Controlling navigation with ActionScript

❑ Importing external images

❑ Adding objects to the display list

❑ Importing an external text file

❑ Importing images using XML file data

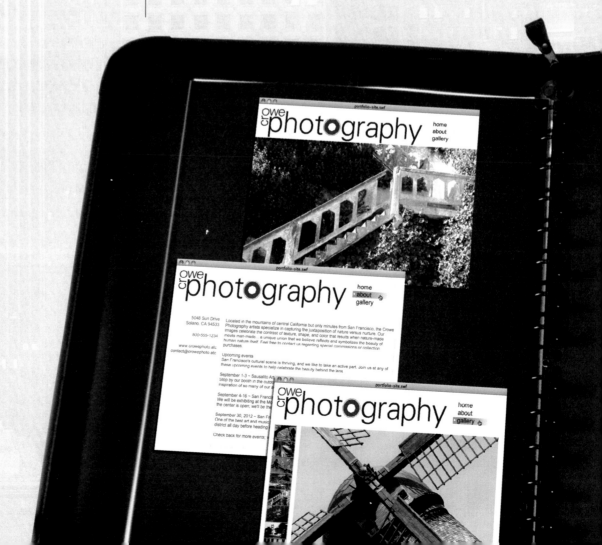

Our site is very basic — only three pages. We don't need to be complicated, we just need to get the information out there. At the same time, we're a photography studio, so we want to include images!

The "about" page is going to include a brief introduction, as well as a list of our upcoming events. New shows are always coming up, so I need to be able to change the text on that page frequently.

I haven't had much time to sift through images to place in the gallery, but I did find a few that we can display for now. I'm hoping to find time soon to gather up more images for that page. Of course, new photos pop in all the time too, so I want to be able to change these images without too much trouble.

There are two stages to this project. First, you need to build the site navigation structure, and then you need to build the individual pages that will feed into the main site.

For the "home" screen, you need to build a script to randomly place one of several images into the page. Make sure it doesn't cover the client's logo and the navigation buttons.

For the "about" screen, you need to include the client's contact information as regular text. You also have to build some mechanism for importing the external text file that the client will change frequently.

For the "gallery" screen, you should build some mechanism that allows the client to keep the image files external to the Flash file. That way she can make whatever changes she wants, as often as necessary, without constantly republishing the Flash file.

To complete this project, you will:

- ❏ Build site navigation using frame labels
- ❏ Create navigation button text
- ❏ Create navigation button symbols
- ❏ Create a site navigation function
- ❏ Define a Loader object for image files
- ❏ Use the addChild function to add to the display list
- ❏ Define a URLLoader object for an external text file
- ❏ Work with XML
- ❏ Conditionally remove a display object

Stage 1 Building Site Navigation

Despite what we are used to as professional designers, many users do not have the luxury of high-speed Internet access. Although high-speed access is certainly more common now than even a year ago, a large part of the world — including many areas of the United States — still relies on dial-up modem connections. And because Flash movies are typically much larger than HTML files, they require longer download times, which can keep users waiting for longer periods of time.

You should also keep in mind that SWF files (published Flash movies) do not work on many mobile devices — specifically iOS devices (iPhones and iPads).

Having issued those warnings, the client in this project asked you to create a Flash version of their site. Their Web developer will later include a simple introduction page that will allow users to select which version they want to use — HTML or Flash.

Note:

When you build a Flash file for Web distribution, always keep your target audience in mind. Research has shown that if a Web page doesn't capture a user's attention within 10 to 15 seconds, the user will abandon the page and move on to another site.

CREATE THE BASE FILE

As with most Flash development projects, this one begins by defining the basic Flash document. In this exercise, you will define the layers and keyframes you need to create site navigation entirely within Flash.

1. **Download FL6_RF_Project8.zip from the Student Files Web page.**

2. **Expand the ZIP archive in your WIP folder (Macintosh) or copy the archive contents into your WIP folder (Windows).**

   This results in a folder named **Portfolio**, which contains the files you need for this project. You should also use this folder to save the files you create in this project.

3. **In Flash, create a new ActionScript 3.0 document that is 800 × 600 px. Leave the frame rate and the Stage color at the default values.**

4. **Save the file as portfolio-site.fla in your WIP>Portfolio folder.**

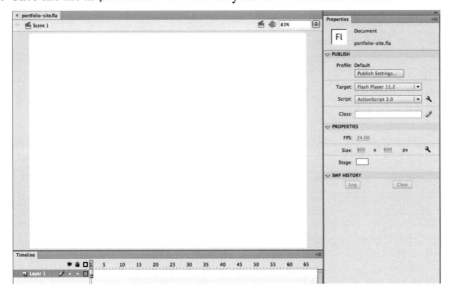

5. **Choose File>Import>Import to Stage. Navigate to and select crowe.ai in your WIP>Portfolio folder, then click Open.**

6. **In the Import to Stage dialog box, make the following choices:**

- **Choose Single Flash Layer in the Convert Layers To menu.**

  This file contains the client's logo, which you will manage as a single object; you do not need to maintain the original layers from the Illustrator file.

- **Uncheck the Set Stage Size ... option.**

  Again, this is simply one component of the overall interface design; you have already defined the correct Stage size.

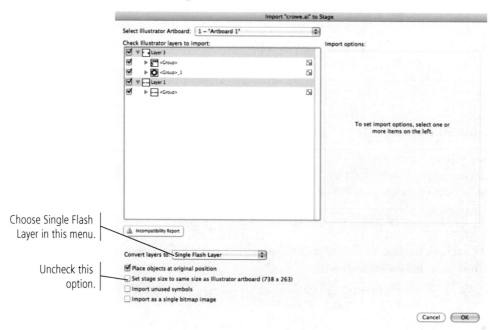

Choose Single Flash Layer in this menu.

Uncheck this option.

7. **Click OK to import the logo onto the Stage.**

8. **Select all the elements of the logo, then choose Modify>Group.**

9. **With the logo group selected on the Stage, use the Properties panel to change the group's position to X: 15, Y: 15.**

Note:

Press Command/Control-G to group selected objects.

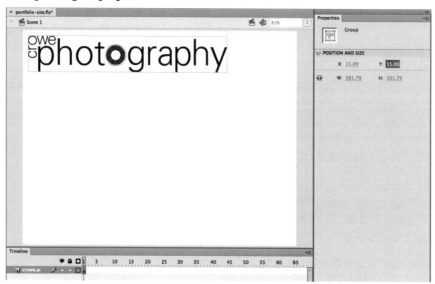

10. In the Timeline panel, change the existing layer name to Logo, then lock the layer.

11. Add four new layers above the Logo layer, using the following names and top-to-bottom stacking order:

> Actions
>
> Labels
>
> Nav Text
>
> Nav Buttons

12. Add one more layer named Content at the bottom of the layer stack.

13. Select Frame 30 on all six layers, and add regular frames to extend the timeline.

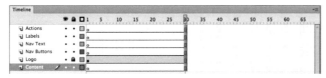

14. Select Frame 1 of the Labels layer. In the Properties panel, define btnHome as the frame label.

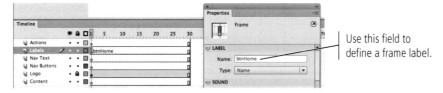

15. Add a new keyframe to Frame 10 of the Labels layer. In the Properties panel, define btnAbout as the frame label.

16. Add a new keyframe to Frame 20 of the Labels layer. In the Properties panel, define btnGallery as the frame label.

In this project you are not using the timeline for animation. Instead, you are using it to define various states for the Flash-based Web site interface. The three keyframes define the three "screens" that make up the overall portfolio site.

These might not be the most intuitive names to give these frames, since the "btn" prefix usually denotes the name of a button. However, you will later define matching instance names to actual buttons, and use an ActionScript shortcut to call the same-named frame when a user clicks a button.

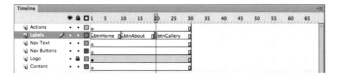

17. Save the file and continue to the next exercise.

CREATE NAVIGATION BUTTON TEXT

In previous projects, you created button symbols to accomplish specific tasks — make a turtle swim across an ocean, start a game, and so on. In this project, you are going to use a different technique to create a single button symbol that can be used in multiple instances. This technique is commonly used by professionals to minimize the amount of repetitive work.

1. **With `portfolio-site.fla` open, select the Nav Text layer to make it active.**

2. **Choose the Text tool in the Tools panel.**

3. **In the Properties panel, choose Classic Text in the Text Engine menu and choose Static Text in the Text Type menu.**

   Later in this exercise, you will add button objects behind the text you are about to create. You are creating this text area as a Static Text object because static-text objects do not recognize mouse interactions, which means the text object will not interfere with the button functionality.

4. **In Properties panel, define the following text formatting options:**

Font:	**Arial**
Style:	**Regular**
Size:	**20 pt.**
Color:	**black (#000000)**
Line Spacing:	**2 pt.**
Format:	**Align Left**

   Note:

   If you don't have Arial, use another sans-serif font such as Helvetica.

5. **Click in the top-right area of the Stage, and drag to create a new text area.**

6. **In the new text object, type:**

   - `home`
   - `about`
   - `gallery`

Note:

To type the bullet character on Macintosh, press Command-8.

On Windows, press the Alt key and type 0149 on your number keypad. If you don't have number keypad access, use the tilde (~) character in place of the bullet.

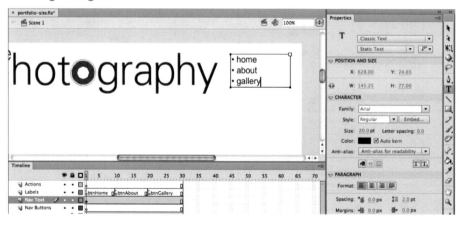

7. **Highlight each of the bullet characters and change their color to white.**

8. **Make the Selection tool active.**

   Unless the insertion point is flashing in the text area, you won't see the white bullet characters against the white Stage background. They will be visible once you create the actual navigation buttons in the next exercise.

9. **With the text object selected, use the Properties panel to define the object's position as X: 620, Y: 25.**

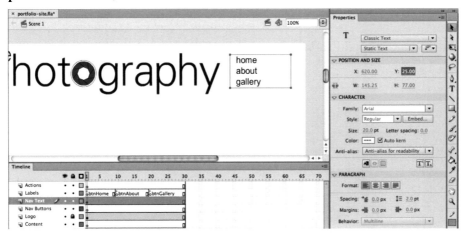

10. **Click the Embed button in the Properties panel.**

11. **In the resulting dialog box, click the Character Ranges:All option, then click OK.**

    Although you only used a few characters in this type object, you will use the same font later in the project. Because you don't know exactly what charcaters will be used, you are simply embedding the entire font.

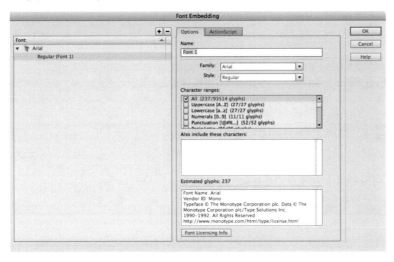

12. **In the Timeline panel, lock the Nav Text layer.**

13. **Save the file and continue to the next exercise.**

Now that you have the navigation text created, you need to create the actual button objects that can be clicked to call a specific screen in the site.

1. **With portfolio-site.fla open, select the Nav Buttons layer.**

2. **Choose the Rectangle Primitive tool. In the Properties panel, change the Stroke color to None.**

3. **Draw a rectangle on the Stage, using the following parameters:**

   X: 620   Y: 78

   W: 120   H: 22

   This creates the rectangle behind the word "gallery", just high enough to contain the ascenders and descenders, starting far enough left to reveal the white bullet, and extending about 50% beyond the right end of the word.

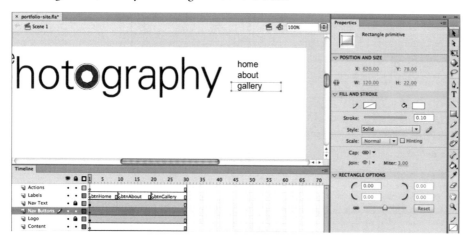

4. **With the rectangle selected, open the Color panel. Click the Fill icon to make that the active attribute, then choose Linear Gradient in the Color Type menu.**

5. **On the gradient ramp at the bottom of the panel, change left color stop to #B00041 with 50% Alpha.**

6. **Change the right color stop to #B00041 with 0% Alpha.**

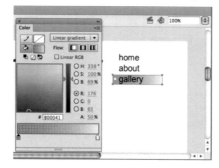

7. **Using the Selection tool, drag the rectangle object to the Library panel.**

   This converts the object to a symbol; it is simply a shortcut to choosing Modify>Convert to Symbol (or Control/right-clicking the object and choosing Convert to Symbol in the contextual menu).

8. **In the Convert to Symbol dialog box, type NavButton in the Name field. Choose Button in the Type menu and select the top-left registration point. Click OK to create the new symbol.**

9. **Double-click the symbol icon in the Library panel to enter into Symbol-Editing mode.**

10. **In the timeline, click the Up frame keyframe, then click it again and drag it to the Over frame.**

    This creates a button that has no visual representation in the Up frame; the gradient will only be visible when the Over state is triggered.

    Because you created the text in a separate text object, rather than as part of the actual button, you can reuse the same button symbol multiple times in the file — a common practice in professional Flash development.

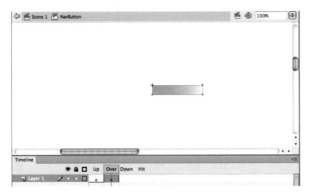

11. **Click Scene 1 in the Edit Bar to return to the main Stage.**

On the Stage, the button instance now appears as a blue rectangle shape. This highlight identifies the invisible hotspot, allowing you to work with the buttons on the Stage even though they won't be seen until the Over state is triggered.

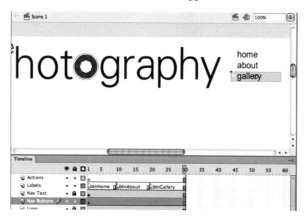

12. **Option/Alt-Shift-click and drag to clone the NavButton instance; place the clone behind the word "about."**

By pressing Shift, you constrain the movement to a 90° angle. The left edges of the original and cloned button instances should be aligned.

13. **Repeat Step 12 to add a third button instance behind the word "home."**

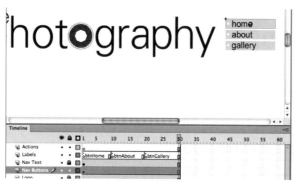

14. **Using the Properties panel, define instance names for each button on the Stage:**

btnHome

btnAbout

btnGallery

Remember, these are the same names that you defined as frame labels for the three keyframes that represent the different site screens.

15. **Select Frame 1 of the Actions layer. Type stop(); on Line 1 of the Actions panel Script pane.**

Because you are using the timeline for site organization rather than animation, you need to add this stop command to prevent the site from automatically cycling through the various "screens" as soon as the file opens.

16. **Add a keyframe to Frame 10 of the Actions layer. In the Script pane of the Actions panel, type** stop();.

In the next exercise, you are going to create a function that moves the playhead to a specific frame based on which button is clicked. You need to add the stop function to each keyframe where the different site contents will be placed.

17. **Add a keyframe to Frame 20 of the Actions layer. In the Script pane of the Actions panel, type** stop();.

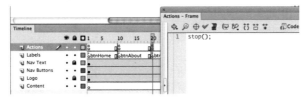

18. **Save the file and continue to the next exercise.**

 ## CREATE A SITE NAVIGATION FUNCTION

Because you used consistent names for each of the button instance names and frame labels, you can now create a simple function that navigates to the correct frame based on which button is clicked.

1. **With** portfolio-site.fla **open, select the Frame 1 keyframe on the Actions layer.**

2. **Place the insertion point at the end of the existing code, then press Return/Enter two times to move to Line 3 of the Script pane.**

3. **Type** btnHome.addEventListener(

4. **After the opening parenthesis, press Control-Space to trigger code hints, then type** mouseev.

Pressing Control-Space in the Script pane triggers the Code Hint menu. Typing automatically scrolls the menu to the first option that matches the characters you type.

Note:

Press Control-Space to trigger code hints when you are typing in the Script pane.

5. With MouseEvent highlighted in the Code Hint menu, press Return/Enter to add it to the script.

The import statement for the MouseEvent class is automatically added to the top of your script.

Note:

You can also double-click an item in the Code Hint menu to add it to the script.

6. Type a period, then choose CLICK in the Code Hint menu.

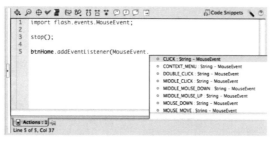

Note:

Since this package path begins with "flash," it is not necessary, but its inclusion will help trigger further code hints for you relating to the MouseEvent class.

7. After the MouseEvent class and method, type

```
, navClickHandler);
```

This is the function that will be called when a user clicks the button instance.

```
1   import flash.events.MouseEvent;
2
3   stop();
4
5   btnHome.addEventListener(MouseEvent.CLICK, navClickHandler);
```

8. Select and copy the statement on Line 5. Beginning on Line 6, paste two copies of the copied code.

```
1   import flash.events.MouseEvent;
2
3   stop();
4
5   btnHome.addEventListener(MouseEvent.CLICK, navClickHandler);
6   btnHome.addEventListener(MouseEvent.CLICK, navClickHandler);
7   btnHome.addEventListener(MouseEvent.CLICK, navClickHandler);
```

9. On Line 6, change the target object to btnAbout.

10. On Line 7, change the target object to btnGallery.

You now have three event listeners, which all call the same function. The next step is to write the navClickHandler() function that defines what happens when one of these buttons is clicked.

```
1   import flash.events.MouseEvent;
2
3   stop();
4
5   btnHome.addEventListener(MouseEvent.CLICK, navClickHandler);
6   btnAbout.addEventListener(MouseEvent.CLICK, navClickHandler);
7   btnGallery.addEventListener(MouseEvent.CLICK, navClickHandler);
```

11. Press Return/Enter to add a line of white space after the third event listener. Beginning on Line 9, type:

```
function navClickHandler (event:MouseEvent):void
{
```

```
5    btnHome.addEventListener(MouseEvent.CLICK, navClickHandler);
6    btnAbout.addEventListener(MouseEvent.CLICK, navClickHandler);
7    btnGallery.addEventListener(MouseEvent.CLICK, navClickHandler);
8
9    function navClickHandler (event:MouseEvent):void
10   {
11
12   }
```

Flash automatically adds the closing brace for you when you press Return/Enter after typing the opening brace.

12. Place the insertion point between the function braces. Type Control-Space to trigger Code Hints, then type `go`.

13. Scroll through the Code Hint menu and select gotoAndStop(), then press Return/Enter to add that method to the script.

```
 5  btnHome.addEventListener(MouseEvent.CLICK, navClickHandler);
 6  btnAbout.addEventListener(MouseEvent.CLICK, navClickHandler);
 7  btnGallery.addEventListener(MouseEvent.CLICK, navClickHandler);
 8
 9  function navClickHandler (event:MouseEvent):void
10  {
11      go
12  }
        globalToLocal3D(point:Point) : Vector3D - DisplayObject
        gotoAndPlay(frame:Object, scene:String=null) : void - MovieC
        gotoAndStop(frame:Object, scene:String=null) : void - Movie
```

Note:

Remember to take advantage of Code Hints whenever possible.

14. After the opening parenthesis, type:

`event.target.name);`

```
 5  btnHome.addEventListener(MouseEvent.CLICK, navClickHandler);
 6  btnAbout.addEventListener(MouseEvent.CLICK, navClickHandler);
 7  btnGallery.addEventListener(MouseEvent.CLICK, navClickHandler);
 8
 9  function navClickHandler (event:MouseEvent):void
10  {
11      gotoAndStop(event.target.name);
12  }
```

Note:

When you use the Code Hint menu to add the gotoAndStop function, the opening parentheses is automatically added for you. You have to manually type the closing parentheses and semicolon.

This argument for the gotoAndStop() method retrieves the name property of the event target — in other words, the <u>name</u> of the <u>target</u> object that triggered the <u>event</u>.

Remember, your frame-label names exactly match the instance names of your navigation buttons. When a user clicks a navigation button, this function will move the playhead to the frame whose label name matches the instance name of the clicked button.

15. Save the file, then press Command-Return/Control-Enter to test the movie.

If you click the navigation buttons, nothing seems to happen because you have not yet created different content on the various keyframes.

16. With the Flash Player window still open, choose View>Bandwidth Profiler.

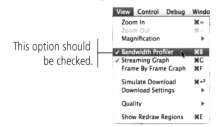

This option should be checked.

17. **Click the navigation buttons and review the top section of the window.**

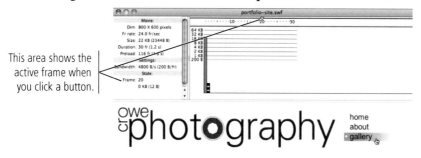

This area shows the active frame when you click a button.

18. **Choose View>Bandwidth Profiler again to toggle that option off.**

19. **Close the Flash Player window and return to Flash.**

20. **Continue to the next stage of the project.**

Stage 2 Loading External Content

Flash movies often require external content. In Project 5: Gator Race Game, you loaded audio files into the Library panel, and then used a variety of methods to incorporate the sounds into your Flash movie. In Project 7: Travel Video Module, you imported video files into the Flash file library, and then used ActionScript to deploy video inside an FLVPlayback component. To complete this project, you need to import a variety of files, including images, text, and other Flash movies.

One important difference exists between this project and the others you already completed — the files you load in this project will not be imported into the Flash Library panel. Instead, you will use ActionScript 3 to dynamically import external files into the movie. Using this method, you can update your movie by simply changing the linked content; there is no need to open, edit, and republish the Flash movie that relies on external content.

DEFINE A LOADER OBJECT FOR IMAGE FILES

Based on the project specifications, the "home" screen of the Crowe Photography Website needs to randomly change each time a user navigates back to the Home page. You need to create a mechanism that loads only one of a variety of images when a user views the page. In ActionScript 3, the Loader object class is perfectly suited to this task.

1. **With portfolio-site.fla open, select Frame 1 of the Content layer.**

2. **In the Actions panel Script pane, type:**

   `var getImage:URLRequest = new URLRequest("home-images/folio1.jpg");`

In ActionScript 3, the URLRequest object identifies the path to a specific image file. This line creates a new URLRequest object named **getImage** with the value **home-images/folio1.jpg**.

The import statement for the URLRequest class is automatically added at the beginning of the script as soon as you define the variable of the URLRequest type (var getImage:URLRequest).

The images for the Home page are in the home-images folder, which is in the same folder where you saved the file you're building now. The path from the portfolio_site.fla Flash file to the image must include the name of the folder that contains the image.

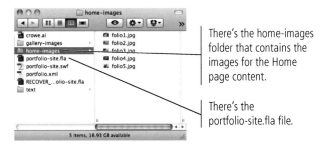

There's the home-images folder that contains the images for the Home page content.

There's the portfolio-site.fla file.

3. **On the next line of the Script pane, type:**

   `var loadImage:Loader = new Loader();`

The Loader class is a partner of the URLRequest object. The URLRequest object simply makes a file available to Flash; the Loader object actually brings the file into the Flash movie.

```
1  import flash.net.URLRequest;
2  import flash.display.Loader;
3
4  var getImage:URLRequest = new URLRequest("home-images/folio1.jpg");
5  var loadImage:Loader = new Loader();
```

The necessary import statement for the Loader class is automatically added at the beginning of the script, after the existing import statement, as soon as you define the variable of the Loader type (var loadImage:Loader).

4. **On the next line of the Script pane, type:**

   `loadImage.load(getImage);`

The load function of the Loader object is the third required step to loading an external file into a Flash file. This statement loads the URLRequest object (**getImage**, which has a value of **home-images/folio1.jpg**) into the Loader object (**loadImage**).

5. On the next line of the Script pane, type `addChild(loadImage);`

The addChild method places the defined Loader object (the external file) in the movie (technically called "adding to the display list").

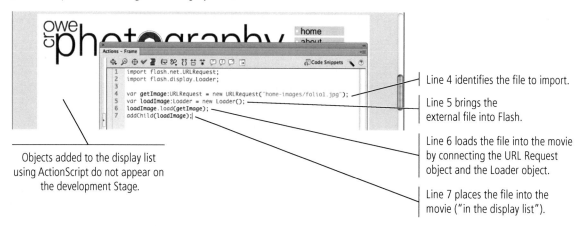

Objects added to the display list using ActionScript do not appear on the development Stage.

Line 4 identifies the file to import.

Line 5 brings the external file into Flash.

Line 6 loads the file into the movie by connecting the URL Request object and the Loader object.

Line 7 places the file into the movie ("in the display list").

6. Press Command-Return/Control-Enter to test the movie.

When you use the addChild method, new objects are added by default at X:0, Y:0 (in the top-left corner of the movie). As you can see, this obscures the logo and buttons at the top of the window.

You might also notice that the image is above the logo and navigation objects, even though you placed the code on the Contents layer (which is at the bottom of the layer stack). Objects added to the display stack are added on top of all other objects in the movie, regardless of the layer position on which you placed the code.

7. Close the Player window and return to Flash.

8. Place the insertion point at the beginning of the addChild statement (on Line 7 of the Script pane) and press Return/Enter to move the statement to Line 8.

9. On the now-empty Line 7, type:

```
loadImage.y=120;
```

If you add this statement after the addChild statement, the image will load first, and then move to the new Y position. By changing the y property of the loadImage object before it's added, the image appears in the correct position as soon as it's visible.

```
2  import flash.display.Loader;
3
4  var getImage:URLRequest = new URLRequest("home-images/folio1.jpg");
5  var loadImage:Loader = new Loader();
6  loadImage.load(getImage);
7  loadImage.y=120;
8  addChild(loadImage);
```

Add this statement to define the Y position of the added display object.

10. Press Command-Return/Control-Enter to test the movie.

The loaded image has now been moved down 120 pixels from the top of the window.

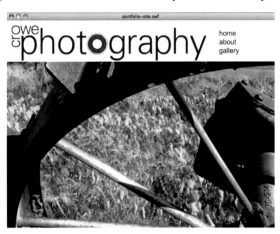

11. Close the Flash Player window and return to Flash.

12. Save the file and continue to the next exercise.

Changing Object Properties in ActionScript

ActionScript 3 allows you to modify a number of properties for objects in the movie (display objects):

alpha	instance.alpha = 0.5;	The object's transparency from 0 (fully transparent) to 1 (fully opaque).
height	instance.height = 250;	The object's height in pixels.
rotation	instance.rotation = 30;	The object's rotation (in degrees) from its original orientation. Positive values (0 to 180) indicate clockwise rotation; negative values (0 to –180) indicate counterclockwise rotation.
scaleX	instance.scaleX = 1.30;	The object's horizontal scale percentage (applied from the object's registration point); 1 equals 100%.
scaleY	instance.scaleY = 1.30;	The object's vertical scale percentage; 1 equals 100%.
visible	instance.visible = true;	Whether the object is visible (true) or not (false).
width	instance.width = 250;	The object's width in pixels.
x	instance.x = 5;	The object's x position relative to the parent object.
y	instance.y = 26;	The object's y position relative to the parent object.

Additional properties can be modified for dynamic text fields, including:

background	instance.background = true;	Whether the text field has a background fill.
backgroundColor	instance.backgroundColor = 0xCC0066;	The text field's background color (defined using hexadecimal values, preceded by "0x").
border	instance.border = true;	Whether the text field has a border.
borderColor	instance.borderColor = 0xFF9966;	The text field's border color (in hexadecimal format).
scrollH	instance.scrollH = 0;	The horizontal scrolling position of text in the field (measured in pixels); 0 represents the left side of the field (no scrolling).
scrollV	instance.scrollV = 1;	The vertical position of text in the field (measured in lines); 1 represents the first line of text at the top of the field.
selectable	instance.selectable = true;	Whether the text field is selectable by the user.
text	instance.text = "This text appears.";	The text that will appear in the field.
textColor	instance.textColor = 0x443724;	The color of text in the field (in hexadecimal format).
wordWrap	instance.wordWrap = true;	Whether the text field has word wrap.

ADD AN OBJECT TO THE DISPLAY LIST

For the site's Home screen, you need to create a mechanism that loads the home images randomly each time the file loads. You can accomplish this task by using a random math variable, similar to the technique you used to move random characters when you completed Project 5: Gator Race Game and Project 6: Gopher Golf Game.

1. **With `portfolio-site.fla` open, select Frame 1 of the Content layer.**

2. **Place the insertion point at the beginning of Line 4 in the Script pane. Press Return/Enter twice to move the existing code down.**

3. **Place the insertion point on Line 4. Insert code to declare a random-number variable named "imageNum", multiplied by 5 and then rounded to the next highest whole number (ceiled):**

   `var imageNum:Number = Math.ceil(Math.random()*5);`

   If you recall from Project 5: Gator Race Game, the Math.random function generates a number from 0 to slightly less than 1. The home-images folder contains 5 images, so you need to multiply the random number by 5.

   ```
   2  import flash.display.Loader;
   3
   4  var imageNum:Number = Math.ceil(Math.random()*5);
   5
   6  var getImage:URLRequest = new URLRequest("home-images/folio1.jpg");
   7  var loadImage:Loader = new Loader();
   8  loadImage.load(getImage);
   9  loadImage.y=120;
   10 addChild(loadImage);
   ```

   You are adding this variable declaration above the existing functions on Frame 1.

4. **On the next line of the Script pane, define a string-type variable named "image" using the following code:**

   `var image:String = "home-images/folio" + imageNum + ".jpg";`

   The right side of the image variable statement appends strings to the imageNum variable, resulting in the full path to one of the five home images in the home-images folder.

   ```
   2  import flash.display.Loader;
   3
   4  var imageNum:Number = Math.ceil(Math.random()*5);
   5  var image:String = "home-images/folio" + imageNum + ".jpg";
   6
   7  var getImage:URLRequest = new URLRequest("home-images/folio1.jpg");
   8  var loadImage:Loader = new Loader();
   9  loadImage.load(getImage);
   ```

   The following table explains what happens in these three lines of code:

Random number	*5	Ceiled	Identifies file
.1	0.5	1	home-images/folio1.jpg
.2	1.0	1	home-images/folio1.jpg
.3	1.5	2	home-images/folio2.jpg
.4	2.0	2	home-images/folio2.jpg
.5	2.5	3	home-images/folio3.jpg
.6	3.0	3	home-images/folio3.jpg
.7	3.5	4	home-images/folio4.jpg
.8	4.0	4	home-images/folio4.jpg
.9	4.5	5	home-images/folio5.jpg

   The final step to randomizing the background image is to replace the URLRequest parameter with the image variable you just defined.

5. **In Line 7 of the script, change the URLRequest object parameter from the specific path name to the image variable (shown here in red):**

   var getImage:URLRequest = new URLRequest(image);

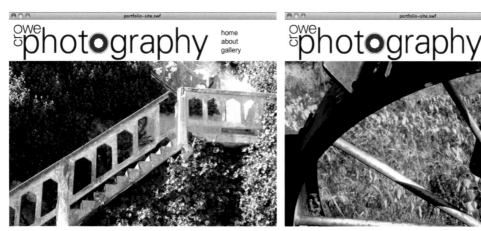

Change this parameter to the image variable.

6. **Press Command-Return/Control-Enter to test the movie.**

7. **Return to Flash, and then press Command-Return/Control-Enter to test the movie again.**

   A different image might appear in the movie background. If the same image appears, return to Flash and continue testing the movie until you see a different background image. (Because the process is random, the same image might appear two or more times in a row.)

8. **Close the Player window and return to Flash.**

9. **Save the file, then continue to the next exercise.**

CREATE THE ABOUT US SCREEN

As we have already explained, you are using different keyframes on the timeline to create the different screens in this site. For the About Us screen, you need a static text area that contains contact information, and a dynamic text area that will display the contents of an external text file.

1. **With portfolio-site.fla open, add a new keyframe to Frame 10 of the Content layer.**

2. **Choose the Text tool in the Tools panel.**

3. In the Properties panel, define the following text formatting options:

Text Engine:	Classic Text
Text Type:	Static Text
Font:	Arial (use the same font that you used for the button text in the top-right corner)
Size:	14 pt.
Color:	black (#000000)
Format:	Align Right
Line Spacing:	3 pt.

4. Click and drag to create a new text object on the left side of the Stage. In that object, type:

5048 Sun Drive
Solano, CA 94533

800-555-1234

www.crowephoto.atc
contact@crowephoto.atc

5. Choose the Selection tool in the Tools panel. With the text object still selected, change the object's parameters to

X: 10   Y: 160
W: 160

Note:

You are not able to set the height of the text object; the height of a classic text object is defined by its content.

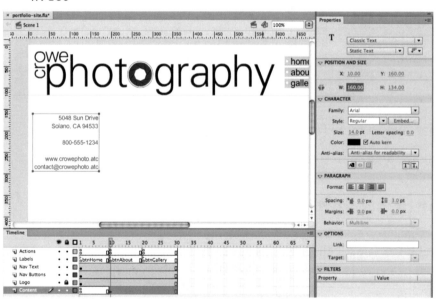

Note:

At the time of writing, the domain name we use here (crowephoto) is not currently registered. However, new domains are registered every day. We use the fictitious ".atc" domain suffix to avoid inadvertently using the domain name of a real company.

When you build sites in a professional environment, you should use the accurate domain (.com, .gov, .edu, etc.) for the job you are building.

6. Using the Type tool, highlight the Website URL.

7. In the Properties panel, expand the Options section. In the Link field, type the full URL (including the http:// prefix).

8. **Highlight the email address text. In the Properties panel, type mailto:contact@crowephoto.atc in the Link field.**

   The mailto: prefix is used to create an email link, which (when clicked) opens a preaddressed email in the user's default email client application.

9. **Open the Components panel and drag a TextArea component (from the User Interface folder) onto the Stage.**

   The TextArea component automatically includes scrolling capability if the contained text is longer than the space allows. If you used a regular dynamic text object here, you would also have to create a scroll bar object to control the area.

10. **Using the Properties panel, define the component's parameters as:**

    Instance Name: txtAbout

    X: 185   **Y: 160**

    W: 595   **H: 420**

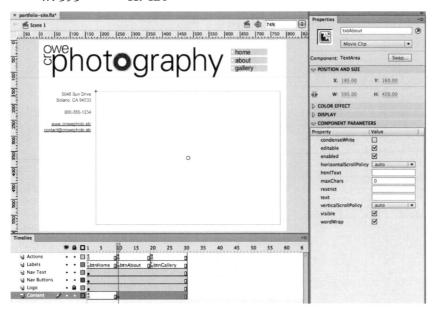

11. **Save the file and continue to the next exercise.**

DEFINE A URLLOADER OBJECT FOR AN EXTERNAL TEXT FILE

Displaying an external text file allows you to create an environment where your client can edit the textual content without touching the Flash files — basically, building an interface that allows easy and frequent updates without editing or republishing the Flash movie file.

In this exercise, you will define code that will import an external text file into the TextArea component that you added in the previous exercise. The advantage of importing text from an external file is that you can edit the text at any time without needing to republish your Flash file.

The URLLoader object class, which you will use to accomplish this goal, is basically the same as the Loader class, except the URLLoader object brings external text or XML files into a Flash movie.

1. With **portfolio-site.fla** open, select the Frame 10 keyframe of the Actions layer.

2. Add a line of white space after the existing code in the Actions panel Script pane.

3. Beginning on Line 3, declare a variable named **txtLoader** of the URLLoader type. Construct a new instance of the URLLoader class for the txtLoader variable to point to. Inside the URLRequest parentheses, construct a new URLRequest object pointing to the "about.txt" file:

   ```
   var txtLoader:URLLoader = new URLLoader(new URLRequest("text/about.txt"));
   ```

   As with the Loader object, you must define the entire path from the active file to the external file that you want to load — including any folders along the way.

   This line performs essentially the same function as the URLRequest/Loader object statements you used in a previous exercise. In this case, you use the URLLoader object instead of the Loader object because you are loading an external text file instead of an image file.

Note:

Two import statements are automatically added to the beginning of the script:

flash.net.URLLoader

flash.net.URLRequest

4. Add a line of white space in the Script pane, then type:

   ```
   txtLoader.addEventListener(Event.COMPLETE, textReady);
   ```

   Like the Loader object class, the URLLoader object brings the text file into Flash. For external text files, however, you must use a custom function that triggers only when the external file is completely loaded.

   This line of code adds an event listener to the external load object (**txtLoader**) that you created in Step 3. The COMPLETE event triggers the custom function (**textReady**) when the external text file is completely loaded.

```
4   stop();
5
6   var txtLoader:URLLoader = new URLLoader(new URLRequest("text/about.txt"));
7
8   txtLoader.addEventListener(Event.COMPLETE, textReady);
```

5. **Add another line of white space in the Script pane, then type the following code to define the custom textReady function:**

```
function textReady(event:Event):void
{
    txtAbout.text = event.target.data;
}
```

The function body defines the text property of the TextArea component named "txtAbout." The right half of the statement (**event.target.data**) calls the target of the COMPLETE event (**txtLoader**) as the data of the defined text area.

```
1   import flash.net.URLLoader;
2   import flash.net.URLRequest;
3   import flash.events.Event;
4
5   stop();
6
7   var txtLoader:URLLoader = new URLLoader(new URLRequest("text/about.txt"));
8
9   txtLoader.addEventListener(Event.COMPLETE, textReady);
10
11  function textReady(event:Event):void
12  {
13      txtAbout.text = event.target.data;
14  }
```

Note:

An import statement for the flash.events.Event package is automatically added to the beginning of the script.

6. **Add a line of white space after the function's closing brace, then declare a new variable named txtAboutFormat of the TextFormat type, instantiated as a new instance of the TextFormat class:**

```
var txtAboutFormat:TextFormat = new TextFormat();
```

If you completed Project 7: Travel Video Module, you learned that you have to use ActionScript to change the appearance of text inside a TextArea component.

7. **Add three more lines to the code to define the size, font, and color of this TextFormat object:**

```
txtAboutFormat.size = 14;
txtAboutFormat.font = "Arial";
txtAboutFormat.color = 0x666666;
```

The "0x" prefix at the beginning of the hexadecimal number is required when designating a color in ActionScript. This replaces the # character that is used in most markup languages.

8. **Add a line of white space at the end of the Script pane. On the next line, use the setStyle() method to apply your new TextFormat object to the TextArea instance.**

```
txtAbout.setStyle("textFormat", txtAboutFormat);
```

Note:

An import statement for the flash.text.TextFormat package is automatically added to the beginning of the script.

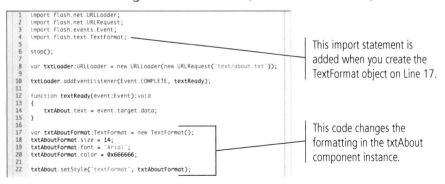

```
1   import flash.net.URLLoader;
2   import flash.net.URLRequest;
3   import flash.events.Event;
4   import flash.text.TextFormat;
5
6   stop();
7
8   var txtLoader:URLLoader = new URLLoader(new URLRequest("text/about.txt"));
9
10  txtLoader.addEventListener(Event.COMPLETE, textReady);
11
12  function textReady(event:Event):void
13  {
14      txtAbout.text = event.target.data;
15  }
16
17  var txtAboutFormat:TextFormat = new TextFormat();
18  txtAboutFormat.size = 14;
19  txtAboutFormat.font = "Arial";
20  txtAboutFormat.color = 0x666666;
21
22  txtAbout.setStyle("textFormat", txtAboutFormat);
```

This import statement is added when you create the TextFormat object on Line 17.

This code changes the formatting in the txtAbout component instance.

9. **Press Command-Return/Control-Enter to test the movie. In the Flash Player window, click the "about" link in the navigation area.**

The image that was added to the display list still appears in the movie, even when you navigate away from the Home screen (Frame 1). Because the image was added with code, you need to use code to remove the image when the timeline moves to the About screen (Frame 10).

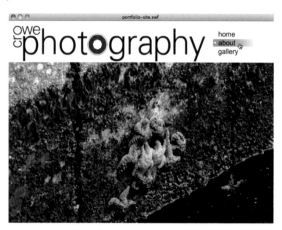

10. **Close the Flash Player window and return to Flash.**

11. **Make sure the Frame 10 keyframe is still selected on the Actions layer.**

12. **Place the insertion point after the last import statement and press Return/ Enter two times.**

13. **On Line 6 of the Script pane, type:**

    removeChild(loadImage);

Add the removeChild statement to remove the loadImage image from the display.

The removeChild function removes an object from the display list. The argument (inside the parentheses) defines the object that you want to remove.

14. **Press Command-Return/Control-Enter to test the movie. In the Flash Player window, click the "about" link in the navigation area.**

15. **Close the Flash Player window and return to Flash.**

16. **Save the file and continue to the next exercise.**

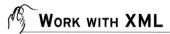 **WORK WITH XML**

Next you will build the actual portfolio portion of the site, including a set of thumbnails that users can click to show different images in the main content area. All of the images exist outside of your Flash movie, and will be described to your movie in an XML document.

XML (eXtensible Markup Language) does not describe display or layout; it simply describes data. XML might seem intimidating at first, but it describes data in a human-readable way.

As with importing an external text file, there is a distinct advantage to using this method for loading images. If you want to add or remove images from the gallery, you can simply place them in the correct folder and modify the content in the XML file. You do not need to republish the Flash movie for the new images to be included.

1. **On your desktop, open and review the portfolio.xml file in the WIP>Portfolio folder.**

    You can open an XML file in most text-editing applications.

    There is a root node named "images," consisting of an opening tag at the beginning and a closing tag at the end. This root node contains multiple *child nodes* named **image**, each of which has a **source** attribute with a value that describes the location of the image.

Note:

The XML for this exercise could have been written in any number of formats, and XML is an entire subject in itself.

2. **In Flash, add a new blank keyframe to Frame 20 of the Content layer in the portfolio-site.fla file.**

3. **With the Frame 20 keyframe of the Contents layer selected, drag an instance of the TileList component from the Components panel onto the Stage.**

4. **Define the properties and parameters of the TileList instance as follows:**

    Instance Name: thumbnails

X: 15	**Y: 135**
W: 75	**H: 450**

    columnCount: 1

    columnWidth: 75

    direction: Vertical

    verticalLineScrollSize: 50

    Leave all other options at their default values.

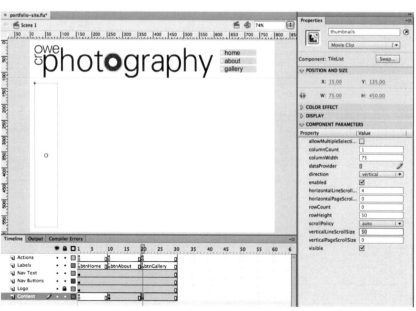

5. **From the Components panel, drag an instance of the UILoader component onto the Stage. Define the properties of the UILoader instance as follows:**

Instance Name: **fullSizeImage**

X: **105**　　　Y: **135**

W: **675**　　　H: **450**

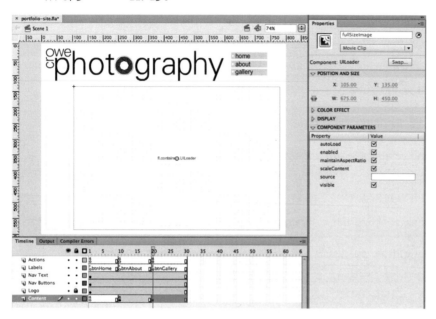

6. **Click Frame 1 of the Actions layer to make the frame the active selection.**

7. **Add a line of white space at the end of the existing code in the Actions panel Script pane.**

8. **Declare a variable named photos of type XML.**

```
var photos:XML;
```

9. **Declare another variable named xmlLoader of type URLLoader. Instantiate this variable as a new URLLoader object.**

```
var xmlLoader:URLLoader = new URLLoader();
```

```
1   import flash.events.MouseEvent;
2   import flash.net.URLLoader;
3
4   stop();
5
6   btnHome.addEventListener(MouseEvent.CLICK, navClickHandler);
7   btnAbout.addEventListener(MouseEvent.CLICK, navClickHandler);
8   btnGallery.addEventListener(MouseEvent.CLICK, navClickHandler);
9
10  function navClickHandler (event:MouseEvent):void
11  {
12      gotoAndStop(event.target.name);
13  }
14
15  var photos:XML;
16  var xmlLoader:URLLoader = new URLLoader();
```

Note:

An import statement for the flash.net.URLLoader package is automatically added to the beginning of the script.

10. Inside the URLLoader constructor function's parentheses, create a URLRequest requesting "portfolio.xml" as its argument.

var xmlLoader:URLLoader = new URLLoader(new URLRequest("portfolio.xml"));

```
1   import flash.events.MouseEvent;
2   import flash.net.URLLoader;
3   import flash.net.URLRequest;
4
5   stop();
6
7   btnHome.addEventListener(MouseEvent.CLICK, navClickHandler);
8   btnAbout.addEventListener(MouseEvent.CLICK, navClickHandler);
9   btnGallery.addEventListener(MouseEvent.CLICK, navClickHandler);
10
11  function navClickHandler (event:MouseEvent):void
12  {
13      gotoAndStop(event.target.name);
14  }
15
16  var photos:XML;
17  var xmlLoader:URLLoader = new URLLoader(new URLLoader(new URLRequest("portfolio.xml"));
```

Note:

An import statement for the flash.net.URLRequest package is automatically added to the beginning of the script.

11. Add an event listener function named xmlReady. This is the same process you used before for txtLoader:

xmlLoader.addEventListener(Event.COMPLETE, xmlReady);

function xmlReady(event:Event):void

{

}

```
1   import flash.events.MouseEvent;
2   import flash.net.URLLoader;
3   import flash.net.URLRequest;
4   import flash.events.Event;
5
6   stop();
7
8   btnHome.addEventListener(MouseEvent.CLICK, navClickHandler);
9   btnAbout.addEventListener(MouseEvent.CLICK, navClickHandler);
10  btnGallery.addEventListener(MouseEvent.CLICK, navClickHandler);
11
12  function navClickHandler (event:MouseEvent):void
13  {
14      gotoAndStop(event.target.name);
15  }
16
17  var photos:XML;
18  var xmlLoader:URLLoader = new URLLoader(new URLRequest("portfolio.xml"));
19
20  xmlLoader.addEventListener(Event.COMPLETE, xmlReady);
21
22  function xmlReady(event:Event):void
23  {
24      |
25  }
```

Note:

An import statement for the flash.events.Event package is automatically added to the beginning of the script.

12. Inside the xmlReady() function, instantiate the *photos* variable as an XML object constructed from the loaded data:

photos = new XML(event.target.data);

Flash does not automatically treat the imported data as an XML object just because it is formatted as XML. You have to explicitly tell Flash to use this data to construct an XML object.

```
17  var photos:XML;
18  var xmlLoader:URLLoader = new URLLoader(new URLRequest("portfolio.xml"));
19
20  xmlLoader.addEventListener(Event.COMPLETE, xmlReady);
21
22  function xmlReady(event:Event):void
23  {
24      photos = new XML(event.target.data);
25  }
```

13. **Select the Frame 20 keyframe on the Actions layer, and add a line of white space after the existing code.**

14. **Set the thumbnails component instance's dataProvider property equal to a new instance of the DataProvider class, constructed from the photos XML object:**

```
thumbnails.dataProvider = new DataProvider(photos);
```

Note:

An import statement for the fl.data.DataProvider package is automatically added to the beginning of the script.

15. **Add an event listener to the thumbnail TileList instance, listening for a click event of the MouseEvent class which will call a function named tileClickHandler, and create the event handler function:**

```
thumbnails.addEventListener( MouseEvent.CLICK, tileClickHandler );

function tileClickHandler( event:MouseEvent ):void
{


}
```

16. **Inside the tileClickHandler() function, set the source property of the fullSizeImage instance equal to event.target.data.source:**

```
fullSizeImage.source = event.target.data.source;
```

Note:

The DataProvider class requires its import statement, because its package path begins with "fl" (not "flash").

```
1   import fl.data.DataProvider;
2   import flash.events.MouseEvent;
3
4   stop();
5
6   thumbnails.dataProvider = new DataProvider(photos);
7   thumbnails.addEventListener(MouseEvent.CLICK, tileClickHandler);
8
9   function tileClickHandler(event:MouseEvent):void
10  {
11      fullSizeImage.source = event.target.data.source;
12  }

   Actions : 20
```

This import statement is added after you define the DataProvider object in Step 14.

This function defines the source of the fullSizeImage instance, based on data in the imported XML file.

17. **Press Command-Return/Control-Enter to test the movie. In the Flash Player window, click the "gallery" link in the navigation area.**

The image that was added to the display list on Frame 1 still appears in the movie, because you only added the removeChild function on Frame 10. If a user clicks straight to the Gallery screen without first visiting the About screen, the removeChild function is never implemented. You need to add the necessary code to Frame 20 as well.

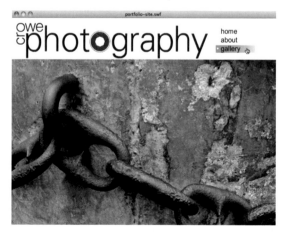

18. **Close the Flash Player window and return to Flash.**

19. **Make sure the Frame 20 keyframe is still selected on the Actions layer.**

20. **Place the insertion point after the last import statement and press Return/Enter two times.**

21. **On Line 4 of the Script pane, type:**

    removeChild(loadImage);

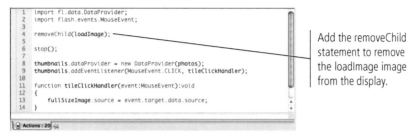

Add the removeChild statement to remove the loadImage image from the display.

22. **Press Command-Return/Control-Enter to test the movie. In the Flash Player window, click the "gallery" link in the navigation area, then click through the thumbnails on the left side of the screen.**

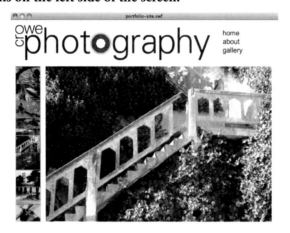

Note:

The fullSizeImage area remains blank until you click a thumbnail.

23. **Close the Flash Player window and return to Flash.**

24. **Save the file and continue to the final exercise.**

The final required step for this project is a slight revision to the existing code, which will prevent errors from occurring if users click through the site screens without first going back to the "home" screen.

1. **With `portfolio-site.fla` open, press Command-Return/Control-Enter to test the movie.**

2. **In the Flash Player window, click the "about" link in the navigation area and then click the "gallery" link in the navigation area.**

   At this point, nothing appears in the TileList object. When things don't appear properly in the Flash Player window, it is a good indication that a problem has occurred somewhere in your code.

3. **Move the Flash Player window so you can see the Timeline panel in the main Flash user interface.**

   ```
   Timeline  Output
   ArgumentError: Error #2025: The supplied DisplayObject must be a child of the caller.
       at flash.display::DisplayObjectContainer/removeChild()
       at portfolio_fla::MainTimeline/frame20()
       at flash.display::MovieClip/gotoAndStop()
       at portfolio_fla::MainTimeline/navClickHandler()
   ```

   If an error occurs in your movie, the Output panel can provide some useful information. In this case, the first line tells you that "The supplied DisplayObject must be a child of the caller." The second and third lines further identify the problem being related to the removeChild command on Frame 20.

   Remember, you added the removeChild(loadImage) method on both Frame 10 and Frame 20 of the Actions layer to remove the loadImage object. The problem is, ActionScript is not capable of ignoring your instructions.

   If the loadImage has already been removed — for example, a user has already clicked the "about" link — the next instance of the removeChild(loadImage) method — say the user clicks the "gallery" link immediately after clicking the "about" link — will cause an error because the loadImage object is not there to be removed.

   Rather than using a simple removeChild statement to remove the loadImage object from the display list, you need to use a conditional statement that says "if the loadImage object is present, remove it."

4. **Close the Flash Player window, then select the Frame 10 keyframe on the Actions layer.**

5. **Delete the removeChild statement on Line 6. In its place, type:**

```
if (this.contains(loadImage))
{
    removeChild(loadImage);
}
```

Replace the removeChild statement with this conditional statement.

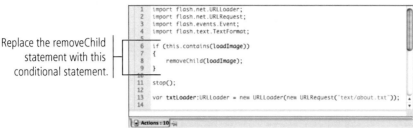

The argument in the if conditional statement tells the script what to look for.

- the target object (**this**) refers to the active timeline.
- The **contains** method determines whether the specific object is a child of the target object.
- The argument inside the contains method parentheses (**loadImage**) defines what object the conditional statement is looking for.

6. **Repeat Step 5 on Frame 20 of the Actions layer.**

Make the same change on Frame 20 of the Actions layer.

Note:

In this case, the original removeChild function is on Line 4 of the script.

7. **Press Command-Return/Control-Enter to test the movie.**

8. **In the Flash Player window, click the "about" link in the navigation area and then click the "gallery" link in the navigation area.**

Because you used the conditional statement to remove the loadImage object, the file now works properly regardless of the order in which users click the navigation links.

9. **Close the Flash Player window and return to Flash.**

10. **Save the file and close it.**

Project Review

fill in the blank

1. The ___ is a text-based way of identifying a specific frame on the timeline.

2. When creating a loader, the _____ object identifies the path to a specific image file.

3. The _____ method places the defined object in the movie (technically called "adding to the display list").

4. The _____ method removes an object from the display list.

5. The _____ prefix at the beginning of the hexadecimal number is required when designating a color in ActionScript.

6. The _____ object class is used to bring external text or XML files into a Flash movie.

7. The _____ event triggers a function when the external file is completely loaded.

8. The _____ class is used to change the appearance of text in a TextArea component.

9. The _____ component can be used to import external content based on parameters defined in the Properties panel.

10. The _____ format describes data; it does not include information about the appearance of that data.

short answer

1. Briefly explain the process of using frames to build site navigation.

2. Briefly explain how matching instance names and frame labels can help in defining navigation.

3. Briefly explain two different methods for adding external images to a Flash movie.

Portfolio Builder Project

Use what you learned in this project to complete the following freeform exercise.
Carefully read the art director and client comments, then create your own design to meet the needs of the project.
Use the space below to sketch ideas; when finished, write a brief explanation of your reasoning behind your final design.

art director comments

Every professional Flash designer or developer needs a portfolio to display their work to prospective clients. If you have completed the projects in this book, you now have a number of different examples to show off your skills using Flash CS6.

The eight projects in this book were specifically designed to include a wide variety of skills and techniques, as well as different types of movies for different types of clients. Your portfolio should follow the same basic principle, offering a variety of samples of both creative and technical skills.

client comments

For this project, you are your own client. Using the following suggestions, gather your work and create your own portfolio.

❏ If possible, get your own domain name to host your portfolio site. If you can't get a personal domain name, use a free subdomain name from an established server company. (If you use a subdomain or a free host, look for one where external ads are not added to your pages.)

❏ Include sample movies that you have created, including those that you completed for this book or for other clients.

❏ If you can't include links to certain files, take screen shots and post those on your own site.

❏ For each sample you include, add a brief description or explanation of your role in the job. (Did you design it? What techniques were used to build the movie?)

❏ Be sure to include full contact information in a prominent location on your site.

project justification

_____

_____

_____

_____

_____

_____

_____

_____

_____

_____

_____

_____

_____

_____

_____

To complete this project, you used named frames to establish the framework of the site navigation, where each button on the page links to a specific frame using the gotoAndStop function. This type of navigation is common for complex files; by naming the individual frames, you don't need to remember which frame number goes with which site page.

This project incorporated a number of techniques for importing external images and text into a Flash movie. You should now understand how to use ActionScript to incorporate external objects, as well as how to use built-in Flash components that simplify the import process.

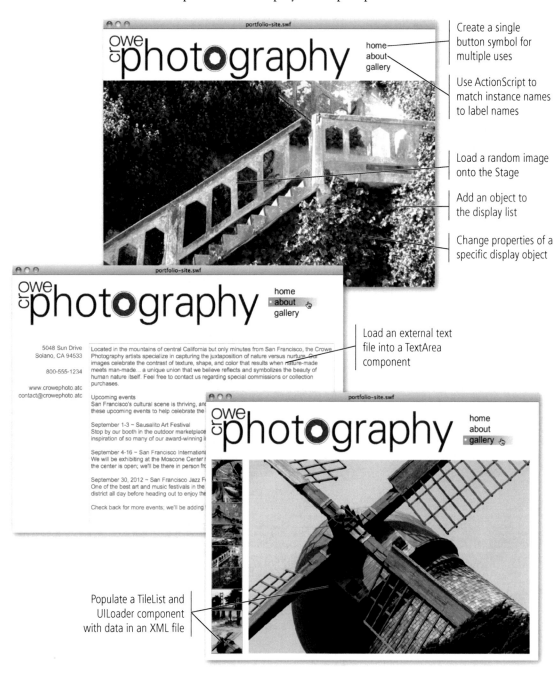

Create a single button symbol for multiple uses

Use ActionScript to match instance names to label names

Load a random image onto the Stage

Add an object to the display list

Change properties of a specific display object

Load an external text file into a TextArea component

Populate a TileList and UILoader component with data in an XML file

Use our portfolio to build yours.

The Against The Clock Professional Portfolio Series walks you step-by-step through the tools and techniques of graphic design professionals.

Order online at www.againsttheclock.com
Use code **PFS712** for a 10% discount

Go to **www.againsttheclock.com** to enter our monthly drawing for a free book of your choice.